FUNDAMENTALS OF
TRANSPORT PHENOMENA IN POROUS MEDIA

Developments in Soil Science 2

FUNDAMENTALS OF TRANSPORT PHENOMENA IN POROUS MEDIA

IAHR

*International Association for
Hydraulic Research*

ELSEVIER PUBLISHING COMPANY
AMSTERDAM / LONDON / NEW YORK

1972

Distribution of this book is being handled by the following publishers:

For the U.S.A. and Canada
American Elsevier Publishing Company, Inc.
52 Vanderbilt Avenue
New York, New York, 10017

For Israel
Jerusalem Academic Press
P.O.Box 2390
Jerusalem, Israel

For all remaining areas
Elsevier Publishing Company
335 Jan van Galenstraat
P.O.Box 211, Amsterdam, The Netherlands

Library of Congress Card Number 71–183910
ISBN 0–444–99897–7

Based on the proceedings of the First International Symposium
on the Fundamentals of Transport Phenomena in Porous Media
Technion City, Haifa, Israel, 23–28 February, 1969

With 109 illustrations and 14 tables

Printed in Israel

PREFACE

Flow through porous materials is encountered in many scientific and engineering disciplines. Although this subject could have been regarded as one of the topics of Fluid Mechanics, there exists sufficient distinction which warrants its treatment as a separate discipline. However, as this subject has been approached by scientists coming from different backgrounds and having different applications, the development of this discipline has progressed along more or less parallel lines by hydrologists, soil physicists, reservoir engineers, chemical engineers, etc. Often a different nomenclature was used to describe the same phenomena.

In 1967 a Section on Flow Through Porous Media was established within the framework of the International Association for Hydraulic Research. Among the aims of this Section are to foster liaisons between those engaged in basic and applied research, as well as in the practical application of the dynamics of fluids in porous materials by encouraging:

(a) Discussion and publication of research information.

(b) Stimulation of basic and applied research and exchange of corresponding information.

(c) Bringing to the attention of members significant progress in the theory of dynamics of fluids in porous media made by research in disciplines such as hydrodynamics, groundwater hydrology, reservoir engineering, soil physics, soil mechanics, chemical engineering, etc.

It is to achieve these aims that the First International Symposium on the Fundamentals of Transport Phenomena in Porous Media has been convened. Special efforts were made by the Section's Committee and by the Organizing Committee to make the Symposium a meeting ground for hydrologists, reservoir engineers, soil physicists and chemical engineers, for an exchange of ideas traditionally derived from different points of view.

As already indicated by the name of this Symposium, it deals with the fundamentals rather than with applications. In addition, a balance was sought in the contributed papers between those which are in nature a review, and those which summarize original studies. Those were the criteria which guided the reviewers of the papers in their decisions.

Financial support from the following organizations is gratefully acknowledged:

- TECHNION — Israel Institute of Technology,
- WATER COMMISSION — Ministry of Agriculture,
- TAHAL — Water Planning for Israel Ltd.,
- MEKOROT — Water Company, Ltd.,
- Ministry of Tourism,

So many other people contributed to the success of the Symposium and the preparation of these Proceedings, that a collective acknowledgment of their help is the only possible way of thanking them.

It is hoped that this Symposium will be the first of a series of similar international Symposia. Whenever possible these Symposia will be held jointly with other organizations dealing with the dynamics of fluids in porous media.

CONTENTS

vii

1

General theory of transport processes

related to fluid flow through porous media

THERMODYNAMIC ANALOGY OF MASS TRANSPORT PROCESSES IN POROUS MEDIA

A. E. Scheidegger and K. H. Liao

ABSTRACT

The paper gives a review of the foundations and justification of an analogy between thermodynamics and mass-transport processes. The stochastic theoretical models of porous media leading up to a general statistical treatment of the mechanics of flow through porous media are described. Then it is shown how the thermodynamic interpretation of large mechanical systems can also be applied to transport processes in general and flow through porous media in particular. Important corollaries bearing upon the stability conditions for displacement processes in porous media are reviewed. It is shown that, based on the theories available to-date, a "front" between displacing and displaced fluid is always unstable (i.e., it grows indefinitely).

1. INTRODUCTION

An important field in water resources research is the study of ground water, which is the subject of the present review. The movement of ground water is a basic part of hydrodynamics and the purpose of this paper is to investigate it, and to pay attention particularly to the spread of pollutants in a ground-water stream.

In ground water problems the formation is considered to be a continuous medium containing many interconnected openings which serve as the fluid carrier. This formation, as a continuous medium of many interconnected openings, is a porous medium.

The problem of fluid flow through a porous medium is difficult to treat by the fundamental Navier-Stokes equations due to the fact that the shape of the internal pore geometry, and therefore the boundary conditions, are unknown. Even if they were known, their complexity would make it difficult to solve the Navier-Stokes equations directly. Since a natural porous medium is so disordered and irregular, it can be best studied by methods of statistical mechanics.

The first attempt along these lines was a random walk model, which led to a diffusivity equation describing the spread of a contaminant or miscibly intruding fluid in a porous medium [6]. Subsequent experiments confirmed the prediction of the theory, notably the existence of a dispersive action in a porous medium. However, the random walk model was unsatisfactory inasmuch as it made very specific assumptions about the microscopic phenomena involved.

It should be noted that the flow of a fluid is a special case of general transport phenomena where the transported quantity is a "fluid." An approach to general transport theory is to apply the methods of Gibbs' thermodynamics which leads to an analogy of energy in energy based statistical mechanics with a positive definite constant of the motion in a system where such a constant exists.

In the case of flow of a fluid through a porous medium, which is a special case of general transport theory, the constant of the motion is the "mass" of a pollutant. It is indeed possible to show that there is a perfect analogy between "mass" transport in a porous medium and "energy" transport in the conduction of heat in solids. Both energy and mass are subject to conservation laws, and both exist only in non-negative amounts. The conservative "mass" system may be defined suitably as to conserve mass within the system (e.g., no mass transfer across the boundary of the system), and thus it is possible to formulate canonical equations of motion. Thus, there is a formal analogy between flow of a contaminant through a porous medium and Gibbsian statistical mechanics, the total mass of contaminant playing the role of the Hamiltonian.

The present paper is devoted to a review of the arguments leading to the above-mentioned analogy between flow in a porous medium and thermodynamics and to a discussion of its implications.

2. MODELS OF POROUS MEDIA

As noted, the structure of porous media is very complicated and disordered. The first attempt to study the flow phenomena of fluids through homogeneous and isotropic porous media was by means of Darcy's law which is an empirical linear relation between specific discharge q and hydraulic gradient J

$$q = KJ$$

where K is a coefficient of proportionality which is called the "permeability." If K is to be replaced by the "specific permeability" k, then according to dimensional analysis, the relation between K and k is

$$k = \frac{\mu}{\rho g} K$$

where μ is the viscosity, ρ the density of the fluid, and g is the gravity acceleration. Darcy's law then can be written as

$$q = \frac{k}{\mu} \rho g J$$

For many years attempts have been made to find the correlation between the specific permeability, or briefly permeability, and other parameters of a porous medium such as porosity, tortuosity, specific surface, pore size distribution (or grain size distribution) and angularity, packing and orientation of the constituent grains. The result is that the correlations are only valid for some specific cases and hardly can be applied to general porous media.

It is from this empirical difficulty that people have begun to set up models to study the correlation between k and other parameters theoretically. The simplest models for this purpose are capillaric models. There are also other kinds of more complicated models, such as statistical models (See e.g., [7]).

(a) *Straight capillaric model*

With this type of model, a porous medium is supposed to be composed of a bundle of straight capillaries entering the porous medium on one face and emerging on the opposite face with uniform pore diameter. The flow in each individual capillary is assumed to be governed by the well-known Hagen-Poiseuille equation. It is clear that, in this fashion, a correlation is obtained between porosity, permeability, internal surface, etc., *in this model*. By extrapolation, these correlations are then postulated to be valid for general porous media. However, it is clear that a model consisting of parallel capillaries gives a permeability in one direction only. Thus, three sets of parallel capillaries (corresponding to the three spacial dimensions) must be employed. However, then the relations between porosity and permeability become patently wrong, and modifications to the simple straight capillaric model have to be made.

(b) *Parallel type model*

Such a modified model is the parallel type model. In this model all the capillaries are parallel in the direction which permits flow of fluid, but the diameters are not uniform. It is clear, however, that this model has not much advantage over the straight capillaric model except that some of the average quantities have a better defined meaning.

(c) *Serial type model*

In straight and parallel type models each capillary is supposed to go through the porous medium without variation of diameter. This is far from reality. Therefore, a serial type model has been proposed in which it is assumed that pores with various diameters are put together going through the porous medium from face to face. Of course in this case the actual length of the flow path σ will be different from the length of the porous medium z. However, this introduces an additional parameter (the "tortuosity" $T = \sigma/z$) into the model which is not very satisfactory.

(d) *Branching type model*

None of the models discussed above take into account the fact that pores may branch and join together again. Thus, branching type models have been postulated. However, assuming branching phenomena introduces a host of specific parameters into the models which does not lend itself to the setting up of a general theory.

(e) *Random walk models*

None of the capillaric models are complicated enough to fairly represent true porous media. It can be said that they are not satisfactory upon general grounds. Therefore one has to try statistical mechanics to solve the problem, which amounts to considering random walk models. There are two fundamental random walk models possible, one of which puts the randomness on to the particles of the fluid while the other puts the randomness onto the porous medium.

Turning to the first possibility, we note that the first attempt [6] was a random walk model. This model is based on the assumption that a fundamental probability distribution function $\psi(r, x, y, z)$ exists which describes the probable position of a fluid particle at certain time steps. The probability distribution is assumed to be Gaussian so that a "dispersion" is introduced. It can be shown that the probability distribution function automatically satisfies a diffusivity equation. This is directly applicable to the description of the spread of a contaminant in a porous medium.

Obviously, the above model is a random walk model referring to the fluid particles. Putting the randomness onto the medium instead of onto the fluid we arrive at a graph theoretical model [4]. In the graph theoretical model, as a unit mass of fluid is injected into a porous medium., it will spread out in a geometrical form. This geometrical form is represented by a bifurcating graph. We then consider not only one graph but the ensemble of all possible graphs with a definite number of free vertices. This is then the basic ensemble for calculating expectation value of observable quantities. The total number N of possible graphs with a definite number n of free vertices can be calculated by the formula

$$N = \frac{1}{2n-1}\binom{2n-1}{n}$$

From the above formula it can be seen that, as n gets large, the total number of possible graphs becomes phenomenal. It would be difficult or completely impossible to find all of the possible graphs. Hence a Monte Carlo technique has to be used to randomly generate a certain number of graphs on a computer from which the required expectation values can be calculated.

Results to-date indicate that a dispersion effect again occurs when a fluid intrudes into another in a porous medium. This dispersion has been verified experimentally.

3. ANALOGY BETWEEN GENERAL TRANSPORT PROCESSES AND GIBBS' SCHEME OF GAS DYNAMICS

(a) *The analogy*

The models of flow through porous media discussed in the last section are not really satisfactory inasmuch as they made very specific ssumptions about the microscopic phenomena involved. Therefore, a much more general approach is desirable in which the property of the amount of tracer being a constant of the motion is exploited. This enables one to make a formal analogy with Gibbsian statistical mechanics, the total mass of contaminant playing the role of the Hamiltonian [8]. In fact, this analogy is valid between Gibbsian statistical mechanics, and *any* transport process involving a non-negative transported quantity.

The statistical mechanics devised by Gibbs has a very general character. It uses the Hamiltonian H as the energy which is a constant of motion. However, it is not necessary in this scheme that H be the energy. The general nature of the scheme demands that H should be a positive definite constant of the motion. In many hydrological cases, for instance, the constant of the motion can be taken as the total mass of a contaminant of flowing fluid.

Thus, let us consider a volume V traveling with the main stream of some fluid in a porous medium, with the assumption that there is no mass transfer across the boundaries of the volume and that a contaminant is introduced therein. Then one gets a system where the mass (of a contaminant) is a constant of the motion, which results from the law of conservation of mass. This mass can then be called H. The volume V is divided into small cells and thus one can write an analogous equation for the Hamiltonian as

$$H = \sum_{i=1}^{N} m_i = \text{const.}$$

where m is the mass in cell i. If one assumes that

$$p_i = +\sqrt{m_i}$$

then the Hamiltonian can be written as

$$H = \sum_{i=1}^{N} p_i^2.$$

The q_i (the conjugates of p_i) in energy-based statistical mechanics are defined through the canonical equations of motion

$$\dot{q}_i = \partial H/\partial p_i \quad \text{and} \quad \dot{p}_i = -\partial H/\partial q_i.$$

Here, p_i and q_i form a phase space to which Liouville's theorem (asserting that any cell in phase space retains a constant phase-volume) can be applied.

An inspection of the last equation shows that it can be interpreted in terms of mass-based statistical mechanics. Then, the system of equations given above describes the behavior of the contaminant in the volume V.

The above Hamiltonian describes static conditions, i.e., the amount of contaminant in each cell remains constant. To obtain a change of the amounts of contaminant in the individual cells, one has to modify the Hamiltonian expression by adding an interaction term U

$$H = \sum_{i=1}^{N} p_i^2 + \varepsilon U(p_i q_i t)$$

where t is the time and ε is a small constant.

After defining the phase space as above, one can define an ensemble in this space. The ensemble then will attain a state of statistical equilibrium because of the presence of the interaction term in the Hamiltonian. The component systems are then distributed canonically, with a probability distribution P_i

$$P_i(m_i) = C' e^{-m_i/kT}$$

where k is a constant, T is the analog of the temperature, and C' is a normalization constant. Thus (leaving out the index i)

$$P = (1/Z) e^{-m/kT}$$

where Z is the "partition function" required for normalization

$$Z = \int e^{-m/kT} \, dp dq = C \int e^{-m/kT} \, dp$$

with $C = \int dq$.

Based upon the partition function, the analogy between a thermal field and the porous medium then maintains the following relations [12]

temperature $(T) \leftrightarrow$ mass per unit cell $(m_0 = 1/k\beta)$,

entropy $(S) \leftrightarrow \ln m_0 + 2k + 1$, where $K = \ln[C\sqrt{(\pi/2)}]$,

internal energy $(U) \leftrightarrow 1/k\beta$,

work done $(W) \leftrightarrow -\alpha \int_v (m_0/V) dV$, ($\alpha$ is a constant)

energy potential (dU) (in differentials) $\leftrightarrow \alpha m_0 - (\alpha m_0/V) dV$,

Helmholtz's free energy $(F) \leftrightarrow -m_0(\ln m_0 + 2K)$,

Gibbs' free energy $(\Phi) \leftrightarrow -m_0(\ln m_0 + 2K) + \alpha m_0 \int dV/V$,

Gibbs' potential $(\psi) \leftrightarrow m_0(1 + \alpha \int dV/V)$,

heat capacity $\leftrightarrow 1$.

Important corrolaries follow immediately from this analogy. The aboves argument refers only to systems under equilibrium conditions. In nature, however, most systems are not at equilibrium. The applications of the principles of irreversible thermodynamics developed by Onsager and others, however, yield corresponding results in flow through porous media. Using the same method as in the deduction of the anisotropic heat-conduction equation in energy-based statistical mechanics, one duly ends up with a diffusivity equation in mass-based statistical mechanics (flow through porous media) as well.

One thus has a complete analogy between energy-based thermodynamics and mass-based statistical mechanics, of which flow through porous media is a special example.

(b) Justification of the analogy

The extent of the analogy between the usual energy-based statistical mechanics and mass dispersion phenomena has to be further explored. It can be shown that in the equilibrium case the interaction function as used in the energy-based statistical mechanics of solid or weakly coupled gases entails a corresponding result in mass-dispersion systems [2]. The analogy can also be extended to irreversible thermodynamics.

Assume that we are dealing with a system which is subject to a positive-definite conservation law so that there are a Hamiltonian and canonical equations of motion. It is well known that the change with time t of any function of the p_i and q_i is given by the corresponding Poisson bracket with the "Hamiltonian" H

$$\frac{df}{dt} = [f, H] = \Sigma \frac{\partial f}{\partial q_i} \frac{\partial H}{\partial p_i} - \frac{\partial f}{\partial p_i} \frac{\partial H}{\partial q_i}$$

In particular, if f represents the density of ensemble points in $p - q$ ("phase") space, normalized to 1, viz.

$$\int dp dq = 1$$

then the well-known theorem of Liouville states that this density behaves as that of an incompressible liquid; hence (Liouville equation)

$$\frac{d\rho}{dt} = 0 \quad \text{or} \quad \frac{\partial \rho}{\partial t} = [H, \rho]$$

Prigogine [5] has shown that the last equation can be written as an operator equation, if one introduces the hermitean operator L

$$L = -i \left(\frac{\partial H}{\partial p} \frac{\partial}{\partial q} - \frac{\partial H}{\partial q} \frac{\partial}{\partial p} \right)$$

Then

$$i \frac{\partial \rho}{\partial t} = L.$$

It is clear that the above formalism is immediately applicable to our transport processes [1]. Let us consider an example. Assuming that the masses at the cells in our transport problem are independent, the Liouville equation becomes

$$i\frac{\partial\rho}{\partial t} = i[H,\rho] = -\Sigma 2p_i\frac{\partial\rho}{\partial q_i}$$

and the Liouville operator is

$$L_0 = -i\Sigma 2p_i\frac{\partial}{\partial q_i}.$$

Of course, no change in ρ is obtained if there is no interaction function present in the Hamiltonian. If it is assumed that such an interaction is present, but small, say, proportional to ε, then it may be noted that the Liouville equation is a differential equation with a Liouville operator $L = L_0 + \delta L$, δL being proportional to ε. The solution for ρ is usually found by expressing ρ in the form of a superposition of eigenfunctions of the Liouville operators L_0. Then the general procedure is to calculate a master equation which has the form

$$\partial\rho\{o\}(t)/\partial t = \varepsilon^2 O_0\rho\{o\}(t)$$

where O_0 is a self-adjoint operator. The master equation represents the first approximation (in ε) to the change of the probability density $\rho\{o\}$ coresponding to L_0. In this, choosing the appropriate interaction function is important. It turns out that the interaction term does not contribute, in the first approximation, to the change of density function for cases where the interaction function is taken as a function of p. On the other hand, if the interaction has terms that are a function of q, it does so contribute. In this case it turns out that the master equations are all similar (for various choices of dependence on q) except that the constant coefficients in each case will be different. The fact that Prigogine's method can be transferred to mass-based statistical mechanics gives the explicit justification of the thermodynamic analogy for transport processes.

4. STEADY STATE CONDITIONS

(a) General theory

Of particular importance are the conditions for the establishment (or prevention) of a steady state in a displacement process in porous media. It is often important (for engineering applications) to know whether a displacing agent forms a "stable front" during a displacement process or whether it will spread rapidly (in an "instable front") through the displaced fluid. The spread may occur actually dispersively on a microscopic scale, or in form of macroscopic "fingers."

In ordinary thermodynamics there is a well-known condition for a system in a steady state to satisfy: the production rate of entropy throughout the system must occur at a minimum rate. To the extent that there is a complete thermodynamic analogy in mass transport, the above principle of "entropy" production must also be satisfied [10].

According to De Groot [3], the entropy production rate σ is in thermodynamics, and hence also in the analogous transport processes, given by

$$\sigma = \sum_i [(\text{grad } T)/T^2] J$$

where the sum is to be taken over all parts of the system and J is the heat (mass) flux. Thus, in the linear continuous case we have as stability conditions

$$-\int_A^B \frac{Jm'}{m^2} dx = \text{a minimum.}$$

(b) *Microscopic spread*

We first turn to the possibility of "microscopic" instability (without fingering), whose conditions correspond to those of "solitary" dispersion [9]. In applying the above minimum-principle, the question is what expression one should take for J. As shown by Scheidegger [9], there are two *a priori* possibilities: either J is a function of m or a function of the gradient of m. Accordingly, let us first set

$$J = J(m)$$

In this case, one ends up sith a Buckley-Leverett equation (c.f. [9]) governing the displacement process

$$\frac{\partial m_1}{\partial t} = r'(m_1) V \frac{\partial m_1}{\partial x} = 0$$

where V is the injection pore velocity and

$$r = m_1 + J(m_1).$$

It becomes evident that it is not possible to find a solution for this case, i.e., the Euler equation corresponding to the minimum entropy principle becomes an identity since the expression under the integral sign is a total differential. This result indicates that, in the case under consideration, the general statistical principles pertaining to the steady state do not lead to any statement of the m-profile

We try next

$$J = a \, dm/dx$$

as a first approximation to J being a function of grad m; in this case, the dynamics of the displacement is governed by a diffusivity equation. It turns out that in this case the minimization can be performed; the solution is

$$m = C_1 e^{C_2 x}.$$

Thus, a solution can be found in this case. We thus have the result that in one of the possible theoretical limits of solitary dispersion [$J = J(m)$; Buckley-Leverett limit] no steady state is possible, in the other ($J = am'$; diffusivity limit), a steady state is possible. However, for the achievement of such a steady state, very special boundary conditions are required: in the *moving* (with pore velocity) coordinate system, a constant m (i.e. saturation) must be maintained at two (moving!) points, say x_1 and x_2. It is clear that, in practical cases, no such conditions can occur.

(c) Fingers

We now turn our attention to instabilities of the type called "fingers." A useful model for treating the overall dynamics of fingers has been suggested by Scheidegger and Johnson [11].

Accordingly, consider a linear displacement experiment parallel, say to the x direction which progresses with time (t). For certain values of x and t, there will be some fingers present. The problem is to set up a determining equation for the function

$$m_i = m_i(x, t)$$

where m_i represents the average relative area (mass!) by the ith fluid at level x.

By drawing an analogy with an immiscible displacement process, and making a rather specific assumption regarding the "fictitious" relative permeability for fluid 1, Scheidegger and Johnson [11] arrived at a Buckley-Leverett equation for m_1 (see above) with

$$r(m_1) = \frac{\mu_2/\mu_1}{(\mu_2/\mu_1 - 1) + 1/m_1}$$

where μ_2/μ_1 is the viscosity ratio of the two fluids involved.

With the thermodynamic analogy at hand, we simply consider the evolution of $m_i(x, t)$, in approximation, as a case of solitary dispersion. One has a complete correspondence between fingering and solitary dispersion theory by setting

$$J = \frac{\mu_2/\mu_1}{(\mu_2/\mu_1 - 1) + 1/m_1} - m_1$$

The above formulas refer to the Buckley-Leverett limit of solitary dispersion theory as applied to fingering. The "diffusivity" limit is reached if we set

$$J = am_1'$$

with some constant. The actual behavior of the fingers may be expected to lie between the two limits given.

As noted above, no stationary state is possible for a displacement process in which the flux is a function of m' (Buckley-Leverett limit) but is possible for a processs in which the flux is a function of m' (diffusivity limit). However, the boundary conditions can evidently not be satisfied for such a case to occur.

Thus, we have the result that the phenomenological-macroscopic theory of fingering in a unidirectional displacement process does not permit a steady state to occur.

REFERENCES

1. CHAUDHARI, N. AND A. E. SCHEIDEGGER (1964), Some statistical properties of certain geophysical transport equations, *Pure and Appl. Geophys.*, *59*, 45–57.
2. CHAUDHARI, N. AND A. E. SCHEIDEGGER (1965), Statistical theory of hydrodynamic dispersion phenomena, *Canad. J. Phys.*, *43*, 1776–1794.
3. DeGROOT, S. R. (1961), *Thermodynamics of Irreversible Processes*, North Holland Publ. Co., Amsterdam, 39–40.
4. LIAO, K. H. AND A. E. SCHEIDEGGER (1968), A computer model for some branching-type phenomena in hydrology, *Bull. Int. Assoc. Scientif. Hydrol.*, *13*, No. 1, 5–13.
5. PRIGOGINE, I. (1962), *Nonequilibrium Statistical Mechanics*, Interscience-Wiley, New York.
6. SCHEIDEGGER, A. E. (1954), Statistical hydrodynamics in porous media, *J. Appl. Phys.*, *25*, 994–1001.
7. SCHEIDEGGER, A. E. (1960), *The Physics of Flow Through Porous Media*, Revised Edition, MacMillan Co., New York.
8. SCHEIDEGGER, A. E. (1961), On the statistical properties of some transport equations *Canad. J. Phys.*, *39*, 1573–1580.
9. SCHEIDEGGER, A. E. (1967), Solitary dispersion, *Canad. J. Phys.*, *45*, 1783–1789.
10. SCHEIDEGGER, A. E. (1969), Stability conditions for displacement processes in porous media *Canad. J. Phys. 47*, 209–214.
11. SCHEIDEGGER, A. E. AND E. F. JOHNSON (1961), The statistical behavior of instabilities in displacement processes in porous media, *Canad. J. Phys.*, *39*, 326–334.
12. TOMKORIA, B. N. AND A. E. SCHEIDEGGER (1967), A complete thermodynamic analogy for transport processes, *Canad. J. Phys.*, *45*, 3569–3587.

DEPARTMENT OF MINING, METALLURGY,
AND PETROLEUM ENGINEERING,
UNIVERSITY OF ILLINOIS,
URBANA, ILLINOIS, 61801, U.S.A.

THE INFLUENCE OF PORE STRUCTURE ON THE PRESSURE AND TEMPERATURE DEPENDENCE OF THE EFFECTIVE DIFFUSION COEFFICIENT

P. Hugo

ABSTRACT

A new method for measuring the effective diffusion coefficient D_{eff} in porous media is described, based upon self-diffusion of hydrogen in o–p–H_2 mixtures. Experiments were carried out with nuclear graphites of different pore size distribution. The correlations between D_{eff} and the porous structure are discussed to derive a general formula for the pressure and temperature dependence of D_{eff}. If there is only one sharp maximum in the pore size distribution from the general equation one obtains the Bosanquet formula. If the porous structure is more complicated a generalized Bosanquet formula has to be used to describe the temperature and pressure dependence of D_{eff}. The theoretical equations are in good agreement with the experimental results.

1. PRINCIPLE OF THE DIFFUSION MEASUREMENTS

Several methods are known to measure the diffusion of gases in porous media. Often the principle of stationary countercurrent diffusion is used, first described by Wicke and Kallenbach [14]. Other methods are of the instationary diffusion type [5], [7] or based upon frequency response or pulse techniques [9], [10]. Whereas these methods are suitable to measure the effective diffusion coefficients for technical purposes [12], [13], they are not very handsome for more fundamental high precision measurements. So for investigations of the pressure and temperature dependence of the effective diffusion coefficient

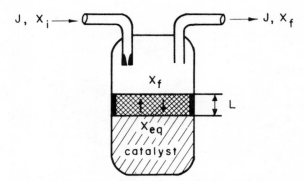

Fig. 1. Diffusion-reaction-cell, schematically

14

a novel method was developed [1], [6]. It is a modified type of the stationary countercurrent method mentioned above.

By this method the countercurrent diffusion of ortho- and para-H_2 across the porous sample is measured in a reaction-diffusion cell: Fig. 1. The stationary countercurrent is maintained by chemical conversion of para-H_2 to ortho-H_2 on a high active catalyst (Pt-alumina) which is filled into the reaction part of the cell. The porous sample separates as a diffusion barrier this part from the upper mixing chamber, which is passed by a stationary hydrogen flow enriched with para-H_2.

The hydrogen flow J (mol/sec) passing the mixing chamber contains the inlet mole fraction x_i of $p-H_2$, it leaves the chamber with x_f. In the reaction part of the cell the equilibrium composition x_q is established. From the stationary mass balance of $p-H_2$ one obtains:

$$(1) \qquad J(x_i - x_f) = QD_{eff}c \cdot \frac{x_f - x_{eq}}{L}$$

Q (cm^2) and L (cm) are the total cross section and the thickness of the porous diffusion barrier, c (mol/cm^3) the total concentration and D_{eff} (cm^2/sec) the effective self-diffusion coefficient of hydrogen. The mole fractions and their differences are measured by thermal conductivity. The details of the measuring device are described elsewhere [1].

The measuring principle has two advantages: the total gas pressures adjust themselves automatically equal on both sides of the diffusion barrier, and the effective diffusion coefficient can be calculated by the simple equation (1), because self-diffusion of hydrogen is measured.

2. EXPERIMENTAL RESULTS

Experiments were carried out with two types of artificial graphites developed specially for nuclear fuel elements. Type G 5 is a high graphitised material with a comparatively homogeneous pore structure, some details of its preparation technique and its properties are described in [4]. The second type HX 12 has due to an additional impregnation [9] a rather inhomogeneous pore structure. Figs. 2 and 3 give the pore size distributions of these materials at different degree of burnoff (see below), measured with a mercury porosimeter.

Using the measuring technique described above the effective diffusion coefficients of these materials were determined. Figs. 4 and 5 give the results for both graphite types at 0% burnoff, covering a range of total gas pressures from 0.02 to about 1 atm and temperatures from $-30°C$ to $+300°C$. The graphites themselves have in this temperature range no catalytic activity on $p-o-H_2$ conversion.

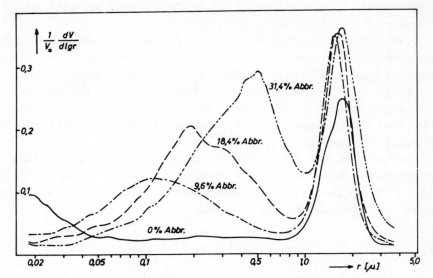

Fig. 2. Pore size distribution of G 5 graphite at different burnoffs

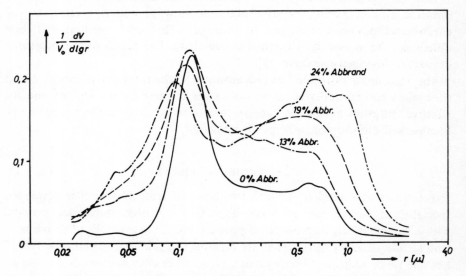

Fig. 3. Pore size distribution of HX 12 graphite at different burnoffs

Further with G 5 graphite the influence of stepwise changes of pore structure on D_{eff} was measured. The stepwise changes were carried out by burning off the material with CO_2 at about 1000°C. Under these conditions all parts within the porous sample react homogeneously. Fig. 6 gives the results of the diffusion measurements at room temperature.

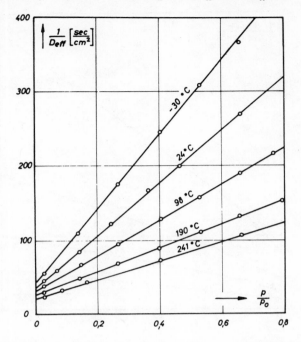

Fig. 4. Dependence of $1/D_{eff}$ on total hydrogen pressure ($p_0 = 1$ atm) for different temperatures, G 5 graphite

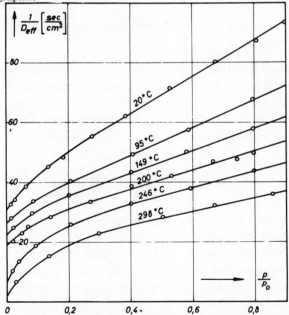

Fig. 5. Dependence of $1/D_{eff}$ on total hydrogen pressure ($p_0 = 1$ atm) for different temperatures, HX 12 graphite

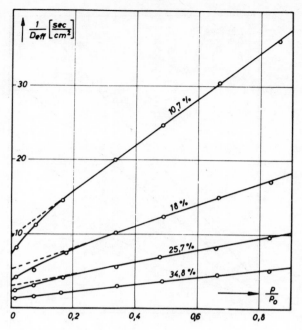

Fig. 6. Dependence of $1/D_{eff}$ on total hydrogen pressure, G 5 graphite at different burnoffs

3. EFFECTIVE DIFFUSION COEFFICIENT AND PORE STRUCTURE

n general porous materials contain a fine-meshed network of pores of different size and shape. So, within a porous sample the total diffusion flow will spread into many small partial flows. If we look at one these partial flows, its diffusion path can be subdivided into length elements l_i with diffusional cross sections q_i; and with local values of the diffusion coefficients D_i. The total diffusional resistance W_j (= flow/concentration difference) of one diffusion path (index j) then results as

$$(2) \qquad W_j = \sum_i \frac{l_{ij}}{q_{ij}D_{ij}} = \frac{L}{Q} \sum_i \frac{x_{ij}}{D_{ij}}$$

However, we must ask how an unique diffusion path can be defined in a network of pores. Each partial flow often will spread among the different pore branches, run along bypasses, and meet again at other branching points to select the favourable path of diffusion. Therefore the main part of a flow may run on a tortuous bypass to avoid straightaway but narrow pore sections with high diffusional resistance.

As far as all bypasses have nearly the same pore size, the local values of the diffusion coefficients are nearly equal. This simple case also can described by an equation of type (2) as only the local values of x_{ij} have to be modified to account for the bypasses. A calculation for bypasses with considerable differences in pore size is more complicated, as the local values of the diffusion coefficients are different. Especially if a bypass contain only a few great pores within a network of small pores, or if great pores in a bypass are nearly blocked by narrow pore sections or dead ends, the diffusion flow may be of the same magnitude within the small and great pores. Such a pore structure cannot be described in details by an equation of the type (2). On the other hand the formal relationship for the pressure and temperature dependence of the overall effective diffusion coefficient is not much changed, if we use the simplified picture of an unique diffusion path without branching of flow. Of course by such a simplification the local values $x_{ij} = Q\, l_{ij}/Lq_{ij}$ lose at least to some extent their meaning of a single pore characteristic and have to be taken rather as statistical weight factors.

The local diffusion coefficients generally lie in the transition range between bulk diffusion and Knudsen diffusion. As was shown by Pollard and Present [8], in this range the self-diffusion coefficient of cylindrical tubes can be calculated from:

$$(3) \qquad \frac{1}{D} = \frac{1}{D_G^0} \cdot \frac{P}{P_0} + \frac{1}{D_k}$$

where D_G^0 is the self-diffusion coefficient in the bulk phase at standard total pressure $p_0 = 1$ atm and $D_K = (2\bar{v}r)/3$ is the Knudsen diffusion coefficient ($\bar{v}$ = mean molecular velocity, r = radius of the tube). By a proper choice of the geometric coefficient x_{ij} the pore sections of a diffusion path can be approximated by cylindrical sections. So, from Eq. (2) and (3) one obtains:

$$(4) \qquad W_j = \frac{L \sum\limits_i x_{ij}}{QD_G^0}\left(\frac{P}{P_0} + \frac{D_G^0}{D_{kj}}\right)$$

The Knudsen diffusion coefficient $D_{Kj} = (2\ \bar{v}r_j)/3$ depends on the effective pore radius r_j of the diffusion path. This value has to be calculated from

$$(5) \qquad r_j = \frac{\sum\limits_i x_{ij}}{\sum\limits_i \dfrac{x_{ij}}{r_{ij}}}$$

where r_{ij} are the radii of the pore sections assuming a cylindrical shape.

What can be measured is the overall effective diffusion coefficient. Summing up all partial flows through the porous sample, one obtains:

$$(6) \qquad D_{eff} = \frac{Q}{L} \sum\limits_j \frac{1}{W_j}$$

or inserting Eq. (4):

$$(7) \qquad D_{eff} = D_G^0 \cdot \sum_j \frac{1}{\left(\dfrac{P}{P_0} + \dfrac{D_G^0}{D_{Kj}}\right) \sum_i x_{ij}}$$

From this equation the pressure and temperature dependence of D_{eff} can be calculated, as the temperature dependence of $D_0 \sim T^{1.65}$ (for hydrogen) and $D_K \sim T^{0.5}$ is known.

As the statistical weight factors x_{ij} of a porous material in general are unknown, a correlation between D_{eff} and the pore structure only can be carried out for some special cases. If there is a porous material in which all pores are of nearly the same size, the mean value of the effective pore radius r_j and so all Knudsen diffusion coefficients D_{Kj} will be nearly the same. Here we obtain from the general equation (7) the well known Bosanquet formula [2]:

$$(7a) \qquad D_{eff} = \frac{\psi D_G^0}{\dfrac{P}{P_0} + \dfrac{D_G^0}{D_K}}$$

where $\psi = \sum (\sum_i x_{ij})^{-1}$ is the permeability of the material. As the mean pore radius can be measured by other methods, e.g., by a mercury porosimeter, in this simple case the temperature and pressure dependence of D_{eff} can be calculated. Of course, the absolute value of D_{eff} must be measured, as the permeability ψ only can be calculated for very simple pore models, which in general the actual behavior of consolidated porous materials can only describe with rather poor accuracy.

Often porous materials contain macropores (index M) and micropores (index m). As far as the diffusion paths of both pore kinds can be handled separately, from Eq. (7) we obtain

$$(7b) \qquad D_{eff} = \frac{\psi_M D_G^0}{\dfrac{P}{P_0} + \dfrac{D_G^0}{D_{KM}}} + \frac{\psi_m D_G^0}{\dfrac{P}{P_0} + \dfrac{D_G^0}{D_{Km}}}$$

Normally the permeability of the micropores ψ_m is small compared with ψ_M so that the second term of Eq. (7b) is a small correction of the Bosanquet formula. If, however, there are a few macropores within a network of micropores or if there is a comparatively broad pore size distribution ψ_M and ψ_m may be of the same magnitude. Of course, in such cases more than two terms of the general equation (7) may be of importance. On the other hand an equation which contains more parameters than Equation (7b) is too complicated for practical purposes. Even with a comparatively high precision technique for diffusion measurements it is often difficult to determine the four parameters ψ_m, ψ_M, D_{Km} and D_{KM} of Eq. (7b) with sufficient security. So, from an experimental point of view Eq. (7b) is the most general form to describe D_{eff} in practical cases. This simplification restrict, however, a more accurate interpretation of the physical significance of the parameters determined from diffusion experiments.

4. COMPARISON WITH EXPERIMENTAL RESULTS

The results of the diffusion measurements were presented in diagrams $1/D_{eff}$ against p/p_0. For G 5 graphite at 0% burnoff straight lines are obtained indicating that the Bosanquet formula (7a) can be applied. Evaluating the $24°C$ line from the slope and the intercept at $p = o$ the values $\psi = 0.0020$ and $D_k = 14$ cm^2/sec are determined ($D_G^0 = 1.43$ cm^2/sec for self diffusion of hydrogen at $24°C$). Applying the Knudsen formula $r = (3D_k)/(2\bar{v})$ a mean pore radius of the diffusion path of $r = 1.2 \cdot 10^{-4}$ cm is obtained ($\bar{v} = 1.77 \cdot 10^5$ cm/sec for hydrogen at $24°C$).

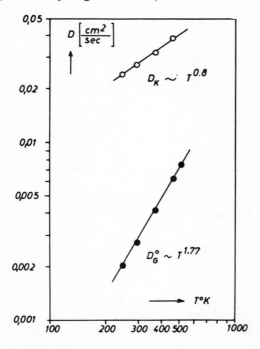

Fig. 7. Temperature dependence of the effective bulk and Knudsen diffusion coefficient, G 5 graphite at 0% burnoff

From Fig. 4 the temperature dependence of the effective bulk diffusion coefficient ψD_G^0 and of the effective Knudsen diffusion coefficient ψD_k can be determined. Fig. 7 gives these values as a function of temperature in logarithmic scale. The bulk diffusion part follows with $T^{1.77}$ within the experimental errors the law to be expected ($D_G^0 \sim T^{1.65}$ for hydrogen). The increase of D_k with temperature $T^{0.8}$ is found somewhat too steep (theoretical $\bar{v} \sim T^{0.5}$). This seems to indicate that there may be some deviations from the Bosanquet formula in the low pressure range.

By burnoff the porous structure of the graphite is changed, Fig. 2. Especially in the low pressure range now deviations from the Bosanquet formula can be detected by the diffusion measurements, Fig. 6. Evaluating these measurements by the generalized Bosanquet formula (7b) the structural parameters given in Table 1 are obtained. As the pressure range between 0.02 and 1 atm. is comparatively small, it is rather difficult to fit four parameters to the experimental values. Especially the values ψ_M and r_M of the macropores contain considerable errors, as these contributions to the total flow are rather small.

TABLE I

Structural parameters of G 5 graphite at different degree of burnoff obtained from diffusion measurements

Burnoff	ψ_M	r_M (cm)	ψ_m	r_m (cm)
0%	0.0020	$1.2 \cdot 10^{-4}$	—	—
10.7%	0.0022	$2.5 \cdot 10^{-4}$	0.021	$0.32 \cdot 10^{-4}$
18.0%	0.0073	$2.0 \cdot 10^{-4}$	0.044	$0.22 \cdot 10^{-4}$
25.7%	0.010	$2.2 \cdot 10^{-4}$	0.079	$0.27 \cdot 10^{-4}$
34.8%	—	—	0.14	$0.50 \cdot 10^{-4}$

Assuming the picture of separate diffusion paths these values can be interpreted. At 0% burnoff there are only a few diffusion paths $\psi_M \ll 1$. The mean pore radius of a path is $1.2 \cdot 10^{-4}$ cm and somewhat smaller than the value expected from the pore size distributuion, Fig. 2, (maximum value at about $1.7 \cdot 10^{-4}$ cm). This indicates that the diffusion resistance of a diffusion path is somewhat influenced by narrow pore sections, which have to be passed by the diffusion flow. So the mean pore radius of a diffusion path, defined by Eq. (5) comes out somewhat smaller.

By burning off the material this picture is changed, now there are two different kinds of diffusion paths. The main part of the diffusion flow runs through micropores ($\psi_m \approx 10 \psi_M$) which had been burnt free by chemical reaction.

The effective pore radii of the micropore diffusion paths up to about 20% burnoff slightly decrease and than increase up to a value of $0.5 \cdot 10^{-4}$ cm at 35% burnoff. This indicates that at low burnoff a diffusion path contains pore sections of micropores and macropores ordered in series. With increasing burnoff the contribution of micropore sections, due to their increasing number, becomes predominant and the effective pore radius decrease. At high degree of burnoff the pore radii increase to the value expected from the pore size distribution.

The number of macropore diffusion paths, already present in the unburnt material, is not much changed by burnoff. Up to 10% burnoff ψ_M remains nearly constant, only the mean pore radius increases indicating that the narrow pore sections between the larger pore sections are burnt off. The increase of ψ_M at higher degree of burnoff seems to be due to new connections between the macropore diffusion paths, so that the diffusion flow runs more straightway through the sample. The number of macropores and the pore radii remain nearly unchanged at higher burnoffs, as can be seen from Fig. 2.

Whereas G 5 graphite has a rather pronounced bimodal pore size distribution for HX 12 a rather broad pore spectrum is obtained, Fig. 3. A separation into macropores and micropores so is more or less arbitrary. Evaluating the diffusion measurements at 20°C, Fig. 5, by Equation (7b) the following structural parameters are obtained for HX 12 graphite:

$$\psi_M = 3.4 \cdot 10^{-4}, \quad r_M = 3.2 \cdot 10^{-4} \text{ cm}$$

$$\psi_m = 1.1 \cdot 10^{-2}, \quad r_m = 0.16 \cdot 10^{-4} \text{ cm}$$

As the calculated contribution of macropore diffusion paths is very small ($\psi_M \ll \psi_m$), the values ψ_M and r_M contain great errors. Further, this graphite seems to be an example of the rather difficult case, mentioned in Section 3, where there are only a few macropores by-passed by a fine meshed network of micropores. In such a case a more detailed interpretation of the physical significance of the "macropore" structural parameters ψ_M and r_M is not possible. The effective micropore radius $r_m = 0.16 \cdot 10^{-4}$ cm, however, is in good agreement with the value expected from the pore size distribution, Fig. 3.

Due to the small contribution of "macropores" the evaluation of the diffusion measurements at different temperatures, shown in Fig. 5, cannot be carried out with sufficient security. The overall temperature dependence of D_{eff} is in the range between $T^{1.0}$ and $T^{1.4}$.

A correlation between the permeabilities ψ and the porous structure, i.e., an evaluation of $\psi = \Sigma_j (\Sigma_i x_{ij})^{-1}$ can be only carried out by assuming simple models for the network of pores within the solid [11]. Unfortunately, the permeability depends sensitively on the details of the network assumed. So simple models can describe the actual ψ-values only with rather poor accuracy.

ACKNOWLEDGEMENT

The author is much indebted to the A. E. E. Dragon, Winfrith, for financial support of these investigations.

REFERENCES

1. BEYER, H.-D. (1966), Diplomarbeit, Münster.
2. BOSANQUET, C. H. (1944), *British TA Report, BR–507.*
3. DEISLER, P. F. AND R. H. WILHELM (1953), *Ind. Engng. Chem., 45*, 1219
4. EVERETT, M. R., D. V. KINSEY AND E. RÖMBERG (1968), in *Chemistry and Physics of Carbon*, Vol. 3, New York.
5. GORRING, L. L. AND A. J. DE ROSSET (1964), *J. Catalysis, 3*, 341.
6. HUGO, P AND A. PARATELLA (1966), *Consiglio Nazionale Delle Ricerche*, Vol. 5, Roma.
7. PIRET, E. L., R. A. EBEL, C. T. KIANG AND W. P. ARMSTRONG (1951), *Chem. Engng. Progr., 47*, 405.
8. POLLARD, W. G. AND R. D. PRESENT (1948), *Phys. Rev., 73*, 762.
9. PRICE, M. S. T. (1959), Paper presented on *O.E.E.C/E.N.E.A. Symposium on Nuclear Graphite*, p. 71.
10. SCHNEIDER, P. AND J. M. SMITH (1968), Paper presented on the *IV. Intern. Congr. on Catalysis*, Novosibirsk.
11. WAKAO, N. AND J. M. SMITH (1962), *Chem. Engng. Sci., 17*, 825.
12. WEISZ, P. B. (1957), *Z. physikal. Chemie N. F., 11*, 1.
13. WEISZ, P. B. AND A. B. SCHWARTZ (1962), *J. Catalysis, 1*, 399.
14. WICKE, E. AND R. KALLENBACH (1941), *Kolloid-Z., 97*, 135. Some earlier measurements by this method were carried out by Buckingham, E., U.S., Dep. of Agriculture, Bureau of Soils, Bull. No. 25 (1904).

INSTITUT F. PHYSIKALISCHE CHEMIE DER UNIVERSITÄT MÜNSTER,
    MÜNSTER, WEST-GERMANY

New Address: P. HUGO, DEOGRANDEL, 2 HAMBURG 11, POSTFAC 1929

STATIONARY HEAT TRANSPORT
BY PLANE GROUNDWATER MOVEMENT
IN A THIN OR A THICK LAYER

A. VERRUIJT

1. INTRODUCTION

For the application of groundwater for cooling purposes use is often made of a system of sources and sinks operating in a confined aquifer. The water being extracted from the soil at the sink is then put back into the soil at the source. It is of interest to investigate by what mechanisms the groundwater flowing from source to sink looses part of its heat.

In the present paper it will be shown that the problem can be treated in a general way, provided that heat transport by conduction or dispersion in the plane of flow is disregarded. Some attention will also be paid to the acceptability of this assumption. As an example the case of a source and a sink in an infinite aquifer is elaborated.

2. BASIC EQUATIONS

Let there be given a two-dimensional stationary groundwater movement, described by a harmonic potential Φ and a harmonic stream function Ψ. The components of the specific discharge vector are denoted by q_x and q_y They are given by

$$q_x = - \partial\Phi/\partial x = - \partial\Psi/\partial y$$

(1)

$$q_y = - \partial\Phi/\partial y = + \partial\Psi/\partial x$$

Let heat be transported through the soil by the water. By assuming that a steady state has been reached and by making use of Fourier's law of heat conduction the basic differential equation is found to be

(2)
$$q_x\frac{\partial T}{\partial x} + q_y\frac{\partial T}{\partial y} - \kappa\left(\frac{\partial^2 T}{\partial x^2} + \frac{\partial^2 T}{\partial y^2} + \frac{\partial^2 T}{\partial z^2}\right) = 0$$

where

(3)
$$\kappa = \frac{n\lambda_f + (1 - n)\lambda_s}{(n/\varepsilon)\rho_f c_f},$$

and where the following notations have been used,

25

T : temperature
n : porosity
ε : effective porosity
λ_f : thermal conductivity of pore fluid
λ_s : thermal conductivity of dry soil
ρ_f : density of pore fluid
c_f : specific heat of pore fluid

Because of the assumption of stationarity T denotes the temperature of both water and soil particles in a representative elementary volume around a point in space.

In case of groundwater movement in a layer of small thickness it may be justified to consider only variations of the temperature in the plane of flow. By integration of Eq. (2) over the thickness H of the layer, and by assuming that the boundary $z = 0$ acts as a thermal isolation and that on the other boundary, $z = H$, heat loss takes place into a confining layer in which the temperature varies linearly with the height, having the value 0 at the height $z = H + d$ one obtains

$$(4) \qquad q_x \frac{\partial T}{\partial x} + q_y \frac{\partial T}{\partial y} - \kappa \left(\frac{\partial^2 T}{\partial x^2} + \frac{\partial^2 T}{\partial y^2} \right) + \beta T = 0$$

where

$$(5) \qquad \beta = \frac{\lambda_0}{(n/\varepsilon)\rho_f c_f H d}$$

with λ_0 and d denoting the thermal conductivity and the thickness of the confining layer. In Eq. (4) T represents the average temperature in the aquifer with the atmospheric temperature as reference temperature.

The coefficients q_x and q_y in Eqs. (2) and (4) may be complicated functions of the coordinates x and y, even for rather simple flow systems, thus prohibiting an analytic solution. For two extremely simple flow systems, namely the cases of uniform rectilinear flow and radial flow, the problem can be solved analytically. In order to enable a quantitative discussion with regard to the influence of convective and conductive heat transport the solution of Eq. (4) for the case of radial flow will be presented in Section 3.

3. SOURCE IN A THIN LAYER

In the case of injection by a single source in a thin confined layer it is appropriate to use polar coordinates r, θ around the source. The specific discharge vector is now directed radially and has a magnitude $q = Q/(2\pi H r)$, where Q is the total discharge in the source. Because of radial symmetry the temperature T is a function of r only, and Eq. (4) reduces to the following ordinary differential equation

(6)
$$\frac{d^2T}{dr^2} + \frac{1-2v}{r}\frac{dT}{dr} - \mu^2 T = 0$$

where

(7)
$$v = Q/4\pi H\kappa, \quad \mu^2 = \beta/\kappa$$

Eq. (6) has solutions expressible in terms of Bessel functions (see [1]). The solution vanishing at infinity and satisfying the condition $T = T_1$ for $r \to 0$ is

(8)
$$T/T_1 = [2^{1-v}/\Gamma(v)](\mu r)^v K_v(\mu r)$$

That for $r \to 0$ the right hand member of Eq. (8) indeed tends to unity is a consequence of the behaviour of $K_v(\mu r)$ near $\mu r = 0$. In practical cases the parameter v will usually be very large. Then the solution (8) reduces to

(9)
$$T/T_1 \approx \exp[-(\mu r)^2/4v] = \exp[-\pi H\beta r^2/Q]$$

provided that T/T_1 is not very small. In Eq. (9) the diffusivity no longer appears. This means that the effect of heat conduction in the plane of flow can be disregarded when $v \gg 1$, or with (7), when

(10)
$$4\pi H\kappa/Q \ll 1$$

It is easily verified that Eq. (9) is the correct solution of the original differential equation for the limiting case $\kappa \to 0$.

The results obtained in this section can be summarized quantitatively by introducing the half-value length L, which is defined as that distance from the source where the temperature difference with the atmosphere has decreased to one-half of its original value. Mathematically speaking: $T/T_1 = 0.5$ for $r = L$. A graphical representation of L, multiplied by $(\pi H\beta/Q)^{1/2}$ to make it dimensionless, as a function of the parameter $1/v = 4\pi H\kappa/Q$ is given in Fig. 1.

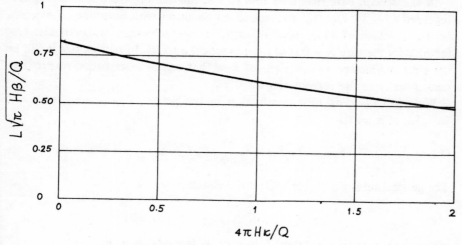

Fig. 1. Half-value length L in the case of radial flow

This figure enables a quantitative evaluation of the influence of heat conduction in the plane of flow. It appears that by disregarding conduction of heat results of sufficient accuracy (within 5% error) are obtained provided that

(11) $4\pi H\kappa/Q < 0.1$

This statement is a more precise version of the requirement (10).

4. THE GENERAL PROBLEM FOR A THIN LAYER

It will appear to be advantageous to rewrite the fundamental differential Eq. (4) with the potential Φ and stream function Ψ as independent variables. This equation then becomes

(12) $(q_x{}^2 + q_y{}^2)\left[\dfrac{\partial T}{\partial \Phi} + \kappa\left(\dfrac{\partial^2 T}{\partial \Phi^2} + \dfrac{\partial^2 T}{\partial \Psi^2}\right)\right] - \beta T = 0$

The introduction of Φ and Ψ as independent variables in problems related to groundwater flow seems to have been proposed first by Muskat [6] for the calculation of residence times. Also in describing the dispersion problem the use of Φ and Ψ as independent variables has been shown to be advantageous [2, 7].

An important simplification obtained in the transition from (4) to (12) is that in the latter equation only one first order partial derivative appears. That $\partial T/\partial \Psi$ does not appear in Eq. (12) corresponds to the physical fact that convective heat transport takes place only in the direction of flow, in which direction Ψ is constant. Another simplification is that the region in which Eq. (12) holds in general is of a simple rectangular form. This circumstance may be of special value for numerical solutions [7].

In the special case that heat loss to the surroundings can entirely be disregarded ($\beta \rightarrow 0$) Eq. (12) reduces to an equation with constant coefficients, for the solution of which several mathematical techniques are available. Unfortunately the case $\beta = 0$ is of little practical interest. In many cases it will be the heat conduction in the plane of flow that can be disregarded rather than heat transfer to the surroundings.

When disregarding heat conduction in the plane of flow, i.e., for $\kappa = 0$ Eq. (12) reduces to

(13) $(q_x{}^2 + q_y{}^2)\dfrac{\partial T}{\partial \Phi} - \beta T = 0$

The general solution of this differential equation is

(14) $T = C(\Psi)\exp\left[\int \{\beta/(q_x{}^2 + q_y{}^2)\}d\Phi\right]$

where the integration constant C in general depends upon Ψ.

The differential Eq. (13) and its solution are analogous to the equations for the calculation of residence times in a two-dimensional flow field (see [6, section 8.5 etc.]). The analogy implies that lines of equal residence times are identical to lines of equal temperature (isotherms). This conclusion has been arrived at earlier, for the more complicated case of non-stationary heat transport in a layer with heat loss to the surroundings by Dietz and Lehner [4], who presented a generalization of the linear flow case studied by Lauwerier [5].

5. SOURCE AND SINK IN A THIN LAYER

In the case of a source at $x = -a$, $y = 0$ and a sink at $x = a$, $y = 0$ (Fig. 2), the flow can be described by the complex potential,

$$(15) \qquad \Omega = \Phi + i\Psi = [Q/2\pi H] \log[(z - a)/(z + a)].$$

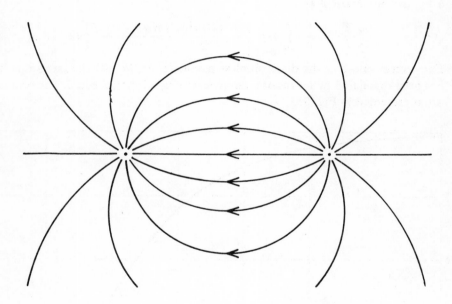

Fig. 2. Source and sink

The region in the Ω-plane corresponding to the entire z-plane is in this case the infinite strip,

$$-\infty < \Phi < +\infty, \quad 0 \leqq \Psi \leqq Q/H$$

The points $z = \pm a$ correspond to $\Phi = \mp \infty$. With the aid of (15) the general solution (14) can now be elaborated. The function C can be determined by requiring that the temperature at $z = -a$ (the sink) be equal to T_1. The temperature in the source $z = +a$ is then found to be

(16) $$T_2/T_1 = \exp\left[-2b\frac{\sin\psi + (\pi - \psi)\cos\psi}{\sin^3\psi}\right]$$

where

$$b = 2\pi Ha^2\beta/Q,$$

$$\psi = 2\pi H\Psi/Q, \qquad 0 \leq \psi \leq 2\pi$$

It appears that the temperature T_2 of the water arriving at the sink depends upon the value of ψ, that is: upon the stream line along which the water particles have travelled. The average temperature $\bar{T}_2$ is found by evaluating the total amount of heat flowing into the sink and dividing this by the total discharge Q. This gives, after some elementary calculations, and after replacing b by its value $2\pi Ha^2\beta/Q$

(17) $$\frac{\bar{T}_2}{T_1} = \frac{1}{\pi}\int_0^\pi \exp\left[-\frac{4\pi Ha^2\beta}{Q}\frac{\sin\psi - \psi\cos\psi}{\sin^3\psi}\right]d\psi.$$

For several values of the dimensionless parameter $4\pi Ha^2\beta/Q$ the integral (17) has been calculated by means of a numerical integration procedure. The results are represented in Fig. 3.

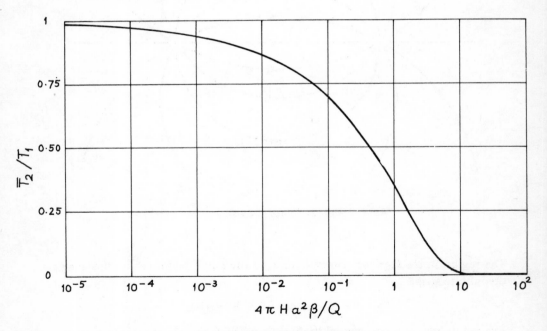

Fig. 3. Source and sink in thin layer.
Temperature in the sink

When the value of $2a$, the mutual distance of source and sink, for which $\bar{T}_2/\bar{T}_1 = 0.5$ is called the half-value length L, then one obtains, with (5),

(18) $$L = 0.183(Q/H\beta)^{1/2} = 0.183[Q(n/\varepsilon)\rho_f c_f d/\lambda_0]^{1/2}$$

One of the conclusions that can be drawn immediately from (17) or (18) is that when a source–sink system is used for cooling purposes, then by taking the distance between source and sink twice as large, the total discharge can be made four times as large without altering the temperature ratio $\bar{T}_2/\bar{T}_1$.

6. SOURCE AND SINK IN A THICK LAYER

In this section the case of a relatively thick layer, in which it is not justified to assume that the temperature distribution is independent of the vertical co-ordinate z, will be investigated. When the terms referring to heat conduction in the x, y-plane in Eq. 2 are disregarded this equation reduces to

(19) $$q_x \frac{\partial T}{\partial x} + q_y \frac{\partial T}{\partial y} - \kappa \frac{\partial^2 T}{\partial z^2} = 0$$

or when the variables x and y are replaced by Φ and Ψ

(20) $$(q_x^2 + q_y^2) \frac{\partial T}{\partial \Phi} + \kappa \frac{\partial^2 T}{\partial z^2} = 0$$

Solutions of Eq. (20) will be sought in the form

(21) $$T(\Phi, \Psi, z) = F(\Phi, \Psi)G(z)$$

and restriction will be made to solutions satisfying the conditions

(22) $$z = 0 \ : \ \partial T/\partial z = 0$$

(23) $$z = H \ : \ \partial T/\partial z = -(H\beta/\kappa)T$$

which express that the surface $z = 0$ is an isolated boundary and that along the surface $z = H$ heat is lost to the surroundings through a thermal resistance of thickness d and conductivity λ_0. In Eq. (23) the parameters β and κ have the same meaning as before, see Eqs. (3) and (5).

One now obtains, by separation of variables, that $G(z)$ must be of the form

(24) $$G(z) = A_j \cos(\gamma_j z/H)$$

where the numbers γ_j ($j = 1, 2, 3, \cdots$) denote solutions of the transcendental equation

(25) $$\cot \gamma_j = \gamma_j(\kappa/\beta H^2)$$

The numbers γ_j have been tabulated (see e.g. [1, p. 225]), and they can thus be considered as known.

On the other hand the function $F(\Phi, \Psi)$ must satisfy the differential equation

(26)
$$\frac{\partial F}{\partial \Phi} = \frac{\gamma_j^2 \kappa / H^2}{q_x^2 + q_y^2} F$$

This differential equation is similar to Eq. (13) and its solution is, analogous to (14),

(27)
$$F(\Phi, \Psi) = C(\Psi) \exp\left[\frac{\kappa \gamma_j^2}{H^2} \int \frac{d\Phi}{q_x^2 + q_y^2}\right]$$

Hence the temperature distribution for the class of problems considered here is, in general

(28)
$$T = \sum_{j=1}^{\infty} A_j(\Psi) \cos\left(\frac{\gamma_j z}{H}\right) \exp\left[\frac{\kappa \gamma_j^2}{H^2} \int_p^{\Phi} \frac{d\Phi}{q_x^2 + q_y^2}\right]$$

where p denotes some arbitrary lower limit of integration and the coefficients $A_j(\Psi)$ are as yet undtermined.

Now further restricting the consideration to problems for which there exists a single point of injection where water of temperature T_1 is injected, it is required that for $\Phi = p$ (where p now is chosen as the value of the potential in the point of injection) and for all values of Ψ and z, the temperature must be equal to the given constant T_1. Then the coefficients A_j must be independent of Ψ and they are to be determined from

$$T_1 = \sum_{j=1}^{\infty} A_j \cos(\gamma_j z / H)$$

which must hold for all z in the interval $0 \le z \le H$. Making use of the orthogonality of the functions $\cos(\gamma_j z / H)$ in this interval and using the characteristic Eq. (25), one obtains

(29)
$$A_j = \frac{2 T_1 \sin \gamma_j}{\gamma_j [1 + (\kappa / \beta H^2) \sin^2 \gamma_j]}$$

and thus the solution becomes

(30)
$$\frac{T}{T_1} = \sum_{j=1}^{\infty} \frac{2 \sin \gamma_j \cos(\gamma_j z / H)}{\gamma_j [1 + (\kappa / \beta H^2) \sin^2 \gamma_j]} \exp\left[\frac{\kappa \gamma_j^2}{H^2} \int_p^{\Phi} \frac{d\Phi}{q_x^2 + q_y^2}\right]$$

It is to be noted that this expression is valid for any flow system satisfying the requirement that the equipotential line $\Phi = p$ is an isotherm, $T = T_1$.

In the case of a source at $x = -a$, $y = 0$ and a sink at $x = a$, $y = 0$ the complex potential is given by Eq. (15). The flow field is then completely defined and the formula (30) can be elaborated. The average temperature $\bar{T}_2$ (averaged over the height of the sink and over the entrance angle) of the water extracted from the soil in the sink is found to be

$$\frac{\bar{T}_2}{\bar{T}_1} = \sum_{j=1}^{\infty} \frac{2\sin^2 \gamma_j}{\pi \gamma_j^2 [1 + (\kappa/\beta H^2)\sin^2 \gamma_j]}$$

(31)
$$\int_0^\pi \exp\left[-\frac{4\pi H a^2 \beta}{Q} \frac{\kappa \gamma_j^2}{\beta H^2} \frac{\sin\psi - \psi\cos\psi}{\sin^3\psi} \right] d\psi$$

This expression contains two dimensionless parameters, namely $\kappa/\beta H^2$ and $4\pi H a^2 \beta/Q$. With (3) and (5) these parameters can be rewritten as

(32)
$$\frac{\kappa}{\beta H^2} = \frac{\lambda d}{\lambda_o H}$$

(33)
$$\frac{4\pi H a^2 \beta}{Q} = \frac{4\pi a^2 \lambda_o}{(n/\varepsilon)\rho_f c_f Q d}$$

where

(34)
$$\lambda = n\lambda_f + (1 - n)\lambda_p$$

the thermal conductivity of the saturated aquifer. For several values of the first parameter, $\lambda d/\lambda_o H$, the results of numerical elaborations of Eq. (31) are represented in Fig. 4. Since λ and λ_0 will be of the same order of magnitude the

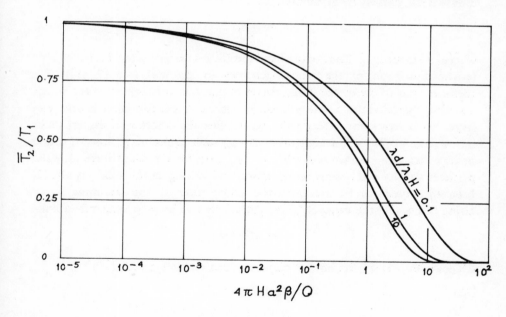

Fig. 4. Source and sink in thick layer
Temperature in the sink

first parameter is almost uniquely determined by the ratio of the thickness d of the confining layer to the thickness H of the aquifer. Since the curve in

Fig. 4 for $\lambda d/\lambda_0 H = 10$ almost completely coincides with the curve in Fig. 3 (which represents the limiting case for $\lambda d/\lambda_0 H \to \infty$) it can be concluded that an aquifer of thickness H, smaller than about one-tenth of d (the thickness of the overlying strata) can be considered as a "thin" layer. For thicker aquifers the assumption of constant temperature over the height may result in a considerable over-estimation of the cooling effect.

7. DISCUSSION

In Sections 4–6 heat conduction in the plane of flow has been disregarded. In Section 3 this effect was not disregarded however, and it can therefore be expected (see Eq. (10)) that the results of the later sections are sufficiently accurate provided that

(35)
$$4\pi H\kappa/Q < 0.1$$

Another effect, altogether disregarded in this paper, is dispersion. The influence of dispersion can be accounted for, in the first instance, by an increase of the diffusivity κ, namely by an amount κ',

(36)
$$\kappa' = \alpha |q| l$$

where $|q|$ is the magnitude of the local specific discharge vector, l is the characteristic dimension of the inhomogeneities in the soil (see [2, 3]). The coefficient α is of the order of magnitude of unity. In the vicinity of the source and sink the velocities are very large, hence the dispersion effect is also very large. Yet it seems not unreasonable to estimate the influence of dispersion by introducing in (36) not a high value of $|q|$ such as occuring close to source or sink, but rather an average value of $|q|$. For, close to the source all water particles will have the same temperature, and mixing in the sink has already been accounted for by averaging the temperature of the incoming water. Using in Eq. (36) the value of $|q|$ in the origin, i.e., $|q| = Q/\pi Ha$, one obtains

$$\kappa' = \alpha Ql/\pi Ha$$

Hence with (35) it is found that dispersion can be disregarded if

(37)
$$4\alpha l/a < 0.1.$$

which expresses that the dimension of the inhomogeneities in the soil must be small compared to the mutual distance of source and sink. The condition (37) will certainly be satisfied when the only type of inhomogeneities in the soil is due to locally unequal pore channels. Then l is approximately equal to the grain size.

REFERENCES

1. ABRAMOWITZ, M. AND I. A. STEGUN (1965), *Handbook of Mathematical Functions*, pp. 1046, Dover, New York.

2. BACHMAT, Y. AND J. BEAR (1964), *J. Geophys. Res.*, *69*, 2561–2567.

3. DE JONG, G. DE JOSSELIN (1958), *Trans. Amer. Geophys. Union*, *39*, 67–74.

4. DIETZ, D. N. AND F. LEHNER (1969), *Appl. Sci. Res.*, *20*, 309–311.

5. LAUWERIER, H. A. (1956), *Appl. Sci. Res.*, *A5*, 145–150.

6. MUSKAT M. (1937), *The Flow of Homogeneous Fluids Through Porous Media*, 763 pp., McGraw-Hill, New York.

7. SHAMIR, U. Y. AND D. R. F. HARLEMAN (1967), *Water Resources Res.*, *3*, 557–581.

TECHNOLOGICAL UNIVERSITY,
    DEPARTMENT OF CIVIL ENGINEERING,
        DELFT, NETHERLANDS

THE SIGNIFICANCE OF THE NET TRANSFER OF VISCOUS STRESS ENERGY AND THE LOCAL PRODUCTION OF KINETIC ENERGY IN STATIONARY SOIL WATER FLOW

P. H. GROENEVELT

ABSTRACT

For stationary saturated flow of water through a homogeneous porous medium non-equilibrium thermodynamics provides a scheme in which measurable fluxes and forces are connected to each other by linear homogeneous relationships. This set of relationships infers the existence of coupling phenomena and constitutes the main virtue of this branche of science, viz. the then proposed equality of "twin" cross-coefficients. Especially for systems with intricate geometry this way of treating cross-phenomena seems to be promising.

To come to a correct and useful set of fluxes and forces one may treat a small subsystem within the liquid phase and then integrate over a certain volume of the porous system. This local treatment brings up the two terms of concern, $-\nabla \cdot \tau \cdot v$ and $-d 1/2 \rho v^2/dt$, in the dissipation function. For stationary flow in a medium with a homogeneous phase distribution these two terms can be shown to disappear upon integration, according to statistical reasonings as used by Mandl [3]. This then leads straightforward to the proposal of a linear relationship between the filter flux and the gradient of the hydraulic head. The existence and the magnitude of the two terms of concern in the dissipation function before integration predicts the lack of a linear relationship between the lScal flow velocity and the local gradient of the hydraulic head.

For non-homogeneous, non-saturated or non-stationary flow, the two terms will not disappear upon integration. A first estimate of the order of magnitude of the two terms relative to the other terms in the dissipation function is made for isothermal stationary saturated flow in a non-homogeneous porous medium by calculations based on the flow pattern between non-parallel plane walls. These calculations show that for systems with a change in thickness of water layers from 100 μ to 50 μ over a distance of 10 cm the ratio of the integrated value of $-\nabla \cdot \tau \cdot v$ to the integrated value of $-v \cdot \nabla p$ (which is the main term of the dissipation function) is only 10^{-7}. The ratio of the integrated value of $-d 1/2 \rho v^2/dt$ to the main term is of the same order of magnitude, assuming a filter flux of 10^{-3} cm/sec to be maintained at a moisture content of 0.50 at the point where the water layer thickness is 100 μ.

INTRODUCTION

The Navier-Stokes equation does not lead to a unique linear relationship between flow velocity and pressure gradient at every point in the system The parabolic velocity profile for flow in a straight tube, which is an exact solution of the Navier-Stokes equation, shows this linear relationship to be dependent on the distance of the point under consideration from the axis of the tube. Upon integration however the linear relationship between the average velocity and the pressure gradient is only dependent on system parameters.

In the teory of irreversible thermodynamics the local dissipation function for Poiseuille flow

(1) $$T\sigma = -\tau : \nabla v = -v \cdot \nabla p - \nabla \cdot \tau \cdot v - v \cdot \rho dv/dt,$$

in which T, σ, τ, v, p, ρ and t denote absolute temperature, entropy production, viscous stress tensor, barycentric velocity, pressure, density and time, respectively, does not suggest a unique linear relationship between flow velocity and pressure gradient locally. The local dissipation is not described by the term $-v \cdot \nabla p$, except at the point $r = (1/2) R\sqrt{2}$. At the central axis ($r = 0$) $- v \cdot \nabla p = R^2(\nabla p)^2/4\mu$, whereas $T\sigma = 0$, and at the wall ($r = R$) $- v \cdot \nabla p = 0$, whereas $T\sigma = R^2(\nabla p)^2/4\mu$. Upon integration however, the linear relationship between the average velocity and the pressure gradient shows up because the last two terms of Eq. (1) vanish (the last already being zero before integration).

Thus for systems with zero net transfer of viscous stress energy to the environment and zero net outflow of kinetic energy, the average velocity will be linearly dependent on the difference in pressure between the entrance and exit side of the chosen system. Such a system is a representative volume element of a porous medium with homogeneous phase distribution in which stationary flow is maintained [cf. 2]. The indicated relationship for this system is well-known as the Darcy law. The purpose of this paper is to estimate the relative magnitude of the "bothersome" terms for a simple system in which they are essentially non-zero. This system will be a porous medium with gradually changing porosity in the direction in which saturated flow is maintained. Calculations will be based upon considerations concerning steady state flow between two non-parallel plane walls.

FORMULATION OF THE HYDRODYNAMICAL PROBLEM

Steady ($\delta/\delta t = 0$) flow of an incompressible (ρ = constant) fluid between non-parallel plane walls is considered to be purely radial ($v_\theta = 0$) [cf. 4].

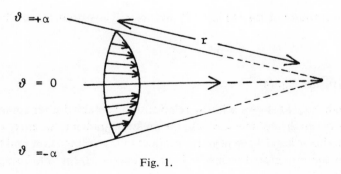

Fig. 1.

In cylindrical co-ordinates (r, θ) the equations for the velocity component v_r are the conservation equation

(2)
$$\frac{\delta}{\delta r}(rv_r) = 0,$$

or

(3)
$$\frac{\delta}{\delta r}v_r = -v_r/r,$$

which makes

(4)
$$v_r = \frac{1}{r} \cdot f(\theta),$$

and the equations of motion

(5) $\qquad \rho v_r \dfrac{\delta v_r}{\delta r} = -\dfrac{\delta p}{\delta r} + \dfrac{\mu}{r^2} \dfrac{\delta^2 v_r}{\delta \theta^2},$ (6) and $0 = -\dfrac{1}{r} \dfrac{\delta p}{\delta \theta} + \dfrac{2\mu}{r^2} \dfrac{\delta v_r}{\delta \theta},$

where μ denotes the viscosity and use have been made of Eq. (2). Solutions of (5) and (6) with the appropriate boundary conditions ($v_r = 0$ at $\theta = \pm \alpha$) shows that for flow with low Reynolds number in channels of small convergence the inertia term of Eq. (5) may be neglected. Using $\delta p/\delta r \neq f(\theta)$ the velocity distribution is then found as

(7)
$$v_r = \frac{r^2}{2\mu} \frac{\delta p}{\delta r}(\theta^2 - \alpha^2),$$

which is the parabolic velocity distribution for plane Poiseuille flow.

The volume of water flowing through the slit per unit time per unit length in the z-direction may be introduced as

$$Q(\text{cm}^2/\text{sec}) = \int_{\theta=-\alpha}^{+\alpha} v_r r d\theta.$$

Using eq. (7)

(8)
$$Q = -\frac{2\alpha^3 r^3}{3\mu} \frac{\delta p}{\delta r}.$$

After substitution of Eq. (8) into (7), the velocity distribution may be written as

(9)
$$v_r = \frac{3Q}{4r\alpha^3}(\alpha^2 - \theta^2),$$

which satisfies Eq. (4).

Though Eq. (8) shows a linear relationship (obtained after some simplifications) between the volume flow and the pressure gradient, the energy dissipation function shows besides the negative product of the volume flow and the pressure gradient the intergrated terms $-\nabla \cdot \tau \cdot v$ and $-d1/2\rho v^2/dt$. Using the above obtained (approximate) flow velocity distribution it is possible to calculate both terms and to consider their relative importance.

CALCULATION OF THE TERMS OF CONCERN

The different terms of the dissipation function, Eq. (1), will be calculated for a volume element (V), stretched out from $\theta = -\alpha$ to $\theta = +\alpha$, from $r = r$ to $r = r + \Delta r$ with $\Delta r \to 0$ and with unit length in the z-direction.

According to the divergence theorem

$$-\frac{1}{V}\int_V \nabla \cdot \tau \cdot v dV = -\frac{1}{V}\left[\int\int_{S_w}[\tau \cdot v]_\theta dS_w - \int_{S_r}[\tau \cdot v]_r dS_r\right.$$

$$\left.+\int_{S_{r+\Delta r}}[\tau \cdot v]_r dS_{r+\Delta_r}\right],$$

in which $\int_{S_w} = 0$ because $[\tau \cdot v]_\theta = 0$ at the wall (w).

In cylindrical co-ordinates one finds [cf. 1]

$$[\tau \cdot v]_r = -2\mu v_r \delta v_r/\delta r.$$

Using Eqs. (3) and (9) one finds

(10)
$$\int_{S_r}[\tau \cdot v]_r rd\theta dz = \frac{6\mu Q^2}{5r^2\alpha}.$$

Applying Eq. (10) for $r = r + \Delta r$, together with $V = \alpha\Delta r(2r + \Delta r)$ and $\Delta r \to 0$ one gets

(11)
$$-\frac{1}{V}\int_V \nabla \cdot \tau \cdot v dV = +\frac{6\mu Q^2}{5r^4\alpha^2}.$$

The barycentric time derivative of the kinetic energy for stationary flow may be written as $v \cdot \rho dv/dt = \rho v^2{}_r \delta v_r/\delta r$.

Using Eqs. (3) and (9) and integrating over the same volume as before it is found that

(12)
$$-\frac{1}{V}\int_V v \cdot \rho dv/dt dV = +\frac{27\rho Q^3}{140r^4\alpha^3}.$$

The main term in the dissipation function, $-v \cdot \nabla p$, using Eqs. (8) and (9) may be calculated as

(13)
$$-\frac{1}{V}\int_V v_r \frac{\delta p}{\delta r} dV = +\frac{3\mu Q^2}{4r^4\alpha^4},$$

which is the same as the negative product of the volume flow per cm^2 $(j^V \equiv Q/2\alpha r)$ and the r-component of the pressure gradient.

Obviously, by dropping the terms $-\nabla \cdot \tau \cdot v$ and $-d1/2\rho v^2/dt$, the errors relative to the main dissipation term are respectively $8\alpha^2/5$ and $12\alpha^2$ Re/35 in which the Reynolds number has been introduced as

$$\text{Re} \equiv \frac{\rho r}{\mu} v_r(\theta = 0) = \frac{3\rho Q}{4\mu\alpha} \quad \text{[cf. 4]}$$

Writing out all terms of the dissipation function, one finds after some substitutions

(14) $$\frac{1}{V} \int_V T\sigma dV = -j^V \left[\frac{\delta p}{\delta r} + \frac{8}{5}\alpha^2 \frac{\delta p}{\delta r} - \frac{6\rho\alpha^4 r^3}{35\mu^2} \left(\frac{\delta p}{\delta r} \right)^2 \right]$$

Considering the *RHS* of Eq. (14) as the product of a flux and a force, one may propose a linear relationship between the two with the coefficient $L = -\alpha^2 r^2/3\mu$ in between.

This leads to the proposal

(15) $$Q = -\frac{2\alpha^3 r^3}{3\mu} \frac{\delta p}{\delta r} - \frac{16\alpha^5 r^3}{15\mu} \frac{\delta p}{\delta r} + \frac{4\rho\alpha^7 r^6}{35\mu^3} \left(\frac{\delta p}{\delta r} \right)^2.$$

Thus irreversible thermodynamic considerations suggest a correction of the linear relationship found by approximation. The non-linear term of Eq. (15) is easily traced back as coming from the inertia phenomena. Thus by substituting Eq. (7) into the inertia term of Eq. (5) and then solving this equation, one finds a non-linear velocity distribution, which gives upon integration an expression for Q with the same non-linear term as which was found in Eq. (15).

SIGNIFICANCE OF THE TERMS $-\nabla \cdot \tau \cdot v$ AND $-d1/2\rho v^2/dt$ FOR WATER FLOW IN A POROUS MEDIUM

To evaluate the calculated expressions for the two terms of concern, saturated stationary flow in a porous medium with non-homogeneous phase distribution will be considered. Choosing the water layer thickness to change from 100 μ to 50 μ over a distance of 10 cm, which makes α to be 2.5×10^{-4} rad., the ratio of the integrated value of $-\nabla \cdot \tau \cdot v$ to the integrated value of $-v \cdot \nabla p$ will be 10^{-7}, which is completely negligible. For a filterflux of 10^{-3} cm/sec to be maintained at a moisture content of 0.50 at the point where the water layer thickness is 100 μ, which makes Q to be 2.10^{-5} cm²/sec per layer, the ratio of the intergrated value of $-d1/2\rho v^2/dt$ to the integrated value of $-v \cdot \nabla p$ will be 1.3×10^{-7}, which also is completely negligible. The conclusion may be that in practice the dissipation of energy in the proces of stationary saturated viscous flow in a porous medium with non-homogeneous phase distribution is sufficiently described by the negative product of the volume flow and the gradient of the pressure.

REFERENCES

1. BIRD, R. B., W. E. STEWART AND E. N. LIGHTFOOT (1960), *Transport Phenomena*, John Wiley, New York.

2. GROENEVELT, P. H. AND G. H. BOLT, Non-equilibrium thermodynamics of the soil-water system, *J. of Hydrology* 7 (1969) 358.

3. MANDL, G. (1964), Change in skeletal volume of a fluid-filled porous body under stress, *J. Mech. Phys. Solids*, 299.

4. MILLSAPS, K. AND K. POHLHAUSEN (1953), Thermal distributions in Jeffery-Hamel flows between nonparallel plane walls, *J. of Aeronautical Sciences*, 187.

LABORATORY OF SOILS AND FERTILIZERS,
    WAGENINGEN, NETHERLANDS

ON THE CORRELATION OF ELECTRICAL CONDUCTIVITY PROPERTIES OF POROUS SYSTEMS WITH VISCOUS FLOW TRANSPORT COEFFICIENTS

HANS — OLAF PFANNKUCH

ABSTRACT

This study emphasizes the role that surface conductivity plays in electrical transport processes in porous media. It shows that this term has to be included in the calculations where phase distributions or porosities are to be deducted from formation factor type measurements even in clean formations when electrolyte concentrations are low. In applying a parallel resistance model of conduction processes through porous media, it can be demonstrated that the pertinent parameters are a conductivity ratio function $k_s/k_f = f(C)$ and the extent of the internal specific pore surface area S_p. The conductivity ratio function will also depend on the nature of the internal surface area which determines the mechanism of charge fixation at the solid-solution interface. Based on the assumption that electrical and viscous flow paths in the porous medium are identical, the electrical transport coefficients can be connected to viscous flow parameters by way of introducing the Kozeny-Carman hydraulic radius concept of porous interstices and linking the internal surface area term to the permeability of the medium.

THEORY OF CONDUCTION IN HETEROGENEOUS SYSTEMS

Conduction of electricity through an electrolyte filled porous system is a case of transport through a heterogeneous system. Bulk properties of heterogeneous systems can be predicted when the properties, the relative distribution, and the governing interrelationships between the pure phases are known. By inverse treatment individual properties and phase distributions in heterogeneous media can be deducted from the bulk measurement. This is of importance in the geophysical context where mixed phase measurements of resistivities and conductivities are relatively easy to perform on otherwise inaccessible samples.

The general form of the relationship is the following:

$$(1) \qquad\qquad K_o = f\,[(1 - w), k_d, k_f]$$

where

K_o = observed conductance of the heterogeneous medium
w = volume fraction of the dispersed phase
k_d = conductivity of dispersed phase (often solid phase)
k_f = conductivity of continuous phase (often fluid phase)

Rigorous treatment of the interrelationships have been carried out for flow of electricity in dilute systems where the dispersed phase is very regularly distributed and where it consists of symmetrically shaped particles. Maxwell's [9] classical problem involves the flow of electricity through a two-phase region of uniform spheres imbedded in a continuum. If a formation factor F is defined as the inverse of K_o, $(F = 1/K_o)$, the results for a suspension of nonconducting spheres $(k_d = 0)$ in a conducting medium of conductivity k_f is as follows:

$$(2) \qquad F = \frac{3 - \phi}{2\phi}$$

where in this special case $F = k_f/k_o$, or r_o/r_f if expressed in terms of the resistivities, and k_o is the conductivity of the bulk phase, and ϕ is the porosity or volume fraction of the continuous phase. In a different derivation, again for spheres, Wagner [21] shows that for very dilute solutions Eq. (2) can be represented by the first term of a series expansion to give a linear form. Dealing with nonconducting spheres one gets:

$$(3) \qquad F = \frac{2}{3\phi - 1}$$

The two methods are based on a solution of Laplace's equation for steady state flow and on simple Ohm-Fourier type laws for the flux-potential relationships. The effective or observed conductance of the mixture is derived by solving Laplace's equation for each phase. The simplifying assumptions hold for very dilute suspensions where no distortion or discontinuities of the field occur.

It should be noted that in both cases the factor F is independent of the size of the spheres and the conductivities, as long as k_d is zero. Rayleigh and later Fricke [7] extended the treatment to dispersed particles of regular but nonspherical shape. Surface conductance is explicitly neglected in all these studies.

Meredith and Tobias [12] have treated the case of composite spheres with an inner core of radius R_1 and conductivity k_d, and an outer shell with R_2 and k_r. For a very thin conductive surface layer (i.e. ratio $R_2/R_1 = 1$) and nonconductive cores $(k_d = 0)$ one obtains:

$$(4) \qquad F = \frac{8 - 3\phi}{6\phi - 1}$$

where again F is only a function of ϕ, since all conductivities are independent of each other.

As pointed out earlier these rigorous treatments can only be applied to relatively dilute dispersions. Long before the solid phase becomes contiguous as in a packed bed, the simplifying assumptions cannot be applied any more and field distortion due to the presence of the particles renders any theoretical treatment extremely difficult.

EMPIRICAL RELATIONS AND THE "CONDUCTIVE SOLID EFFECT"

At this stage empirical or semi-empirical relationships are introduced. Slawinski [16] presented a treatment of the bulk conductivity of an electrolyte containing non-conducting spheres by using a tortuosity concept and neglecting any possible interaction of fields. His tortuosity factor T is an indication to what degree the actual flow path L_f of the quantity under consideration (mass, electricity, etc.) differs from the overall bulk length measure L of the sample. It is defined as:

$$(5) \qquad T = \left(\frac{L_f}{L}\right)^2$$

The formation factor concept involving only the geometry of the flow paths leads to the formulation of an intrinsic or pure formation factor F_i:

$$(6) \qquad F_i = \frac{T}{\phi}$$

His treatment is based on geometric reasoning for regular spherical packings where the volume of the unit cell and hence its overall length depend on the mode of packing only. In this case there exists for each porosity a corresponding tortuosity so that T can be replaced by ϕ in the following way:

$$(7) \qquad F_i = \frac{(1.3219 - 0.3219\phi)^2}{\phi}$$

This expression is valid as long as the spheres are in contact and has been verified experimentally in the range between 0.29 and 0.52.

From a purely empirical point of view Archie [2] has proposed the following relationship both for unconsolidated and consolidated porous media:

$$(8) \qquad F = \frac{k_f}{k_o} = \phi^{-m}$$

Here m is a cementation factor describing the degree of consolidation and the textural characteristics of a rock sample. It ranges from 1.3 for unconsolidated packings of spheres to about 2 for consolidated rocks. Relationships of this kind are of importance in electric logging where porosities of subsurface formations can be evaluated by remote resistivity measurements.

Archie's and Slawinski's expressions compare reasonably well for unconsolidated spheres in the range of porosities between 0.25 to 0.40. In the case of consolidated media Slawinski's relation has no theoretical basis, and Archie's cementation factor is no longer independent of ϕ, becoming a function of other non-specified factors. More refined expressions have been proposed [Winsauer et al. (25)], but they only introduce additional empirical factors.

The above mentioned approximations seem to break down whenever clays or "conductive solids" are present in the sample. These questions have been treated by Wyllie and co-workers [26], [27], De Witte [6], Hill and Milburn [8], and recently by Waxman [23]. Consideration of conductive solids content leads to the introduction of an apparent formation factor F_a, and Eq. (8) takes the form:

$$(9) \qquad \frac{1}{r_0} = \frac{1}{F_a} \left(b + \frac{c}{r_f} \right)$$

where b and c are constants depending on the shale content and texture of the rock. Wyllie and Southwick [28] propose a more complicated model for conduction in porous media in form of three parallel conductances. There is conduction through the electrolyte, through the contiguous solid matrix and a contribution from the effect of particles suspended in the electrolyte within the pore space as well as that of conducting particles dispersed in the matrix. Still the overall effects of dispersion are lumped into dimensionless geometrical parameters that describe the arrangement and distribution of the interstitial solutions. The nature of the conductivity of the solids and its possible dependence on the electrolyte concentration of the saturating liquid has not been clarified explicitly.

For shaly reservoir rock Winsauer and McCardell [24] have proposed an ionic double layer conductivity as a possible contributing mechanism to the overall bulk conduction. It enters their formulation in terms of an excess double layer conductivity $1/Z$.

One has:

$$(10) \qquad \frac{1}{r_0} = \frac{1}{F} \left(\frac{1}{r_f} + \frac{1}{Z} \right)$$

Eq. (10) does not show the interrelations between the extent of internal surface area S and the formation factor or the importance of bulk solution concentration on the excess conductivity term in an explicit way.

Recent publications seem to indicate that abnormal values for F are obtained in field measures even in clean sands with no clay content where the saturating fluid is of low conductivity [clean water aquifers, Alger [1]]. This is of importance in water well logging.

The purpose of the following considerations is to study the effects of conductance in clean, non-shaly granular media, to point out the parameters that effect surface conductance, and finally to use this information in a functional model approach that permits an explicit incorporation of surface conductance phenomena into the formation factor concept.

ELECTRICAL TRANSPORT PROCESSES IN POROUS MEDIA

Transport of electricity through a heterogeneous system can be achieved through electronic or ionic conduction. In the special case of a porous medium saturated with an electrolyte, conduction will be ionic through the liquid and electronic through the solid matrix if it is conductive at all. Surface conduction is a special ionic form of electricity transport occuring at the interface between the solid and liquid electrolyte phase. The most general model of electric transport phenomena in a porous medium is that of resistances in parallel, with electronic conduction through the contiguous matrix (K_d), ionic conduction through the saturating bulk fluid (K_f), through the surface or electrical double layer (K_s), at the surface by exchange mechanisms (K_{ex}), if applicable, with the added Maxwellian effects of either solids of different conductivity dispersed within the matrix material or suspended in the pore fluid (McEuen et al. [10] McKelvey et al. [11]). The combination of all gives a combined bulk or observable conductivity of the mixture K_o. The last two mechanisms and the possibility of a base exchange mechanism will not be treated further in this study. The contribution and nature of the surface conduction to the formation factor concept seem to be not well understood. The following is an attempt to clarify its role.

THE ELECTRICAL DOUBLE LAYER AT THE SOLID-LIQUID INTERFACE

The seat of surface conduction phenomena is the electrical double layer that forms at the solid liquid interface. It is characterized by a higher ion concentration than the bulk phase. Hence, it will also show different ionic conductance properties since these are a function of the ion concentration among other things.

The double layer is due to the existence of surface charges at the solid liquid interface. As a result of electrostatic attraction these are compensated by migration of counterions from the bulk solution to the interface. This results in the accumulation of an ionic atmosphere near the surface. The distribution of counterions and charges can be described in different ways by the classical theories of Helmholtz, Smoluchowski, Gouy, Chapman or combinations thereof as the one carried out by Stern.

The two main mechanism of charge formation and counterion attraction at the surface are those directly linked to lattice effects of the solid or those connected with the existence of active adsorption sites on the surface. Lattice effects are independent of the concentration of ions in the bulk solution. Therefore, the specific surface charge is independent of the concentration, provided that the minimum number of counterions is present to satisfy the surface charges. In this case the thermodynamic or surface potential ψ_o changes inversely with the bulk solution concentration. The other principal mechanism of surface charge acquisition is by way of occupying active adsorption sites on the solid surface. Here the valencies of the lattice atoms are not completely compensated and give rise to chemisorption or physical adsorption. This mechanism is called potential determining since the surface potential remains constant whereas the specific surface charge is a direct function of the bulk solution concentration. Glass or silica surfaces are affected by this type of mechanism; clay surfaces, however, belong to the first group (Van Olphen [20]). In both cases the counterion atmosphere, and with it the electrical double layer, is compressed when the bulk solution concentration rises. These are important points since surface conductance is intimately linked to the structure and the properties of the electrical double layer.

SURFACE CONDUCTION DUE TO THE ELECTRICAL DOUBLE LAYER

Surface conduction is due to the displacement of ions along the solid-liquid interface in an electrolyte filled system under the influence of an electric field. The particular and different values for surface conductivity are attributed to the fact that ion concentration at the interface is different from that in the bulk solution. The theoretical expression for the surface conductance term has been derived by Bikerman [3], [4] [also by Urban et al. (19)]. It is based on the concentration and mobilities of the excessions in the double layer which are distributed in a diffuse fashion according to the Gouy-Chapman model. Bikerman's expression is written to show the contribution of an excess ion term (E) and of an electro-osmotic term (O). The first one describes the contribution of excess ion concentration near the interface to the surface conductance effect. It depends mainly on the mobility of the ions and on the zeta potential. The second term shows the influence of electro osmotic phenomena based on the Smoluchowski concept of electrokinetics.

For the surface conduction or specific surface conductivity k_s (Ω^{-1}) the equation has the following form:

$$(11) \qquad k_s = \sqrt{\frac{2\,DRT}{\pi}}\sqrt{C}\left[\underbrace{\frac{v_c}{A-1} + \frac{v_a}{A+1}}_{E} + \underbrace{\frac{DRT}{\pi\mu v F}\frac{1}{A^2-1}}_{O}\right]$$

where:

$$A = \frac{\exp\left(\dfrac{vF\zeta}{2\,RT}\right)+1}{\exp\left(\dfrac{vF\zeta}{2\,RT}\right)-1}$$

$$E = \tfrac{1}{2}v_c[(\exp\,(vF\zeta/2\,RT)-1) + v_a\,(\exp\,(vF\zeta/RT)-21))]$$

$$O = \frac{1}{2}\frac{DRT}{\pi\mu F}\left(\cos h.\left[\exp\left(\frac{vF\zeta}{2\,RT}\right)\right]-1\right)$$

It is D = dielectric constant of the solvent

R = gas constant

T = absolute temperature

C = bulk solution concentration (mole/cm^3)

μ = viscosity

v = valency of excess ions

ζ = zeta potential

F = Faraday constant

v_a, v_c = mobility of anions, cations, respectively

Other phenomena may influence surface conductivity. Bikerman [3] shows that the added viscous flow resistance of ions moving near a solid wall may give a negative contribution. It can, however, be neglected when C is low and k_s is the predominant factor. Urban et al. [19] consider a positive contribution due to ion or charge migration in the outer Helmholtz layer depending this time on C, and again on some exponential function of ζ. For further discussion only the simpler form of Eq. (11) is considered. As indicated, k_s is a function of $(C)^{-1/2}$, and that changes rapidly with ζ. ζ depends on the amount and way surface charges appear at the interface. Indirectly it is linked to the concentration of the solution since the surface charge density depends on it in the case of a self determining potential mechanism of electrical double layer formation.

It is important to know how much of a contribution to the overall conduction is given by surface conductance. In this case the ratio k_s/k_f and its dependence on C has to be studied. The electrolyte is considered to be ideal, i.e., $k_f = f(C)$. Two different situations have to be considered. In the first, potential determining case, ζ is independent of concentration. Hence the E and O terms of Eq. (11) are independent of C and the terms in the bracket can be considered a constant B (Bikerman [5]). Then:

(12)
$$\frac{k_s}{k_f} = \frac{\sqrt{C}B}{C} = f\left(\frac{1}{\sqrt{C}}\right)$$

This relationship can be found in some experimental observations made by Watillon [22] and Rutgers and de Smet [15] over limited concentration ranges (k_f from 10^{-7} to $10^{-8}\Omega^{-1}$ cm^{-1}) for certain univalent and symmetrical electrolyte glass systems (KCl, KI, KNO$_3$). Eq. (12) also indicates that at high concentrations the surface conductance contribution to the overall conductance can be neglected:

(13) $k_s/k_f \rightarrow 0$ as $C \rightarrow \infty$

This had been suggested by Smoluchowski and supported by numerical examples of Bikerman [4].

The other case where lattice effects cause surface charge densities to remain constant as potentials vary with C is more complex. No simple dependence on concentration as that in Eq. (12) can be given. It can be shown, however, that the E and O terms of Eq. (11) tend toward zero when concentration tends toward infinity (Mossman and Mason [13]). In this case, the same conditions exist as in Eq. (13) for high bulk solution concentrations. The great number of different and sometimes contradicting experimental data found in the literature gives an indication of the complexity of the problem.

In summarizing the findings of the literature research for the particular problem under study, it suffices to say that k_s/k_f is a function of C, in glass bead and sand packs most likely of $C^{-1/2}$ over a certain range of concentrations. It has an upper limit of zero when $C \to \infty$, and possibly a lower limit of zero when $C \to 0$. The latter would be the case when no ions are present to form a double layer. For all practical purposes, the upper limit of C where $k_s/k_f = 0$ has to be determined experimentally for each system in question. The assumption that k_s/k_f passes through zero twice would suggest that the functions have a maximum between $C = 0$ and $C = \infty$. A definite extremum effect is displayed when there is reversal in sign of the zeta potential. Furthermore, it has to be pointed out that the expression of Eq. (12) becomes indeterminate for $C = 0$, and that no experimental data exist near this point. To show the order of magnitude of the effect some examples for typical porous media (ceramic membranes) are given as quoted in Bikerman [4]. Here the contribution of surface conductance to the everalll conductance can be in the order of 15% to 30% of the bulk conductivity at low concentrations, the data were for KCl-solutions of $10^{-3}n$ to $10^{-3}n$.

THE PARALLEL RESISTANCE MODEL OF CONDUCTION IN A POROUS MEDIUM

Since it has been shown that the surface conduction can be an appreciable part of the total conduction of an electrolyte in a porous system, k_s will be explicitly introduced into the resistance network model of conduction through a porous medium.

In the absence of solid suspensions the passage of electrical current through an electrolyte saturated porous medium can best be represented by a network of parallel resistances. One would have to consider three resistances in the particular case under study: R_f, the resistance due to the bulk fluid phase, R_d, the resistance of the solid matrix, and R_s, the resistance due to the electrical double layer at the solid liquid interface. The resultant and observable resistance of the combined system, R_o, can then be written:

(14a)
$$\frac{1}{R_o} = \frac{1}{R_f} + \frac{1}{R_d} + \frac{1}{R_s}$$

or in conductivity connotations:

(14b) $$K_o = K_f + K_d + K_s$$

Physically speaking, these conductivities can be represented by the sample in Figure 1. It has a cross-sectional area A (cm^2) and an overall length L (cm) in the direction of general flow.

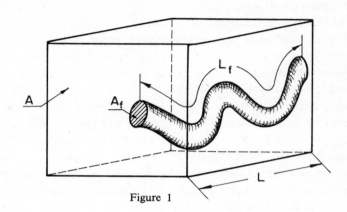

Figure 1

The length of a tortuous pore channel is L_f, where $L_f > L$. The free cross-sectional area available to flow is according to Street [17]:

(15a) $$A_f = A\phi T^{-1/2}$$

By a very loose analogy the cross-sectional area occupied by the solid matrix material could be defined as:

(15b) $$A_d = A(1 - \phi)(L/L_d)$$

where L_d is the flow path through the solid material or the length of a portion of the matrix material as bounded by the pore walls. The conductances would then be as follows:

(16)
$$
\begin{aligned}
K_o &= k_o\,(A/L) \\
K_f &= k_f\,(A_f/L_f) \\
K_d &= k_d\,(A_d/L_d) \\
K_s &= k_s\,(A_f/L_f)\,S_p
\end{aligned}
$$

where: k_o, k_f, k_d = specific conductivity for the mixed, fluid and solid phase, respectively in (Ω^{-1} cm^{-1})

k_s = specific surface conductivity in (Ω^{-1})

S_p = specific internal pore surface in (cm^{-1}), as defined by the ratio of the total internal surface area S over the pore volume of the sample. Also inverse of the hydraulic radius.

Introducing Eq. (16) into Eq. (14b), rearranging and making use of the definitions in Eq. (15) one obtains:

$$(17) \qquad k_o = k_f \frac{\phi}{T} + k_d(1 - \phi)(L/L_d)^2 + k_s(\phi/T)S_p$$

When the definition of the intrinsic formation factor of Eq. (6) is recalled and an apparent formation factor F_a is defined as the ratio of the observed bulk conductivity measurement k_o which contains contributions from all conduction mechanisms and their interrelation effects to the bulk conductivity of the free electrolyte such that $F_a = k_f/k_o$, it is:

$$(18) \qquad F_i = F_a\left(1 + \frac{k_d}{k_f} \frac{(1 - \phi)}{\phi} \left[\frac{L_f}{L_m}\right]^2 + \frac{k_s}{k_f} S_n\right)$$

All other factors being equal, Eq. (18) shows that the ratios k_d/k_f and k_s/k_f determine the magnitude of the correction to be introduced in order to transform the apparent formation factor F_a to the intrinsic formation factor F_i. In the case of a porous medium constituted of sand grains or glass spheres, the matrix conductivity becomes very low ($k_d = 10^{-12}\Omega^{-1}$ cm^{-1}). Then the second term in the brackets of Eq. (18) can be neglected because of $k_d/k_f = 0$. Eq. (18) also shows explicitly how the specific internal pore surface area S_p influences the magnitude of the last correction term. This term contains the ratio k_s/k_f which was shown to be of importance in surface conduction.

CORRELATION BETWEEN VISCOUS AND ELECTRICAL TRANSPORT PHENOMENA
IN POROUS MEDIA

It is difficult to establish values for S_p in a direct fashion. For unconsolidated porous aggregates, of which many aquifers consist, the Kozeny-Carman concept of relating internal surface area to permeability k_p has proved to be a good working hypothesis (Wyllie and Spangler [27], Rose [14]). One can write:

$$(19) \qquad S_p = \left[\frac{\phi}{Tk_p t_s}\right]^{1/2}$$

with: k_p = permeability in cm^2

t_s = shape factor related to the geometry of the cross-sectional area of the individual pore channel normal to the flow direction. (Most values fall between 1.7 and 2.5.)

When Eq. (19) is introduced into Eq. (18) with $k_d/k_f = 0$ and $t_s = 2.25$, one obtains:

$$(20) \qquad F_a = \frac{F_i}{1 + \dfrac{k_s}{k_f}\dfrac{2}{3}\left[\dfrac{1}{F_i k_p}\right]^{1/2}}$$

If the Kozeny-Carman equation were to be modified in a way consistent with the treatment of Street [17], the bracketed expression in the denominator of Eq. 20 would take the form $(F^{1.5}\phi^{0.5}k_p)^{-1/2}$.

Equation (20) shows the apparent formation factor to be an implicit function of the intrinsic formation factor, depending on k_s/k_f which again is a function of the bulk electrolyte concentration, and finally on $k_p^{1/2}$.

The relation is based on the assumption that viscous flow paths and paths of electricity transport are identical and that the hydraulic radius model of Kozeny-Carman holds. This is the case for unconsolidated porous media. Only when $k_s/k_f = 0$ can one assume that $F_a = F_i$. This may well be the case in clean oilfield sands saturated with brine for which the early contributions were made. Even in so-called clean sands with no clay content, $F_a = F_i$ cannot be postulated when the total dissolved solid content is low and when one deals with low permeability systems. It has been pointed out that the surface conduction contribution to the total conduction can be in the order of 15 to 30% for representative porous media. If F_a is used under these circumstances without correction in Eq. (8) to obtain porosities, errors of 3 to 6 porosity percent or results that are 11 to 20% higher than the actual values will occur in the range of normally encountered porosities ($\phi = 0.25$ to 0.35). Errors of this order of magnitude cannot be tolerated when introduced into the specific storage or storage coefficient term as it is used in the interpretation of hydrological pumping tests.

Eq. (20) also indicates that viscous transport coefficients (k_p) can be deducted from electrical resistivity measurements. This, however, implies that the intrinsic formation factor and the surface conduction function would have to be established independently. F_i could be obtained by injecting a high salinity solution into the porous environment of interest. The surface conduction function $k_s/k_f = f(C)$ would have to be determined in the laboratory on a representative sample.

The phenomena in the electric double layer depend on the nature of the surfaces involved since they determine charge densities, surface potential, and their variation with the concentration of the saturating electrolyte solution. The correction factor involving the k_s/k_f function will therefore also depend on the nature of the porous medium. Hence, the lumped "conductive solid effect" Street [18] has to be broken down into the contribution due to clay surfaces with lattice effects and those due to potential determining surfaces such as displayed by quartz grains. Only then can electrical transport phenomena in porous media be completely understood for a full utilization of resistivity measurements in the quantitative interpretation of hydrologic properties of porous systems such as aquifers.

ACKNOWLEDGMENT

The work upon which this study is based was supported by the Water Resources Center at the University of Illinois, by the Illinois Mining Institute, and by Geo-Engineering Laboratories, Mt. Vernon, Illinois.

REFERENCES

1. ALGER, R. P. (1966), Interpretation of electric logs in fresh water wells in unconsolidated formations, Soc. of Prof. Well Log Analyst Trans., Art. CC, 1–25.

2. ARCHIE, G. E. (1942), The electrical resistivity log as an aid in determining some reservoir characteristics, *Trans. AIME, 146,* 54–67.

3. BIKERMAN, J. J., (1933), Ionentheorie der Elektroosmose, der Strömungsströme und der Oberflächenleitfähigkeit, *Z. Phys. Chem.,* A, *163,* 378–394.

4. BIKERMAN, J. J. (1935), Die Oberflächenleitfähigkeit und ihre Bedeutung, *Kolloid Z., 72,* 100–108.

5. BIKERMAN, J. J. (1940), Electrokinetic equations and surface conductance. A survey of the diffuse double layer theory of colloidal solutions, *Farad. Soc. Trans., 36,* 154–160.

6. DEWITTE, L. (1955), A study of electric log interpretation methods in shaly formations, *Trans. AIME, 204,* 103–110.

7. FRICKE, H. (1931), Electric conductivity and capacity of disperse systems, *Physics,* **1,** 106–115.

8. HILL, H. F. AND J. D. MILBURN (1956), Effect of clay and water salinity on electrochemical behavior of reservoir rocks, *Trans. AIME, 207,* 65–72.

9. MAXWELL, J. C., *A Treatise on Electricity and Magnetism* (1881), 2nd Ed., Vol. I, Oxford, p. 435.

10. MCEUEN, R. B., J. W. BERG, JR. AND K. L. COOK (1959), Electrical properties of synthetic metalliferous ore, *Geophysics, 24,* No. 3, 510–530.

11. MCKELVEY, J. G., JR, P. F., SOUTHWICK, K. S. SPIEGLER, AND M. R. J., WYLLIE, (1955), The application of a three-element model to the S. P. and resistivity phenomena evinced by dirty sands, *Geophysics, 20,* 913–931.

12. MEREDITH, R. E. AND C. W. TOBIAS (1962), Conduction in heterogeneous systems (pp. 15–47), in *Advances in Electrochemistry and Electrochemical Engineering,* Vol. 2 (C. W. Tobias, Ed.), Interscience, New-York-London p. 300.

13. MOSSMAN, C. E. AND S. G. MASON (1959), Surface electrical conductance and electrokinetic potentials in networks of fibrous materials, *Can. J. Chem., 37,* 1153–1164.

14. ROSE, W. D. (1966), Transport through interstitial paths of porous solids, Princeton Seminar on *Hydrology and Flow through Porous Materials* (R. J. M. DeWiest, director), July, 1965, Urbana, 225.

15. RUTGERS, A. J. AND M. DE SMET (1953), Electrokinetic researches in capillary systems and in colloidal solutions, *Natl. Bur. Stand.* (J. S.), Circ. No. 524, 263.

16. SLAWINSKI, A. (1926), Conductibilité d'un électrolyte contenant des sphères diélectriques, *J. Chimie Phys., 23,* 710–727.

17. STREET, N. (1958), Tortuosity concepts (short communication), *Austr. J. Chem., 11,* 4, 607–609.

18. STREET, N. (1961), Electrokinetics, III., Surface conductance and the conductive solid effect, *Ill. State Geol. Survey Circular, 315,* 16.

19. URBAN, F., H. L. WHITE AND E. A. STRASSNER (1935), Contribution to the theory of surface conductivity at solid-solution interfaces, *Jour. Phys. Chem.*, *39*, 311–330.

20. VAN OLPHEN, H. (1963), *An Introduction to Clay Colloid Chemistry*, Interscience, New York-London, 301.

21. WAGNER, K. W. (1914), Erklärung der dielektrischen Nachwirkungsvorgänge auf Grund Maxwellscher Vorstellungen, *Arch. Elektrotech.*, *2*, 371–387.

22. WATILLON, A. (1956), Contribution à l'étude de la conductibilité de surface et du potentiel électrocinétique aux interfaces continus. I.—Interface verre-solution aqueuse d'électrolyte, *J. Chim. Phys.*, *54*, 130–145.

23. WAXMAN, M. H. AND SMITS (1968), Electrical conductivities in oil-bearing shaly sands, *SPE Journal*, *8*, No. 2, 107–122.

24. WINSAUER, W. O. AND W. M. McCARDELL (1953), Ionic double-layer conductivity in reservoir rock, *Trans. AIME*, *198*, 129–134.

25. WINSAUER, W. O., A. M. SHEARIN, JR., P. H. MASSON AND M. WILLIAMS (1952), Resistivity of brine-saturated sands in relation to pore geometry, *Bull.*, *AAPG*, *36*, 253.

26. WYLLIE, M. R. J. ANY W. D. ROSE (1950), Some theoretical considerations related to the quantitatitative evaluation of the physical characteristics of reservoir rock from electrical log data, *Trans. AIME*, *189*, 105–118.

27. WYLLIE, M. R. J. AND M. B. SPANGLER (1952), Application of electrical resistivity measurements to problem of fluid flow in porous media, *Bull. AAPG*, *36*, No. 2, 359–403.

28. WYLLIE, M. R. J. AND P. F. SOUTHWICK (1954), Experimental investigation of the S. P. and resistivity phenomena in dirty sands, *Trans. AIME*, *201*, 43–56.

SOUTHERN ILLINOIS UNIVERSITY,
    CARBONDALE, ILLINOIS

presently:

DEPARTMENT OF GEOLOGY AND GEOPHYSICS,
    UNIVERSITY OF MINNESOTA,
        MINNEAPOLIS, MINNESOTA, 55455.

SOME ASPECTS OF HEAT
AND MASS TRANSFER IN POROUS MEDIA

G. DAGAN

LIST OF SYMBOLS

Dotted symbols represent microscopical entities (at the pore scale). Undotted variables are macroscopical.

A_D, A_S	—	coefficients representing Dufour and Soret effects
c	—	specific heat of fluid
c_s	—	specific heat of solid
C	—	concentration of solute (mass solute/total mass)
d	—	pore size
D'_m	—	coefficient of molecular diffusion
D_m	—	coefficient of molecular diffusion at the macroscopical scale (bulk coefficient)
D_{if}	—	dispersion tensor
e	—	internal energy
g	—	gravity acceleration
h	—	coefficient of heat transfer between solid matrix and fluid
J_i	—	vector of solute flux
J^q	—	vector of heat flux
j^q	—	thermal equivalent of mechanical work
k	—	intrinsic permeability
L	—	macroscopical length scale
n	—	porosity
p, P	—	pressure
$Pe = ud/D'_m$	—	Peclet number
q_i	—	vector of specific discharge
t	—	time
T	—	fluid temperature
T_s	—	solid temperature
u_i	—	velocity vector
u_{if}	—	heat front velocity
U	—	macroscopical velocity scale
x_i	—	cartesian coordinates
z	—	vertical coordinate
a	—	coefficient of thermal expansion
β	—	coefficient relating the density of solution to concentration
$\Delta\rho$	—	scale of fluid density variation
ΔT	—	scale of temperature variation
ε	—	rate of energy dissipation by viscous stresses
λ	—	fluid heat conductivity coefficient
λ_s	—	solid heat conductivity coefficient
λ_{if}	—	tensor of heat dispersion
λ_t	—	coefficient of heat conductivity of the saturated porous medium
μ	—	dynamic viscosity
ρ_o	—	average fluid density
ρ	—	fluid density
ρ_s	—	solid density

1. INTRODUCTION

The transport of heat and solutes by a fluid moving through a porous matrix is a phenomenon of great interest from both application and theory points of view. Mass transfer in isothermic conditions has been studied under the generic name of "Hydrodynamic Dispersion," with applications to problems of mixing of fresh and salt waters in aquifers, miscible displacements in oil reservoirs, spreading of solutes in fluidized beds and crystal washers, salt leaching in soils, etc. Heat transfer, in the case of a homogeneous fluid, has been studied less systematically, but rather extensively, with relation to different applications, like: dynamics of hot underground springs, terrestrial heat flow through aquifer, hot fluid and ignition front displacements in reservoir engineering, heat exchange (with evaporation and condensation) between surface soil and atmosphere, flow of moisture through porous industrial materials, heat exchanges with fluidized beds etc.

Here, at the Hydraulics Laboratory of the Technion, we have been attracted to this field by studying a problem in which the two transport phenomena are combined, namely the case of the hot and salty springs of the Sea of Galilee. Our research work is only at its beginning. The purpose of this paper is to present a systematical derivation of the equations of flow in porous media with heat and mass transfer and of the different types of approximations used in applications. Physical interpretation and estimates of order of magnitude, rather than intricate mathematical derivations, are emphasized in the following sections.

The considerations are limited to the cases of an immobile and inert solid matrix and a slightly compressible liquid undergoing a slight density variation, i.e., having a low solute concentration and a moderate temperature drop. The flow is assumed to be in the Darcian regime. The Schmidt number (viscosity over diffusion coefficients) is of order 10^3.

2. THE MICROSCOPICAL EQUATIONS

First, the equations at the pore scale are considered in the most general manner, for both fluid and solid (De Groot and Mazur, (1963)).

For the fluid we have*:

Conservation of mass

(1)
$$\frac{d\rho'}{dt} + \rho'\frac{\partial u'}{\partial x_i} = 0 .$$

Conservation of solute mass

(2)
$$\rho'\frac{dC'}{dt} + \frac{\partial J_i'}{\partial x_i} = 0 ,$$

Equation of momentum (Navier-Stokes)

(3)
$$\rho'\frac{du_i'}{dt} = -\frac{\partial p'}{\partial x_i} + \mu\frac{\partial^2 u_i'}{\partial x_j \partial x_j} - \rho_g'\frac{\partial z}{\partial x_i}$$

---

* Symbols are defined in List of Symbols. The tensorial summation convention is adopted.

where the terms containing the derivatives of μ are neglected.

Energy

(4) $$\rho' \frac{de'}{dt} = - \frac{\partial J'_{q,i}}{\partial x_i} + \frac{\mu}{2} \left(\frac{\partial u'_i}{\partial x_j} + \frac{\partial u'_j}{\partial x_i} \right) \left(\frac{\partial u'_i}{\partial x_j} + \frac{\partial u'_j}{\partial x_i} \right)$$

where the radiative transfer of heat is neglected.

Equation of state

$$\rho' = \rho(T', C').$$

This equation becomes, in the case of small density variations

(5) $$\rho' = \rho_0[1 - \alpha(T' - T_0)] + \beta C'$$

For the case of liquids and in the range of the parameters considered here the internal energies, of both fluid and solid matrix are given by

(6)
$$e' = cT'$$
$$e'_s = c_s T'_s$$

the specific heats being practically constant and equal to their values for constant volume.

The thermodynamical fluxes

(7) $$J'_i = -\rho' D'_m \frac{\partial C'}{\partial x_i} - A_s \frac{\partial T'}{\partial x_i}$$

(8) $$J'_{q,i} = -\lambda' \frac{\partial T'}{\partial x_i} - A'_D \frac{\partial C'}{\partial x_i}.$$

For the solid, through which only heat is transferred, we have

Energy

(9) $$\rho'_s \frac{\partial e'_s}{\partial t} = - \frac{\partial J'_{qs,i}}{\partial x_i}.$$

Flux

(10) $$J'_{qs,i} = - \lambda'_s \frac{\partial T'_s}{\partial x_i}.$$

Our next step is to rewrite Eqs. (1)–(10) in a macroscopical form, i.e., at a scale much larger than the pore scale, but small when compared with the characteristic length of the flow domain. This crucial step is a matter of study and debate and the present Symposium is in part dedicated to this problem. Here, generally accepted concepts will be used without proof or reference.

3. THE MACROSCOPICAL EQUATIONS

Macroscopical variables are defined as averages over macroscopical volumes.

For instance

(11) $$\rho = \frac{\iiint_{voias} \rho' dv}{\iiint_{voids} dv}, \qquad T_s = \frac{\iiint_{solid} T_s' dv}{\iiint_{solid} dv}.$$

We have now for the different equations of Section 2 the following set:
Conservation of mass (from Eq. (1))

(12) $$\frac{\partial \rho}{\partial t} + u_i \frac{\partial}{\partial x} + \rho \frac{\partial u_i}{\partial x_i} = 0$$

where u_i is the average mass velocity, related to the specific discharge through $q_i = nu_i$.

Conservation of solute mass (from Eq. (2))

(13) $$\rho \frac{\partial C}{\partial t} + \rho u_i \frac{\partial C}{\partial x_i} = -\frac{\partial J_i}{\partial x_i} + \frac{\partial}{\partial x_i}\left(\rho D_{ij}\frac{\partial C}{\partial x_{kj}}\right).$$

Eq. (13) is the dispersion equation in which D_{ij} is the tensor of hydrodynamic dispersion which appears in the process of averaging of Eq. (2), from the advective term $u_i'(\partial C'/\partial x_i)$;
Momentum (from Eq. (3))

(14) $$\rho \frac{\partial u_i}{\partial t} = -\frac{\partial p}{\partial x_i} - \frac{\mu}{k} q_i - \rho \frac{\partial z}{\partial x_i}.$$

Here, the main point is the representation of the viscous term of Eq. (3) by its Darcian counterpart. Generally k is a tensor. For simplicity we will consider an isotropic porous medium and a scalar k. The inertial advective term is neglected, the flow being at small Reynolds number.
Energy (from Eqs. (4) and (6))

(15) $$\rho c \frac{\partial T}{\partial t} + \rho c u_i \frac{\partial T}{\partial x_i} = -\frac{\partial J_{q,i}}{\partial x} + \varepsilon + \frac{\partial}{\partial x_i}\left(\lambda_{ij}\frac{\partial T}{\partial x}\right) + h(T_s - T).$$

Here ε is the rate of dissipation by viscous stresses per unit volume and λ_{ij} is the tensor of temperature dispersion, similar to D_{ij} of Eq. (13), h is a coefficient of heat transfer between the solid matrix and the fluid.
Equation of state (from Eq. (5))

(16) $$\rho = \rho_0[1 - \alpha(T - T_0)] + \beta C$$

which remains unchanged.

The thermodynamical fluxes (from Eqs. (7) and (8))

$$(17) \qquad J_i = -\rho D_m \frac{\partial C}{\partial x_i} - A_S \frac{\partial T}{\partial x_i}$$

$$(18) \qquad J_{q,i} = -\lambda \frac{\partial T}{\partial x_i} - A_D \frac{\partial C}{\partial x_i}.$$

Here D_m, λ, A_S, A_D are macroscopical coefficients of transfer for a statical fluid ($u_i = 0$). Their relationship with the original microscopical coefficients will be discussed later.

For the solid matrix we have:
Energy (from Eqs. (9) and (16))

$$(19) \qquad \rho_s c \frac{\partial T_s}{\partial t} = \frac{\partial J_{qs,i}}{\partial x_i} + h(T - T_s).$$

Heat flux

$$(20) \qquad J_{qs,i} = -\lambda_s \frac{\partial T_s}{\partial x_i}.$$

Again λ_s is a "statistical" macroscopic coefficient.

4. A DISCUSSION OF THE COEFFICIENTS APPEARING IN THE MACROSCOPICAL EQUATIONS (11)–(19)

(a) The dispersion tensor D_{ij} has been studied theoretically and experimentally (see, for instance, Saffman (1960), Pfankuch (1963)) The state-of-the-art may be summarized as following: in an isotropic medium D_{ij} has the principal axes along the velocity vector and normal to it. The ratio D_I/D'_0 between the longitudinal dispersion coefficient and the microscopical diffusion coefficient is a function of $Pe = u_i d/D'_0$. For small Pe numbers ($Pe < 1$) $D_I/D'_0 \gg 1$, while at large Pe ($Pe > 100$), $D_I/D'_0 \approx 1.4\,Pe$, or $D_I = 0(ud)$. The coefficient of lateral dispersion has the same trend, but is sensibly smaller than D_I.

(b) The significance of k, the intrinsic permeability, is well known. Its order of magnitude is d^2.

(c) The rate of dissipation ε, in Eq. (15), may be estimated from Darcy's law for a homogeneous fluid

$$(21) \qquad \varepsilon = \rho n u_i \frac{\partial}{\partial x_i} \left(\frac{p}{\rho} + gz \right).$$

(d) The tensor of heat dispersion λ_{ij} should be similar to D_{ij}, if concentration and temperature are regarded as scalars transported by the fluid.

These are, however, two meaningful differences: (i) heat is transferred through the solid matrix, and not only by fluid. As effect one can expect a smaller heat dispersion. (ii) the heat diffusivity is much larger, in the case of liquids, than mass diffusivity, the Schmidt number being of order 10^3. As an effect λ_{ij} is negligible when compared with λ in a much larger range of velocities than D_{ij}.

Green (1963) has analyzed different existent data and carried out experiments which led to the conclusion that up to $Pe = 10^4$ λ_{ij} and D_{ij} are practically identical, the influence of heat flow through the solid being immaterial. Moreover, the longitudinal heat dispersivity is negligible as compared with heat diffusivity at $Pe < 3000$, which corresponds roughly to the Darcian regime ($Pe < 1$). At this stage, therefore, one may assume $\lambda_{ij} \approx 0$.

(e) The statical or "bulk" coefficients of diffusion (D_m) and heat conductivity (λ) of Eqs. (17) and (18), as well as λ_s from Eq. (17) are equal to the molecular diffusivities times a correction factor which depends on porosity and the geometry of pores. For D_m, and λ, it is generally assumed, that this factor is roughly equal to 2/3.

The bulk heat conductivity of the solid-liquid complex (for standing fluid and common temperature) is of practical relevance. If may be found experimentally much easier that λ and λ_s separately. For a combination in parallel the total heat conductivity λ_t may be related to λ and λ_s by

(22) $\lambda_t = n\lambda + (1-n)\lambda_s.$

(f) The cross coefficients representing the Soret and Dufour effects, A_S and A_D of Eqs. (17) and (18) are interrelated by Onsager relations (De Groot and Mazur (1963)). Both coefficients are negligible in liquids and very small in gases, when compared with the "direct" coefficients (the ratio is of order 10^{-3} for liquids). Hence, we can assume $A_S \approx A_D \approx 0$.

(g) The coefficient of heat transfer h (Eqs. (15) and (16) has been discussed by Green (1963). It depends on geometry of pores, relative velocity, viscosity and coefficient of heat transfer.

For Reynolds number up to 1 the effect of heat transfer due the difference between the solid and liquid temperatures is negligible when compared with the thermal conduction (less than 10%). Hence h may be neglected excepting boundary layers and short time intervals when $T_S - T$ is high.

Summarizing this discussion, we may take $\lambda_{ij} = 0$, $h = 0$, $A_S = A_D = 0$ in the macroscopical equations, as a valid approximation in the conditions considered here.

5. FURTHER SIMPLIFICATIONS OF THE MACROSCOPICAL EQUATIONS: BONSSINESQ APPROXIMATION

Additional simplfications may be found from the examination of the order of magnitude of different terms of the macroscopical equations.

In Eq. (12) the terms $u_i(\partial\rho/\partial x_i)$ and $\partial\rho/\partial t$ are of order $U\Delta\rho/L$, while $\rho(\partial u_i/\partial x_i)$ is of $O\,(U\rho_0/L)$. Hence, their ratio is of order $\Delta\rho/\rho_0$ and the above terms may be neglected (Boussinesq approximation).

In Eq. (14) the inertial term $\rho(\partial u_i/\partial t) = O(\rho_0 U^2/L)$, while the frictional term $(\mu/k)q_i = O(\mu U/d^2)$. Their ratio is of order $O[(\rho_0 ud/\mu)(d/L)] \ll 1$ and the inertial term may be neglected.

In Eq. (15) the dissipation term ε is, by Eq. (18) of order $j(\rho_0{}^2 d^2 g^2/\mu)$ while other typical terms are of order $\rho_0 C\Delta TU/L$. Their ratio is of order $(\rho_0 dU/\mu)(jLg^2 d/U^2 C\Delta T)$ which is exceedingly small, i.e. the equivalent thermal energy of dissipation is small when compared with the charges in thermal energy due to heat transfer. Hence, ε may be neglected.

Eq. (16)links the effects of temperature and concentration. It may be written in the form

$$\frac{\Delta\rho}{\rho_0} = -\alpha\Delta T + \beta\frac{\Delta C}{\rho_0}$$

where α/ρ and β are of order 4×10^{-4}/grade and 0.7 respectively.

Hence, inorder that thermal and concentration effects on density should be of the same order of magnitude, $\Delta c/\rho_0$ must be 5×10^{-3} for ΔT say $100°$.

Summarizing Sections 4 and 5, the macroscopical equations become in their simplified version

$$(23) \qquad \frac{\partial u_i}{\partial x_i} = 0$$

$$(24) \qquad nu_i = \frac{k}{\mu}\left(\frac{\partial p}{\partial x_i} + \rho g\frac{\partial z}{\partial x_i}\right)$$

$$(25) \qquad \frac{\partial c}{\partial t} + u_i\frac{\partial c}{\partial x_i} = \frac{\partial}{\partial x_i}\left((D_m + D_{ij})\frac{\partial c}{\partial x_j}\right)$$

$$(26) \qquad \frac{\partial T}{\partial t} + u_{if}\frac{\partial T}{\partial x_i} = \frac{\lambda_t}{n\rho_0 c + (1-n)\rho_s c_s}\frac{\partial^2 T}{\partial x_i{}^2}$$

$$(27) \qquad \rho = \rho_0[1 - \alpha(T - T_0)] + \beta c.$$

The system (23)–(27) is a complete system for the functions u_i, ρ, c, T, p. There are no simple means of reducing it by elimination because of the presence of the variable ρ in Eq. (24). Eqs. (25), (26), (27) cannot be combined either into a single equation in ρ because of the different mass and thermal dispersion coefficients appearing in Eqs. (25) and (26). The mathematical problem seems very difficult and no analytical solution is known to have been found so far.

The last sections are dedicated to the review of typical problems which permit further simplifications.

6. HORIZONTAL FLOW, STABILITY

In a cartesian system with one of the axes vertical (say $x_3 \equiv z$), ρ appears only in one of the three equations of (24). If the flow is purely horizontal, i.e. $u_3 \equiv 0$, the Eqs. (23) and (24) separate and permit the determination of u_1 and u_2 as solution of a potential flow. The vertical pressure distribution becomes hydrostatical and Eqs. (25) and (26) allow us to determine separately T and C distributions.

Such as solution is possible only if the flow is stable, i.e. $\partial \rho / \partial z < 0$. Otherwise, a simple check shows that fingering instability may occur locally, since fluid of larger density and viscosity overtops lighter fluid. In the unstable case u_3 is different from zero and again the equations of flow are coupled.

Unstable flows have been considered by Wooding (1962) (isothermal uniform vertical flow) and by Lapwood (1948) Wooding (1960) and Elder (1966) (homogeneous fluid, horizontal layer with vertical temperature gradient, linearized equations). Only recently Nield (1968) has considered linear instability in presence of both thermal and concentration vertical gradients.

7. QUASI-HORIZONTAL STABLE FLOW, STRATIFIED FLOW, INTERFACE

In the case of a stable horizontal flow the pressure distribution is hydrostatical and given by $p_h = -\int \rho \, dz$. Since we have assumed that $\Delta \rho / \rho_0 \ll 1$, p_h is approximately given by $p_h = -\rho_0 Z + P(x_1, x_2)$. Substituting in Eqs. (24) one may write again, similarly to Bonssinesq approximation for stratified ocean flows (Phillips (1966)),

$$nu_i = -\frac{k}{\mu}\frac{\partial P}{\partial x_i} \quad (i = 1, 2)$$

(28)

$$nu_3 = -\frac{k}{\mu}g(\rho - \rho_0).$$

If the flow is almost horizontal, P becomes a function of $x_1, x_2, x_3 = z$ and the last equation becomes

(29)
$$nu_3 = -\frac{k}{\mu}\left[\frac{\partial P}{\partial x_3} + g(\rho - \rho_0)\right].$$

If, now, $g(\rho - \rho_0)/(\partial P/\partial x_3) \ll 1$ buoyancy term may be neglected and again the flow may be regarded from a dynamical point of view as homogeneous and u_i, P may be found from Eqs. (23), (24) with $\rho = \rho_0$, while C and T may be found separately. The whole procedure may be presented systematically by a perturbation expansion.

Another type of approximations stem from Eqs. (25) and (26). The order of magnitude of the diffusive terms in these equations is DC/L^2 while the adjective are $O(UC/L)$ and the ratio DU/L is generally small. If one neglects the diffusive terms in Eqs. (25) and (26), heat and solute are transferred only by advection. In this case one deals with stratified flow (discussed by Yih (1965)) or, in the case of two fluids, with interface flows. Both cases are much simpler than the original general case. A similar case is that of a "thermal" jet (Wooding (1951)), where a hot fluid flows upwards, due to buoyancy, in a cold environment.

And, finally, the above simplifications become invalid in boundary layers. For instance the neglection of dispension at an interface is not justified since the local scale for the concntration. gradient is very high. The same is true the for thermal jet. In both cases a boundary layer solution valid locally, may improve the overall solution of the problem.

8. CONCLUSION

The purpose of the presentation is to discuss the different aspects encountered in the derivation and solution of the equations of flow through porous media with heat and mass transfer.

The ideas are derived mainly from the existent knowledge on flow of a homogeneous fluid or isothermal flow. The problem of combined solute and heat transfer was not yet considered thoroughly and the field is open for research. Such a research will contribute to the solution of problems encountered in applications—like that of hot salty springs— and also to a better under-standing of the fundamentals of flows through porous media.

9. ACKNOWLEDGMENT

The present paper is partly based on a report prepared in the frame of a research project carried out at the Hydraulic Laboratory of the Technion and sponsored by Tahal—Water Planning for Israel. Mr. A. Kahanowitz, from Tahal, has participated in the preparation of the report.

REFERENCES

1. DE GROOT, S. R. AND P. MAZUR (1963), *Non-Equilibrium Thermodynamics*, North Holland Publ. Co., Amsterdam, 510 pp.

2. ELDER, J. W. (March 1966), Steady free convection in a porous medium heated from below, *Journ. Fluid Mech.*, *27*, part 1, 29–48.

3. GREEN D. W., (1963), Heat transfer with a flowing fluid through porous media, Ph. D. thesis, University of Oklahoma, 251 pp.

4. LAPWOOD, E. R. (1948), Convection ot a Fluid in Porous Media, *Proc. Cambridge Phil. Soc. 44*, 508–520.

5. NIELD, D. A. (1968), Onset of thermoline convection in a porous medium, Water Resources Research, Vol. 4, No. 3, pp. 553–559.

6. PFANKUCH, H. (1963), *Contribution a l'étude des deplacements de fluides miscibles dans un milieu poreaux*, Ins. Francais du Petrole, Paris, 54 pp.

7. PHILLIPS, O. M. (1966), *The Dynamics of the Upper Ocean*, Cambridge University Press, 261 p.

8. SAFFMAN, P. A. (1960), Dispersion due to molecular diffusion and macroscopic mixing in flow through a network of capillaries, *Journ. Fluid Mech.*, *9* 194–208.

9. WOODING, R. A. (1957), Mixing — layer flows in a saturated porous medium, *Journ. Fluid Mech.*, *2* 273–285.

10. WOODING, R. A. (1960), Rayleigh instability of a thermal boundary layer in flow through a porous medium, *Journ. Fluid Mech.*, 183–192.

11. WOODING, R. A. (1962), The stability of an interface betwwen miscible fluids in a porous medium, *ZAMP*, *13* 255–265.

12. YIH, C. S. (1965), *Dynamics of Non Homogeneous Fluids* McMillan Co., New York Ch. 5, 306 pp.

TECHNION ISRAEL INSTITUTE
OF TECHNOLOGY, HAIFA

2

Deterministic and statistical characterization of porous media, and computational methods of analysis

STUDY OF ACCESSIBILITY OF PORES
IN POROUS MEDIA

F. A. L. Dullien

INTRODUCTION

One obstacle in the way of an improved understanding of the capillary and flow behaviour of porous media probably is the scant statistical information we have on the geometry of the pore structure.

A number of authors have pointed out [1, 2, 3, 4, 5] that pores in porous media can be expected to be 'wavy', that is to say, to contain bulges alternating with constrictions in series. Up to the present time, however, no attempt appears to have been made to extract some statistical information experimentally on the manner in which pores of different size may alternate in a porous medium.

Work has been started recently in the Department of Chemical Engineering of the University of Waterloo to collect this type of data on porous media.

THEORY

Consider a pore segment of 'radius' r in the porous medium. The purpose of our present work is to determine the 'radius' r_e ($r_e \leq r$) of the largest constriction through which the pore segment in question is accessible.

The physical picture associated with this concept is as follows. There are, in general, several pore paths connecting any pore segment to the outside surface of the sample. Each of these approaches contains at some point a segment which is the narrowest constriction in that path. Of all these necks the largest is the one through which a non-wetting fluid will be able to pass under the influence of the least pressure differential and invade the pore segment in question. In case the neck is the pore segment itself, $r_e = r$, otherwise $r_e < r$. Conversely, for given values of the interfacial tension and the contact angle the 'radius' r_e will determine the least pressure difference across the interface which is necessary to dislodge an isolated drop of the non-wetting fluid in the pore segment of 'radius' r and move it to the outside of the sample.

This type of information on porous media is being obtained by us at the present time by the combined use of "Wood's Metal Intrusion Porosimetry" and micrographic methods.

Wood's metal is injected into a sample at the least penetrant pressure and subsequently the metal is allowed to solidify. The size distribution $\alpha(r, r_e)\,dr$ of the pores containing metal is determined micrographically using polished sections of the sample. Here r_e is the 'radius' of the narrowest pore penetrated by the metal in the sample. More metal is injected into another, statistically identical sample at higher pressure and the above sequence of operations is repeated. The injection pressure is raised from sample to sample until the final sample is "saturated" with the metal.

In the presence of pores of limited accessibility in the sample the distribution function $\alpha(r, r_e)\,dr$ is a function of both r and r_e, the latter of which is determined by the pressure of injection into the sample. The expression $-[\partial\alpha(r, r_e)/\partial r_e]_r = \beta(r, r_e)$ is the "accessibility distribution". $\beta(r, r_e)\,dr\,dr_e$ gives the frequency of the pores in the interval r and $r + dr$ that are accessible through pores between r_e and $r_e + dr_e$.

It is evident that if similar experiments are carried out with a wetting fluid the accessibility will have an inverted geometrical meaning. Considering, as before the various paths connecting any pore segment to the outside surface of the sample each of these will contain at some point a segment which is its widest part. In the process of imbibition of a wetting fluid into the porous medium the widest part of an approach to the pore segment determines the capillary pressure at which the pore segment in question can be invaded. The narrower that widest part of the approach, the higher the capillary pressure and the more readily accessible is the pore segment to the wetting fluid. Hence, for the case of a wetting fluid the accessibility of a pore of 'radius' r will be determined by the 'radius' r_e of the narrowest pore leading to it such that $r_e \geq r$. More pictorially one could say that r_e is the 'radius' of the smallest bulb in all paths of access to the pore in question. This "inverse accessibility" could be determined experimentally by letting some polymerizable liquid imbibe into a set of statistically identical (dry) samples at different capillary pressures and, after reaching equilibrium, polymerizing the liquid. Micrographic analysis of each sample would yield the size distribution $\alpha'(r, r_e)\,dr$ of the pores containing the polymer. Here r_e is the 'radius' of the widest pore penetrated by the wetting liquid. The inverse accessibility distribution function $\beta'(r, r_e)$ could be calculated in an analogous fashion as in the case of the non-wetting liquid.

EXAMPLES OF APPLICATIONS

Knowledge of the accessibility distribution functions can be expected to be useful in understanding the movement of fluids in porous media. Two examples of possible applications are discussed below. It is quite likely that there will be many more of them once accessibilities have been determined for a number of porous materials.

a) *Hysteresis of capillary pressure*

Knowledge of the accessibility distribution functions $\beta(r, r_e)$ and $\beta'(r, r_e)$ will permit a prediction of the hysteresis of the capillary pressure expected on the basis of accessibility of pores. Comparison of the predictions with actually observed hysteresis loops might shed more light on the importance of alternating bulges and constrictions in determining the level of capillary hysteresis.

Prediction of capillary hysteresis using the accessibility distribution functions is based on the following reasoning.

Consider any pore segment of 'radius' r and volume ∂v in the sample. Let us assume that of all the paths connecting this pore with the surface of the sample there is none in which the narrowest constriction would be bigger than or equal to the pore under consideration. In this case the pore segment will be invaded by a non-wetting fluid (drainage curve) only at a pressure differential that is greater than the one predicted by Laplace's equation for 'radius' r. As a matter of fact the pressure differential will correspond to the 'radius' $r_e \leqq r$ of the largest constriction in all paths of access to the pore:

$$\text{(1)} \qquad\qquad P_{nw} - P_w = \frac{2\gamma}{r_e} \cos \theta_w,$$

where P_{nw} and P_w are the pressures in the non-wetting phase and the wetting phase, respectively, γ is the interfacial tension and θ_w is the contact angle measured in the wetting phase.

When the same pore is invaded by a wetting fluid (imbibition curve) the pressure differential at which this will take place can be calculated by Eq. (1) if the 'radius' $r_e \geqq r$ of the smallest bulb in all paths of access to the pore is used. It is evident that, for the same value of $\gamma \cos \theta_w$, the pressure differential at which this pore segment is invaded by the wetting fluid is less than in the case of the non-wetting fluid. It is also clear that a difference of the same sign between the drainage and imbibition capillary pressures must exist for all pores except those that are connected to the surface of the sample by a pore having exactly the same 'radius' all the way as that of the pore in question.

In order to make a quantitative prediction, based on accessibilities, of the magnitude of capillary hysteresis in a sample one must calculate the volume of all those pore segments that are accessible to a non-wetting fluid through pores between r_e and $r_e + dr_e (r_e \leqq r)$

$$\text{(2)} \qquad\qquad s(r_e)dr_e = dr_e \int_{r=r_e}^{\infty} \beta(r, r_e)dr$$

This volume has to be compared with the volume of all those pore segments that are accessible to a wetting fluid through pores between r_e and $r_e + dr_e$ $(r_e \geqq r)$ at the same value of $\gamma \cos \theta_w$

(3) $$s'(r_e)dr_e = dr_e \int_0^{r=r^e} \beta'(r, r_e)dr.$$

The capillary pressure is the same value in both cases and is calculated by Eq. (1).

Experimental values of $s(r_e)$ and $s'(r_e)$ can be calculated from the drainage and imbibition curves, respectively, of a capillary pressure test and comparison can be made with the values calculated from accessibilities by Eqs. (2) and (3).

b) *Recovery of petroleum by waterflooding*

It is common knowledge in the field of petroleum recovery research that after waterflooding much of the residual oil is present in the form of variously shaped droplets that are completely surrounded by water. Usually it is also taken for granted that these oil droplets are cut off from an initially continuous oil mass by the floodwater which is advancing more rapidly in some pores than in others.

According to Poiseuille's equation for laminar flow in a cylindrical capillary the average velocity of the liquid is proportional to the pressure difference driving the liquid and the square of the radius of the tube. Hence, for a given pressure difference the fluid will advance more rapidly in a larger tube everything else being equal. If the pressure difference driving the liquid is the capillary pressure which, according to Laplace's equation is inversely proportional to the radius of the tube, then the average velocity of the liquid becomes proportional to the radius of the tube. Hence, on this basis, it might be expected that the invading water should be advancing faster in longer capillaries. This, in all likelihood will be the case if the capillaries of different radii running side-by-side are isolated from each other. In porous media, however, the capillaries are interconnected with each other laterally. Therefore there is a constant flow of oil from the narrower capillaries into the wider ones. As a result of this phenomenon it may happen that the water will advance in narrow capillaries faster than in wider ones. As a result it may well happen that the oil in some pore segments with narrow necks gets cut off by the flood water advancing faster in parallel narrow capillaries. The oil thus cut off is difficult to recover because the available pressure differential across the oil drop is not usually sufficient to squeeze it through the narrow necks.

It is probable that it will be possible to make certain prediction regarding the expected residual oil saturations after waterflooding based on the accessibility of the pores in the sample. For example one would expect relatively poor oil recovery from a rock containing a large number of bulges alternating with constrictions in the presence of a network of narrow pores. On the other hand relatively uniform pores should promise good recoveries regardless of whether the pores are all very narrow or all very wide.

The pressure differential necessary to squeeze an oil drop into a capillary of radius r_e from a capillary of radius $r \geq r_e$ can be shown to be

$$(4) \qquad \Delta p = 2\gamma \cos \theta_w \left(\frac{1}{r_e} - \frac{1}{r} \right)$$

If one starts with a physical situation where there are isolated oil drops in the capillary bulges surrounded completely by water the Eq. (4) gives the pressure drop necessary to dislodge the oil from a bulge characterized by the pair of parameters r and r_e. The relative number of bulges characterized by (r, r_e), however, is given by $\beta(r, r_e) dr dr_e$. (E.g. if there are a relatively large number of pores with small r_e and big r then β will have a relatively large value at this point). Therefore it appears logical to weigh Δp with β and integrate over all values of r and r_e

$$(5) \qquad D = \gamma \cos \theta_w \int\int \alpha(r, r_e) \left(\frac{1}{r_e} - \frac{1}{r} \right) dr dr_e.$$

The greater the value of D (for "difficulty") the higher residual oil should be expected on the basis of pore-accessibility in the sample.

Since the physical situation where isolated oil drops are surrounded by water certainly does exist at the end of a waterflood Eq. (5) may predict best the changes of improving the oil recovery by lowering the surface tension and/or increasing the contact angle.

DISCUSSION

The success of determining accessibilities seems to depend to a considerable degree on the accuracy with pore size distributions can be determined micrographically using polished sections of the sample. Therefore, a detailed study of these techniques has been undertaken by us.

As a first step the known methods of calculating the distributions of sphere diameters were studied. These calculations are based either on the distribution of circle diameters obtained by intersecting the spheres with random planes or on the chord length distribution obtained by intersecting the spheres with random straight lines. This work was done on a high speed digital computer. The results indicate that the sphere diameter distribution can be calculated with a very high degree of accuracy on samples that can be easily handled by present-day automatic microscopes.

As a second step in testing the micrographic methods the size distributions of granular materials consisting of non-spherical particles have been determined both micrographically and by standard sieve analysis. So far the results indicate reasonable agreement with such materials as common salt.

The idea underlying the micrographic determination of pore size distributions assumes that the pores can be thought of being put together from segments of low geometric anisotropy and that the statistical information obtained by scanning the polished sections of the sample of the porous solid with an automatic microscope is essentially the same as one would obtain if the pore segments were separate particles. According to this viewpoint, the "particle size distribution" of the pore segments is regarded as the pore size distribution of the sample.

In order to test this assumption packed beds have been made from granular solids such as common salt, the pore space saturated with Wood's metal, the salt dissolved, and polished sections analyzed micrographically. The pore size distribution obtained in this way will be compared with the particle size distribution of the salt.

ACKNOWLEDGMENTS

Acknowledgment is made to the donors of the Petrolem Research Fund, administered by the American Chemical Society, for partial support of this research. Partial support by Imperial Oil, Ltd. is also gratefully acknowledged.

REFERENCES

1. BARRER, R. M., N. McKENZIE AND J. S. S. REAY (1956), *J. Coll. Sc. 11*, 479.
2. McBAIN, J. W., (1935), *J. Am. Chem. Soc. 57*, 699.
3. LE FOURNIER, M. J., (1966), *Bull. Ass. Franc. Tech. Petrol. 175*, New Ser., *17*.
4. PETERSEN, E. E. (1958), *A.I.Ch.E. Journal, 4*, 343.
5. EVERETT, D. H. AND F. S. STONE, (1958), *Structure and Porous Properties of Materials*, Butterworths Scientific Publications, London.

DEPT. OF CHEMICAL ENGINEERING,
    UNIVERSITY OF WATERLOO,
        WATERLOO, ONTARIO.

ON THE PLANE STEADY FLOW
THROUGH INHOMOGENEOUS POROUS MEDIA

St. I. Gheorghitza

1. Let us assume that in an isotropic porous medium there is a **Newtonian** fluid in linear steady motion, i.e., when the Darcy law is valid. If k is the filtration coefficient,

γ — the specific weight of the fluid,

p — the pressure,

z — the height above an arbitrary chosen horizontal plane, and

v — the filtration velocity, then

$$(1) \qquad v = -k \operatorname{grad}(p\gamma^{-1} + z).$$

If $k \neq$ const. in the domain D filled by the porous medium, this is inhomogeneous, and the inhomogeneities can be classified in several principal types and the corresponding derived types. Thus we have first the type I, when D is built up of a homogeneous medium and an impervious medium whose frontier S has no common point with the exterior frontier S of the porous medium (i.e. an impervious body is included in a porous medium). When $k(M) = k_j \, (= \text{const})$ for $M \in D_j$ but $D_i \cap D_j = \varnothing$, $\bigcup_1^n D_j = D$, the medium is of the type II. In general k is variable with the point M, and then the medium is said to be of the type III. There are two special categories of media of the last type, namely the "Helmholtz inhomogeneous media," when $k^{1/2}$ satisfies the Helmholtz equation

$$(2) \qquad \Delta k^{1/2} + \alpha^2 k^{1/2} = 0,$$

where α is a real constant and the "harmonically inhomogeneous media," when the equation

$$(3) \qquad \Delta k^{1/2} = 0$$

is satisfied, Δ being the Laplace operator ($\Delta = \partial^2/\partial x^2 + \partial^2/\partial y^2 + \partial^2/\partial z^2$); these media could be designated media of the type III_H and the type III_a, respectively.

There is another principal type of inhomogeneity, characterized by the absence of the medium in a certain domain $D^*(\subset D)$, i.e. when the porous medium contains a cavity. When the porous medium in $D - D^*$ is homogeneous, then the porous medium it is said to be of the type 0 [1]. The derived types of inhomogeneities will of the type 0–I, I–III$_a$, II–III, 0–I–II etc.

In the last case we could consider the fluid in the cavity as a perfect fluid or as a viscous fluid, and so we have two important problems: the "reduced" problem, when the fluid is considered to be perfect in the cavity, and the "restricted" problem when in the cavity is Newtonian. Because the motion in the cavity is very slow in general, we may use in D^* the linearized Stokes equation

(4) $\operatorname{grad} p^* = \mu \Delta v^*,$

where p^* and v^* are, respectively, the pressure and the velocity in the cavity; the corresponding motion problem is named the "second restricted" problem [2]. When not all the frontier S^* of the cavity is in contact with the porous medium, we say that we have a partially obturated cavity. Here we shall deal briefly with problems concerning the steady motion of incompressible fluids, therefore the equations

(5) $\operatorname{div} v = 0, \quad \operatorname{div} v^* = 0$

must be satisfied in the domains of definition of the functions v and, respectively, v^*, the motion taking place in domains of the type 0, I or II.

2.   First let us assume that we deal with a reduced problem. A problem which has a solution in closed form is that of a spherical cavity in an unbounded homogeneous porous medium. Using the spherical coordinates (r, θ, λ) with the origin in the centre of the sphere and denoting by $\phi_0(r)$ the velocity potential in the whole space, the singularities of motion being in the exterior of the sphere $r = R$, for the motion in the presence of the spherical cavity $r < R$ we find for the velocity potentials ϕ in the exterior of the cavity and ϕ^* in its interior [3]

$$\phi = \phi_0(r) - Rr^{-1}\phi_0(R^2 r^{-1}), \qquad r > R,$$

$$\phi^* = 2\phi_0(r) + \int_0^1 \phi_0(sr)s^{-1}\,ds, \qquad r < R.$$

It is interesting to compare the plane fundamental problem for a cavity with the fundamental problem of the classical plane hydrodynamics, where the bodies are supposed to be impervious. In the plane of motion let C^* be the frontier of the cavity, the arc $\tilde{C}^*$ being in contact with an impervious body in the cavity and the porous medium being unbounded. For the complex potential $f(z) = \phi(x, y) + i\psi(x, y)$, $z = x + iy$ then we have the following mixed boundary value problem: $\text{Re}\{f(z)\} = 0$ on $C^* - \tilde{C}^*$ and $\text{Im}\{f(z)\} = 0$ on $\tilde{C}^*$. In particular if we have a uniform stream of speed V_0, then the problem is to determine a function $f(z)$ holomorphic in each point at finite distance in the exterior of C^*, when in the neighborhood of the point at infinity the expansion

$$f(z) = V_0 e^{-i\alpha} z + a_0 + a_1 z^{-1} + a_2 z^{-2} + \cdots$$

is valid, where α is the angle between the stream and the axis Ox, but $\phi = \text{Re}\{f(z)\}$ on $C^* - \tilde{C}^*$ and $\psi = \text{Im}\{f(z)\}$ on $\tilde{C}^*$. If $\tilde{C}^* \equiv C^*$ we have a problem relative to a medium of the type I, if $\tilde{C}^*$ does not exist we have a problem relative to a medium of type 0, but when $C^* - \tilde{C}^* \neq \emptyset$ the medium is of the type 0–I. Let us observe that only the first problem has a sense in the classical hydrodynamics [4].

For a cavity the following theorems can be proved [5]:

a) The discharge of the fluid passing through the cavity does not depend on the direction of velocity at infinity;

b) If two cavities fill the bounded domains D^* and $\tilde{D}^*(\tilde{D}^* \supset D^*)$, whose frontiers are the simple closed curves C^* and $\tilde{C}^*$, respectively, then the discharge of the fluid passing through the second cavity is greater than the discharge of the fluid passing through the first cavity, if at infinity the fluid has the same uniform motion.

A simple case is that of a cylindrical cavity of radius O. When $\alpha = 0$, then the complex potentials will be

$$(6) \qquad f(z) = V_0(z - R^2 z^{-1}), \; f^*(z) = 2V_0 z.$$

For a crack in the form of a cross, $y = 0, -L < x < L$ and $x = 0, -L < y < L$, applying the first theorem we find that

$$(7) \qquad Q_c = 2\sqrt{2} V_0 L.$$

As an example of an obturated cavity, let us consider the cavity $|z| < R$, obturated along the arc $|z| = R$, $2\pi - \theta > \arg z > \theta$ and a uniform stream at great distances, parallel to Ox. Using a Joukowsky transformation

$$2Z = z + R^2 z^{-1}, \; z = Z + \sqrt{Z^2 - R^e},$$

in the plane $Z = X + iY$ we have to solve the following problem: to determine a function $f(Z)$, holomorphic in each point from the half-plane $Y > 0$, having the expansion in the neighborhood of the point at infinity in the form $K_1 Z + K_0 + K_{-1} Z^{-1} + \cdots$ where $K_j (j = 1, 0, -1, \cdots)$ are constants, when on the real axis

$$\text{Re}\{f(Z)\} = 0, \; R\cos\theta < X < R,$$

$$\text{Im}\{f(Z)\} = 0, \; X < R\cos\theta \text{ and } X > R.$$

In the physical plane the solution is

$$(8) \qquad f(z) = V_0(1 - Rz^{-1})(z^2 + R^2 - 2Rz\cos\theta)^{1/2}$$

and the discharge Q_c is

$$(9) \qquad Q_c = 4V_0R(\cos^2\frac{\theta}{2} - \cos\theta)^{1/2}\sin\frac{\theta}{2}.$$

If $\theta = 0$, i.e. there is no obturation, $Q_c = 4V_0R$, but if $\theta = 0$ we find $Q_c = 0$, as it was to be expected. If the stream is parallel to $0y$, solving the corresponding boundary value problem we find $f(z)$, which is not transcribed here, but the discharge is

$$(10) \qquad Q_c = 4V_0R\sin(\theta/2).$$

The semi-inverse method ([6], [7]) can be applied to our problems. As a first example let us consider a cavity, symmetrical with respect the $0x$-axis, obtained in the following manner (having in view the symmetry, we can deal only with the points in the upper half-plane): by the Joukowsky transformation the points $A(-R)$ and $B(R)$ remain unchanged, but the point $C(iR)$ correspond to the origin in the Z-plane; in the Z-plane the image of the arc $B\gamma C$ from the physical plane is the half-circle of radius $R/2$ and the centre in the point $Z = R/2$. Solving the boundary value problem for a uniform stream at great distances we find that the discharge through the cavity is ([1], p. 286)

$$(11) \qquad Q_c = 16V_0R/3.$$

The semi-inverse method has been applied also to certain problems concerning the media of type II [8].

The phenomenon of the interference of cavities, analogous to the phenomenon of the interference of wells from the underground hydrodynamics can be made manifest [4]. If we have only two cavities D_1^* and D_2^*, let Q_{c10} and Q_{c1} the discharge through the first cavity when the second one does not exist and, respectively, the discharge through the first cavity when the two cavities coexist; what does Q_{c20} and Q_{c2} mean is evident. Then, denoting by Q_{ct} the sum $Q_{c1} + Q_{c2}$, we have the inequalities:

$$Q_{ct} > Q_{c10}, \; Q_{ct} > Q_{c20}, \; Q_{ct} < Q_{c10} + Q_{c20}.$$

There are certain interference effects between a cavity and an impervious inclusion, between a cavity and an inhomogeneity of the type II etc. Considerations about these interferences allow us sometimes to solve approximately certain problems. We notice that from (7) we deduce effects of interference in the form of the same inequalities as above.

3. Let us assume that in the physical plane $z = x + iy$ the porous homogeneous medium fills the domain D bounded by a curve L, this curve being

closed, or open and unbounded in both directions; the angle $\theta(\zeta)$ between the tangent in a point $\zeta \in L$ with a fixed directions is supposed to satisfy a Hölder condition. Let us choose a sense on L and let A_j and B_j $(j = 1, 2, \cdots, N)$ be $2N$ points on L, numbered in that sense; the arcs $A_j B_j$ will be denoted by C_j, their union by C, but the arcs $B_j A_{j+1}$ $(A_{N+1} \equiv A_1)$ are denoted by C'_j and their union by C'. We assume that C are feeding contours but C are impervious contours. In D there are several perfect wells represented by the circles γ_s of radius δ_s and centers z_s $(s = 1, 2, \cdots, P)$. As usual, we continue the motion in the interior of these circles, so that the porous medium fills the simple connected domain D, but in this case we must introduce singularities in the points z_s. If q_s is the discharge of fluid passing through γ_s (in order to fix the ideas we consider all $q_s < 0$), K_j are constants (only one can be chosen arbitrarily) and $H_s (>0)$ is the difference between the level of the free fluid in contact with C and the level of the fluid in the well having its centre in $z_{s.}$, then the problem can be stated thus: to determine an analytical function $f(z) = \phi(x, y) + i\psi(x, y)$, defined in D, such that:

a) the difference $f_0(z) = f(z) - (1/2\pi) \sum_{s=1}^{P} |q_s| \ln (z - z_s)^{-1}$ is holomorphic in D,

b) $\phi(M) = 0$ when $M \in C_j$,

c) $\psi(M) = K_j$ when $M \in C'_j$,

d) $\phi(M) = kH_s$ when $M \in \gamma_s$.

After the complex potential is obtained from the conditions a), b) and c), the discharges q_s are determined using the condition d). This problem is the general problem of the plane motion of the artesian waters, but the results, for certain boundary curves L, can be applied immediately to problems relative to inhomogeneous media of the type 0, I, or 0–I.

Introducing the complex velocity, $w(z)$, we can write

$$w_0(z) = w(z) - \frac{1}{2\pi} \sum_{s=1}^{P} \frac{|q_s|}{z - z_s},$$

where $w_0(z)$ is holomorphic in D. Because from b) $d\phi/ds = 0$ on C_j and from c) $d\phi/dn = 0$ on C'_j, denoting by $a(M)$ and $b(M)$ the functions defined by the relations

$$a(M) = \begin{cases} \dfrac{dx}{ds}, & P \in C \\[2mm] \dfrac{dx}{dn}, & P \in C' \end{cases} ; \quad b(M) = \begin{cases} \dfrac{dy}{ds}, & P \in C \\[2mm] \dfrac{dy}{dn}, & P \in C', \end{cases}$$

we can then write

(*) $$a(M)u(M) + b(M)v(M) = 0.$$

Now the problem can be stated thus, denoting by $g_q = x_q + iy_q$ a point A_j or B_j: to determine a function $w(z) = u - iv$, analytic in D, having polar singularities in a finite number of points at finite distances, continuously prolongable on L save for the points A_j and B_j $(j = 1, 2, \cdots, N)$ in whose neighborhood

$$\left| w(z) \right| < \frac{\text{const.}}{\left| z - g_q \right|^\alpha}, \qquad \alpha \in (0, 1),$$

which satisfies the condition (*) on L minus the points g_q, the functions a and b having discontinuities of the first kind in the points g_q; if the point at infinity belongs to D, then we must have

$$w(z) = O\left(\frac{1}{z^2}\right).$$

Therefore we have to solve the Hilbert problem with given singularities and discontinuous coefficients [9].

From the general solution of this problem, given by S. Gogonea, several cases considered previously [10], [11], [12] can immediately be obtained as particular cases, some previous results are generalized, and innumerable problems can be solved in closed form. Consider, for instance, the case of the partially obturated cavity, but where the complex velocity at great distances is w_0; in this case it is found for the complex potential

$$(12) \qquad f(z) = (\bar{w}_0 - w_0 R z^{-1})(z^2 + R^2 - 2Rz\cos\theta)^{1/2}.$$

If we have the same cavity and the uniform stream at great distances but also a source of intensity q in the point z_0, denoting by $Q(z) = (z^2 + R^2 - 2Rz\cos\theta)^{1/2}$, we obtain in the end

$$
\begin{aligned}
(13) \quad f(z) &= (w_0 - \bar{w}_0 R z^{-1})Q(z) - \frac{|q|}{2\pi}\ln\left[\frac{z}{R}\cdot\frac{Q(z) + z - R\cos\theta}{Q(z) - z\cos\theta + R}\right] \\
&+ \frac{q}{2\pi}\ln\frac{[zz_0 + R^2 - (z + z_0)R\cos\theta + Q(z_0)Q(z)](zz_0 - R^2)}{[(z + \bar{z}_0)R - (z\bar{z}_0 + R^2)\cos\theta + Q(z_0)\overline{Q(z)}](z - z_0)R}.
\end{aligned}
$$

Thus Eq. (12) generalizes the result (8) and in its turn Eq. (13) generalizes Eq (12). When $w_0 = 0$, from (13) it follows a generalization of a result obtained by Pilatowsky [13].

4. When we have two homogeneous media separated by a simple closed curve, a new method founded on the theory of conformal mapping, the theory of boundary value problems and the theory of singular integral equations has been proposed by Elena Ungureanu [14]. Let C be a simple closed curve with all its points at finite distances, D_1 the domain exterior to this curve, D_2 the

domain interior to the same curve, in D_j the filtration coefficient being k_j ($= $const; $j = 1, 2$). Denoting by $f_0(z)$ the complex potential due to the singularities of motion, $f_0(z)$ being defined in the whole plane, we can write

$$f_j(z) = f_0(z) + F_j(z) \qquad (j = 1, 2).$$

Denoting by $k_0 = k_2/k_1$, by β the angle between the $0x$-axis and the tangent to C, by z^* a point C and by s a parameter on the same curve, from the matching conditions

$$\operatorname{Re}\{k_0 f_1 - f_2\} = 0, \ \operatorname{Im}\{f_1 - f_2\} = 0$$

it ollows that

$$(14) \qquad 2k_0 w_1(z^*)\frac{dz^*}{ds} = (1+k_0)w_2(z^*)\frac{dz^*}{ds} + (1-k_0)\overline{w_2(z^*)}\frac{\overline{dz^*}}{ds}.$$

Let us map conformally D_1 onto the exterior of the circle $\tilde{C}$ ($|Z| = 1$) by the transformation $z = f_1(Z)$ and D_2 onto the interior of the same curve by the function $z = f_2(Z)$. Because in the new plane Z we can write for the complex velocity $\tilde{w}_j(Z_j) = \tilde{w}_0(Z_j) + \tilde{W}_j(Z_j)$, $(j = 1, 2)$ and denoting by t_1 and t_2 the images of z^* by the two transformations, instead of (14) we have

$$(1 + k_0)\tilde{W}_2(t_2) = (1 - k_0)t_2^{-2}\overline{\tilde{W}_2(t_2)} + 2k_0\tilde{W}_1(t_1)(dt_1/dt_2)$$

or

$$+ (k_0 - 1)\tilde{w}_0(t_2) - (k_0 - 1)\overline{\tilde{w}_0(t_2)}t_2^{-2},$$

$$(k_0 - 1)\overline{\tilde{W}_1(t_1)} = (k_0 + 1)t_1^2\tilde{W}_1(t_1) - 2\tilde{W}_2(t_2)t_1^2(dt_2/dt_1)$$

$$+ (k_0 - 1)\tilde{w}_0(t_1)t_1^2 - (k_0 - 1)\tilde{w}_0(t_1).$$

Introducing the holomorphic functions

$$G_1(Z_1) = \begin{cases} \tilde{W}_1(Z_1) & \text{in } D_1 \\[2mm] \overline{\tilde{W}_1\left(\dfrac{1}{Z_1}\right)} & \text{in } D_2 \end{cases}, \qquad G_2(Z_2) = \begin{cases} W_2\left(\dfrac{1}{Z_2}\right) & \text{in } D_1 \\[2mm] \tilde{W}_2(Z_2) & \text{in } D_2 \end{cases},$$

denoting by $G_j^-(t_j)$ and $G_j^+(t_j)$ $(j = 1, 2)$ the boundary values of the functions $G_j(Z_j)$ when Z_j reaches the circle C coming from D_1 or from D_2, and by $t_2(t_1)$ the connection between t_1 and t_2, E. Ungureanu obtains for $G_2(t_2^+)$ an integral equation. In particular for the circular contour the circle theorem can be written in the more general form

$$f_1(z) = f_0(z) + \frac{1 - k_0}{1 + k_0}\overline{f_0\left(\frac{R^2}{\bar{z}}\right)}, \qquad |z| > R,$$

$$f_2(z) = \frac{2k_0}{1 + k_0}f_0(z), \qquad |z| < R,$$

obtained in works concerning the motions in media of type II.

Elena Ungureanu has been able to show by this method that only the circle and the ellipse are the curves of separation of the two media allowing a uniform stream in D_2 when in D_1 at great distance there is a uniform stream.

5. The above mentioned results concerning the motion in media of the type 0 are unsatisfactory because the fluid has been considered to be perfect. When in D^* the fluid is real we have to solve the Navier-Stokes equations or the Eq. (4) and to apply the matching conditions on S^*. Neglecting the influence of the gravity, denoting by n the exterior normal to S^* and by ϕ the velocity potential in the porous medium we can write when $M \in S^*$

(15)
$$\left.\begin{array}{l} p^*(M) = -\gamma k^{-1}\phi(M), \\ v^*(M) = v(M)n, \\ v^*(M) = (\partial\phi/\partial n)_M \end{array}\right\}.$$

The first condition has been generalized in 1956 ([15]; [1], p. 272), but here we shall use only the conditions (15). When the motion in the porous medium, which contains the point at infinity is linear, the solution of Equations (4) and (5) has been obtained in closed form in several cases when at great distances we have a uniform stream parallel to 0x. An interesting case is that of a cylindrical cavity, and using the polar coordinates (r, θ), we can write

$$\lim_{r\to\infty} v_r(r,\theta) = V_0 \cos\theta, \quad \lim_{r\to\infty} v_\theta(r,\theta) = -V_0 \sin\theta.$$

The Equations (4) and (5) in polar coordinates can be written

$$\mu^{-1}\frac{\partial p^*}{\partial r} = \frac{\partial^2 v_r^*}{\partial r^2} + \frac{1}{r^2}\frac{\partial^2 v_r^*}{\partial\theta^2} + \frac{1}{r}\frac{\partial v_r^*}{\partial r} - \frac{2}{r^2}\frac{\partial v_\theta^*}{\partial\theta} - \frac{v_r^*}{r^2},$$

$$(\mu r)^{-1}\frac{\partial p^*}{\partial\theta} = \frac{\partial^2 v_\theta^*}{\partial r^2} + \frac{1}{r^2}\frac{\partial^2 v_\theta^*}{\partial\theta^2} + \frac{1}{r}\frac{\partial v_\theta^*}{\partial r} + \frac{2}{r^2}\frac{\partial v_r^*}{\partial\theta} - \frac{v_\theta^*}{r},$$

$$r\frac{\partial v_r^*}{\partial r} + \frac{\partial v_\theta^*}{\partial\theta} + v_r^* = 0.$$

For a cylindrical cavity of radius R, denoting by K the dimensionless number

$$K = \frac{k\mu}{\rho g R^2},$$

which appears in the theory of motion of Newtonian fluids in the presence of porous bodies [1], the solution of these equations satisfying the conditions (15) is ([17]; [1], p. 289)

$$\phi(r,\theta) = V_0\left(r - \frac{1-4K}{1+4K}\frac{R^2}{r}\right)\cos\theta,$$

$$v_r^* = \frac{V_0}{1+4K}\left(3 - \frac{r^2}{R^2}\right)\cos\theta, \quad v_\theta^* = \frac{3V_0}{1+4K}\left(\frac{r^2}{R^2} - 1\right)\sin\theta,$$

$$p^* = \frac{-8\mu V_0 r\cos\theta}{(1+4K)R^2}.$$

The speed in the porous medium and the cavity are, respectively,

$$V = V_0\left[1 + \left(\frac{1-4K}{1+4K}\right)^2\left(\frac{R}{r}\right)^4 + 2\frac{1-4K}{1+4K}\left(\frac{R}{r}\right)^2\cos 2\theta\right]^{1/2},$$

$$V^* = \frac{V_0}{1+4K}\left\{9 - 6\left(\frac{r}{R}\right)^2 + \left(\frac{r}{R}\right)^4 + 4\left(\frac{r}{R}\right)^2\left[2\left(\frac{r}{R}\right)^2 - 3\right]\sin^2\theta\right\}^{1/2}.$$

For $K = 1/4$ the motion in the porous medium is imperturbed by the presence of the cavity, the stream lines having the equation $y = $ const., but

$$v_r^*(r,\theta) = \frac{V_0}{2}\left(3 - \frac{r^2}{R^2}\right)\cos\theta, \quad v_\theta^*(r,\theta) = \frac{3V_0}{2}\left(\frac{r^2}{R^2} - 1\right)\sin\theta,$$

$$p^*(r,\theta) = -4\mu V_0 R^{-2} r\cos\theta.$$

The stream lines in the two domains are in general given by

$$\text{const.} = \psi(r,\theta) = V_0\left(r + \frac{1-4K}{1+4K}\frac{R^2}{r}\right)\sin\theta, \qquad r > R,$$

$$\text{const.} = \psi^*(r,\theta) = (1+4K)^{-1}V_0(3 - r^2 R^{-2})r\sin\theta, \qquad r < R,$$

and the discharge of the fluid passing through the cavity is

$$(17) \qquad\qquad q_c = 4V_0 R(1+4K)^{-1}.$$

Taking into account that $4V_0 R$ is just the discharge of a perfect fluid, we could write instead of (17)

$$(18) \qquad\qquad q_c = Q_c/(1+4K).$$

In general, for a certain cavity, the discharge of a Newtonian fluid may can written

$$(19) \qquad\qquad q_c = Q_c F(K),$$

where K is given by (16) with R replaced by a certain characteristic length of the cavity. The function F must satisfy the condition $F(0) = 1$, because for $\mu = 0$ we must find that $q_c = Q_c$ and certainly F must be a positive monotonically decreasing function of K, such that F tends towards zero for K tending towards infinity. Therefore we will have always that $q_c < Q_c$. For a spherical cavity and a uniform motion at great distances, writing the equations

(4) and (5) in polar coordinates in a meridian plane, we have to satisfy the conditions (15), which are now written:

$$p^*(R, \theta) = -\gamma k^{-1}\phi(R, \theta),$$

$$v_\theta^*(R, \theta) = 0,$$

$$v_r^*(R, \theta) = (\partial\phi/\partial r)_{r=R}.$$

The solution in this case is, with K given equally by (16),

$$\phi = \left(1 + \frac{R^2}{r}\frac{10K-1}{20K+1}\right)V_0 r\cos\theta,$$

$$v_r^* = \frac{3V_0}{20K+1}\left(2 - \frac{r^2}{R^2}\right)\cos\theta, \quad v_\theta^* = \frac{-6V_0}{20K+1}\left(1 - \frac{r^2}{R^2}\right)\sin\theta,$$

$$p^* = -\mu\frac{30V_0 r\cos\theta}{(20K+1)R^2}.$$

When $\mu \to 0$, then we obtain ([3]; [1], p. 383)

$$\phi(r, \theta) \to V_0(r - R^3 r^{-2})\cos\theta, \quad p^* \to 0,$$

$$v_r^* \to 3V_0(2 - r^2 R^{-2})\cos\theta, \quad v_\theta^* \to -6V_0(1 - r^2 R^{-2})\sin\theta$$

$$V \to V_0\left[\left(1 - \frac{R^2}{r^2}\right)^2 + \frac{R^2}{r^2}\left(2 + 4\frac{R}{r} - \frac{R^2}{r^2} + 4\frac{R^4}{r^4}\right)\cos^2\theta\right]^{1/2},$$

$$V^* \to 3V_0[4 - 4r^2 R^{-2}(1 + \sin^2\theta) + r^2 R^{-4}(1 + 3\sin^2\theta)]^{1/2}.$$

We deduce that in the porous medium the speed cannot exceed the value $3V_0$, but in the cavity the maximum speed is obtained in its centre, where the speed is less than $6V_0$ for a real fluid.

The above results have been generalized for inhomogeneous media when there is a constant filtration coefficient k_1 for $R_1 > r > R$ and another constant filtration coefficient k_2 for R_1 [18]. The solutions for the cylindrical and spherical cavities will contain, beside the dimensionless number (16), two dimensionless numbers, namely the ratios R_1/R and k_2/k_1. Denoting them by r_0 and k_0, respectively, it follows that instead (19) we have to write

(20) $q_c = Q_c F(r_0, k_0, K),$

where F is a positively definite function, monotonically decreasing with K, and always $F < 1$.

There is no difficulty in considering media of the type III when the Equations (2) or (3) are satisfied but k depends only on r. For harmonically inhomogeneous media we will have

$$k^{1/2} = A + Br^{-1},$$

where A and B are constants, and the corresponding solution can be obtained without essential difficulties for the cylindrical and spherical cavities.

6. We point out two cases of inhomogeneous media of the type II when the solution can be obtained in closed form. In [19], using the theory of conformal mapping and solving the corresponding matching conditions, we have obtained the solution for the case when in the plane of motion the frontier between D_1 and D_2 is the ellipse of half-axes a and b and at great distances we have a uniform stream. Elena Ungureanu has obtained by her method the same result, and, after [14], we can write it in the form

$$f_1(z) = w_0 z + (k_0 - 1) \frac{ab}{a - b} \left(\frac{w_0 + \bar{w}_0}{2(a + bk_0)} - \frac{w_0 - \bar{w}_0}{2(ak_0 + b)} \right) \left(\sqrt{z^2 - c^2} - z \right) \text{ in } D_1$$

$$f_2(z) = w_0 z + (k_0 - 1) \left(\frac{w_0 + \bar{w}_0}{2(1 + ba^{-1}k_0)} + \frac{w_0 - \bar{w}_0}{2(1 + ab^{-1}k_0)} \right) z \text{ in } D_2,$$

where $k_0 = k_2/k_1$, $c^2 = a^2 - b^2$, but w_0 is the complex velocity of the unperturbed stream. When $k_0 \to \infty$ we obtain the solution for the case of an elliptical cavity [20] in the reduced problem.

The solution for the case of the sphere $r = R$ separating the domain D_1 ($r > R$) with the filtration coefficient k_1 from the domain D_2 ($r < R$) with the filtration coefficient k_2 and a uniform stream at great distances can be obtained from the general solution when we know the velocity potential $\phi_0(r)$ in the whole space and all the singularities of motion are in D_1; here $\phi_0(r)$ is an abbreviation for $\phi_0(r, \theta, \lambda)$, as in the second paragraph, ([20]; [1], p. 377). The general solution is

$$\phi_1(r) = \phi_0(r) + Rr^{-1}(k_1 - k_2)(k_1 + k_2)^{-1}[\phi_0(R^2 r^{-1}) - k_1(k_1 + k_2)^{-1} \cdot$$
$$\cdot \int_0^1 \phi_0(sR^2 r^{-1}) s^{-k_2/(k_1 + k_2)} ds],$$

$$\phi_2(r) = k_2(k_1 + k_2)^{-1} \left[2\phi_0(r) + (k_2 - k_1)(k_2 + k_1)^{-1} \cdot \right.$$
$$\left. \cdot \int_0^1 \phi_0(sr) s^{-k_2/(k_1 + k_2)} ds \right].$$

This result can be considered as a "sphere theorem", analogous to the "circle theorem" of the classical hydrodynamics or of the filtration theory. For a uniform stream at great distances parallel to $0x$ we obtain from the above formulae:

$$\phi_1(r, \theta) = V_0 \left(r + \frac{R^3}{r^2} \frac{k_1 - k_2}{k_2 + 2k_1} \right) \cos \theta, \quad r > R,$$

$$\phi_2(r, \theta) = \frac{3V_0 k_2 r \cos \theta}{k_2 + 2k_1}, \quad r < R.$$

The discharge of the fluid passing through the sphere will be

(21) $Q = 3\pi V_0 R^2 k_2/(k_2 + 2k_1)$

and when $k_2 \to \infty$ we obtain the discharge in the case of the reduced problem:

$$Q_c = 3\pi V_0 R^2 \ .$$

The discharge q_c through a spherical cavity in the second restricted problem it is found easily,

$$q_c = 3\pi V_0 R^2/(1 + 20K),$$

that is here the function $F(K)$ pointed out in (19) is

$$F(K) = 1/(1 + 20K).$$

7. The motion of fluids through various inhomogeneous porous media is a vast field of research which present practical interest. Even the theory of the reduced problem has some importance, because from this theory we can obtain, for example, the *maximum* discharge of a fluid passing through a cavity. Beside the problems outlined there are still numerous problems which must be solved in the future. If the theory of the reduced problem for a cavity is well developed, we are lacking a similar theory for the restricted problem. For instance, the interaction of cavities in the restricted problem was not tackled till now. Equally, we need a satisfactory theory for the motion in porous media of the type *II*, when three or more different homogeneous media are in presence.

The problem becomes much more intricate if we use other relationships between the filtration velocity and the pressure, i.e. if we replace (1) by another law. It would be interesting to consider the case of rapid motions ([1], p. 65), the case of very slow motions ([1], p. 63), or the case which contain these two extreme cases [21].

REFERENCES

1. St. I. Gheorghitza, (1966) Mathematical Methods in Underground Hydrogasdynamics, *Ed. Acad. R.S.R.*, (in Rumanian).

2. St. I. Gheorghita: (1964), On the motion in porous media having cavities, *Com. Ac. R.P.R.*, *13*, 789.

3. St. I. Gheorghita (1964), On the motion of fluids in an unbounded porous medium having a spherical cavity, *Rev. Roum. Math. Pures et Appl.*, *9*, 425.

4. St. I. Gheorghitza (1967), Sur le mouvement dans les milieux poreux ayant des cavités, Fluid Dynamics Transactions, Pergamon Press.

5. St. I. Gheorghitza (1965), Sur le mouvement plan dans les milieux poreux homogènes ayant une cavité, *Comptes Rendus*, *260*, 5457.

6. St. I. Gheorghitza (1959), A semi-inverse method in plane hydrodynamics, *Arch. Mech. Stos.*, *9*, 681.

7. St. I. Gheorghitza (1965), On a semi-inverse method in plane hydrodynamics, *Rozprawy Inzy..erskie*, vol. *13*, 447 (in Polish).

8. St. I. Gheorghita (1967), On the analytical solution of some problems for inhomogeneous media of the second type, *Rev. Roun. Math. Pures Appl.*, *12*, 71.

9. Sorin Gogonea (1968), Contributions to the study of certain motions in porous media, Doctoral Disertation, Bucharest University, (in Rumanian).

10. Sorin Gogonea (1964), Motions in porous media in the prescnce of some finite feeding contours, *St. Cerc. Mat.*, tom *15*, 825.

11. Sorin Gogonea (1966), Sur les mouvements plans dans les milieux poreux en presence d'une classe de frontières d'alimentation, *J. de Mécanique*, 309.

12. Sorin Gogonea (1967), Sur le mouvement des eaux artesiennes en présence de quelques contours d'alimentation partiellement obturés, *Rev. Roum. Math. Pures Appl.*, *12*, 955.

13. V. P. Pilatowsky (1954), The filtration in an imperfect layer, I. A. N. SSSR, OTN, 121 (in Russian).

14. Elena Ungureanu (1968) On some boundary value problemes in the fluid motion through inhomogeneous porous media, *Doctoral Dissertation, Bucharest University*, (in Rumanian).

15. St. I. Gheorghita (1956), On the generalizstion cf the Stokes' law, *Com. Ac. RPR*, *6*, 763 (in Rumanian).

16. St. I. Gheorghita (1955), The generalization of the Stokes' law, *Bul. St. Sect. St. Mat. Fiz.*, *7*, 751 (in Rumanian).

17. St. I. Gheorghitza (1965), On the steady flow through porous media in presence of a cylindrical cavity, *Ganita*, *16*, 7.

18. St. I. Gheorghitza (1966), On the steady motion of incompressible fluids in porous inhomogeneous media with cavities, *Proc. Eleventh Int. Congress Appl. Mech.*, 847.

19. St. I. Gheorghita (1954), Some motions in porous inhomogeneous media, *Bul. St. Sect. St. Mat. Fiz.*, *6*, 823 (in Rumanian).

20. St. I. Gheorghitza (1964), Sur le mouvement dans un milieu poreux homogène ayant une cavité elliptique, *Comptes Rendus*, *259*, 2779.

21. St. I. Gheorghita (1959), On the nonlinear motions with initial gradient, *Analele Univ. Bucuresti*, *22*, 39 (in Rumanian).

Institute of Mathematics,
    Calea Grivitei 21,
        Bucharest 12,
            Rumania

A NUMERICAL STUDY OF THE
NONLINEAR LAMINAR REGIME OF FLOW
IN AN IDEALISED POROUS MEDIUM

K. P. STARK

INTRODUCTION

The study of flow through porous materials presents many complex problems. As a result of these complexities, present day knowledge of the behaviour of flow through such materials relies almost completely on empirical approaches. Literally, thousands of experiments have been performed since Darcy's historical tests performed in 1856. The variety of empirical relations developed from these tests gives quite conflicting, and often, milseading results. Scheidegger (1960) points out that a general relation for flow through porous materials can be developed only if "one is able to understand exactly how all these properties (of the porous media) are conditioned by the geometrical properties of the pore system." Such an understanding will be possible when a variety of solutions of the fundamental equations of flow in porous materials have been obtained. This paper outlines a technique for obtaining such solutions by solving the Navier-Stokes laminar flow equations for flow through a number of simplified and idealised porous materials.

Previously, some solutions of the equation for creeping flow in idealised mcdia have been available and most of these are summarised by Happel and Brenner (1965). The solutions presented here extend the treatment into the nonlinear (non-creeping) flow range by retaining the inertia terms in the Navier-Stokes equations. Irmay (1956) and others have shown that retention of these inertia terms is required if a study of the non-creeping range of flow is contemplated and experimental evidence exists (e.g. Wright, 1968) to illustrate that these effects become important before the onset of turbulence.

A few other solutions are available in this nonlinear laminar regime, however, the solutions presented here are believed to be the first attempt at solving the full nonlinear laminar flow equations for a variety of idealised porous materials and relating the results to the conventional empirical approaches. Flows in the nonlinear regime are associated with non-zero Reynolds numbers and high velocities or pressure gradients. Thus in the design of coarse filters, self-spillway rockfill dams, pumpwells, etc., the nonlinear flow regime must be considered. Further, the method of analysis can be used to illustrate the magnitude of the nonlinear effects, which are generally neglected, at very low Reynolds numbers. Although the retention of the inertia terms complicates the numerical analysis, this complication is offset by the detailed description of the flow characteristics which results—thus, pressure, shear, velocity, streamline and vorticity values at every pount in the flow are automatically obtained; the form and shear drag on any particle, the resultant vortex pattern in the wake region and its growth with varying Reynolds numbers, the non linear inertia components of the flow and the hydrodynamic force distribution on any element of the medium can all be evaluated; finally, the empirical relations of Darcy, Forchheimer, Kozeny and others, as presently used in the variety of problems associated with media flow can all be tested.

The particular idealised models used in the numerical analysis are shown in Figures 1a and b, where the two dimensional media are assumed to extend to infinity in all directions. Watson (1964) considered the model of Figure 1a with $a = b$, however, his calculations were invalidated, except at zero Reynolds numbers, by an incorrect boundary assumption. The simplified boundary geometries, representing the idealisations incorporated in the models, were adopted so that the mathematical model would become tractable. The model of Figure 1.1b involves two particle sizes where the smaller particles are placed symmetrically around the larger ones. The porosity of Figure 1a was varied by altering the ratio a/b. In this way porosities from 0.36 to 0.97 were analysed. However, only one arrangement of Figure 1b with $a = b$, a porosity of 0.56, was considered.

The analysis of a two dimensional idealised media cannot be expected to give quantitatively useful results for application to real three dimensional media. The two dimensional model has considerably greater mathematical simplicity than a three dimensional one and is particularly useful in illustrating the analytical approach adopted. Further, if the flow characteristics of a simplified model cannot be clearly defined it is unlikely that the behaviour of a complex real material will be understood. Finally, the two dimensional model used acquires added significance when it is realised that flow through fibres and textiles is essentially two dimensional. Such flows, perpendicular to the longitudinal axis of the fibres, are considered by Lord (1955) for measuring fibre fineness.

THE DIFFERENTIAL RELATIONS

For steady, incompressible viscous flow in the $x - y$ plane the Navier-Stokes equations may be written

$$(1) \qquad \rho \frac{dV}{dt} = -\operatorname{grad}(\gamma h) + \mu \nabla^2 V$$

and continuity gives

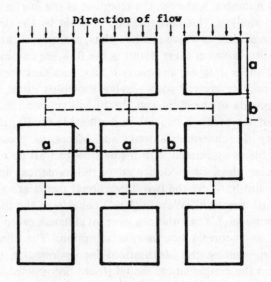

Figure 1a

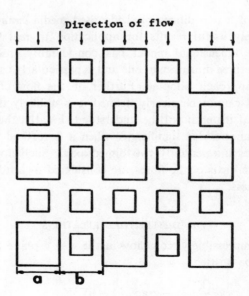

Figure 1b

Fig. 1. The two-dimensional models used in the numerical analysis

(2) $$\text{div } V = 0$$

where the velocity $V = u_i + v_j$, the piezometric head is h and the fluid has density ρ, specific weight γ, dynamic viscosity μ, and kinematic viscosity v.

If a stream function ψ is defined by

(3) $$u = -\frac{\partial \psi}{\partial y} \quad \text{and} \quad v = \frac{\partial \psi}{\partial x}$$

and the vorticity, ε, is given by

(4) $$\varepsilon = \frac{\partial v}{\partial x} - \frac{\partial u}{\partial y}$$

then a dimensionless form of the vorticity-transport equations (5) is obtained by eliminating the pressure terms from (1) and introducing representative parameters U (velocity) and L (distance) such that

$$\bar{u} = \frac{u}{U}, \quad \bar{v} = \frac{v}{U}, \quad \bar{x} = \frac{x}{L}, \quad \bar{\psi} = \frac{\psi}{UL}, \quad \bar{\varepsilon} = \frac{\varepsilon L}{U},$$

$$\bar{h} = \frac{h}{L}, \quad e \quad .$$

where the bar indicates a dimensionless quantity.

$$\nabla^2 \bar{\psi} = \bar{\varepsilon}$$

(5)

$$\nabla^2 \bar{\varepsilon} = RN \left(\frac{\partial \bar{\psi}}{\partial \bar{x}} \cdot \frac{\partial \bar{\varepsilon}}{\partial \bar{y}} - \frac{\partial \bar{\psi}}{\partial \bar{y}} \cdot \frac{\partial \bar{\varepsilon}}{\partial \bar{x}} \right)$$

and RN = Reynolds number = UL/v.

Further, the change in piezometric head between any two points is obtained by integration of the dimensionless form of (1), Stark and Volker (1967).

Thus, integrating along a line of constant value of '$\bar{y}$'

(6a) $$-\frac{\gamma L^2}{\mu U} h = RN \frac{\bar{V}^2}{2} - RN \int \bar{v}\bar{\varepsilon} d\bar{x} + \int \frac{\partial \bar{\varepsilon}}{\partial \bar{y}} d\bar{x} + f(y)$$

and along a line of constant value of $\bar{x}$

(6b) $$-\frac{\gamma L^2}{\mu U} h = RN \frac{\bar{V}^2}{2} + RN \int u\bar{\varepsilon} d\bar{y} - \int \frac{\partial \bar{\varepsilon}}{\partial \bar{x}} d\bar{y} + f(x) .$$

It is stressed that RN in the above relations is defined in terms of the representative values of U and L. These values may be selected as any velocity and length in the flow, e.g. the seepage velocity and a particular particle (or pore) size could be chosen. Any two flow patterns will be similar if (i) their pore geometries are similar and (ii) RN is identical for each and has been defined using corresponding values of U and L.

THE NUMERICAL MODEL

The numerical technique used to solve the dimensionless relations (5) follows the squaring method outlined by Thom and Apelt (1961). Finite difference approximations of these equations were solved iteratively at each node of a rectangular Eulerian mesh which was set up to cover one complete bay of the flow field, e.g. Figure 2.

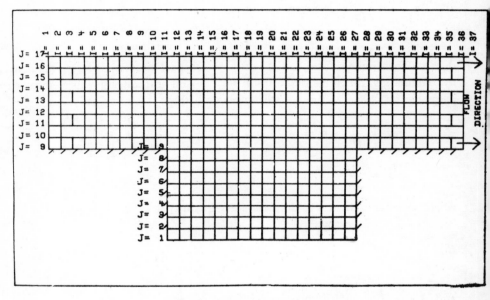

Fig. 2. Grid pattern for arrangement 1

The appropriate boundary conditions involving no-slip on the rigid boundaries, symmetry on fluid boundaries parallel to the flow direction and similarity of flow patterns at the entrance and exit of each bay were included to complete the field equations. The discontinuities at the corners of the squares required particular attention.

A detailed description of the numerical procedures, including stability and convergence considerations is given by Stark (1968). As the equations involved are nonlinear, a finite difference solution is required for each value of RN and successive solutions were obtained for each pattern of blocks by using the settled solution of a lower RN as the initial set of values for a higher flow.

Some 65 solutions were obtained for the patterns of Figure 1 with porosities ranging from 0.36 to 0.972, for the ratio $a : b$ from 0.2 to 4.0 and for Reynolds numbers from zero to 500 where $U = $ seepage velocity and $L = $ the side of the (larger) block.

Considerable difficulty was encountered in obtaining convergence at the higher flows and approximately 200 iterations were required for each solution to meet the convergence criterion.

For each solution, equations (5) were used to give values of $\bar{\psi}$ and $\bar{\varepsilon}$ at each grid point, the velocity values then follow from (3) and piezometric heads are obtained from (6). Contour plots of ψ and ε for each solution were obtained using a Calcomp incremental plotter to facilitate study of the flow profiles. Typical polts of ψ and ε at $RN = 50.0$ for the arrangement of Figure 1a with $a = b$ are shown as Figure 3.

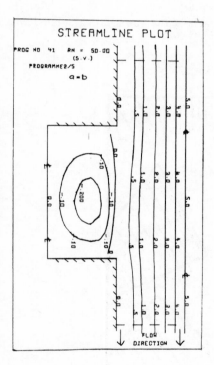
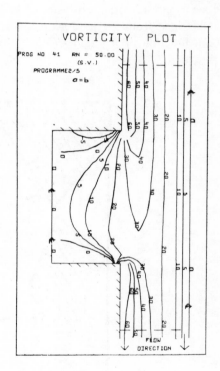

Fig. 3. Typical solution of Model 1a at $RN = 50$

MACROSCOPIC FLOW EQUATION

The dimensionless equations used in the numerical analysis illustrate very clearly the assumptions underlying the conventional formulae for flow in porous materials and it will be helpful to consider these developments before discussion of the numerical results. Although Irmay (1956) has shown the derivation of a Darcy and a Forchheimer relation, the following treatment underlies the rationale of these laws more closely.

If Figure 4 represents a typical domain of porous material, then theoretically the Navier-Stokes equations can be solved for flow in this domain. The solution obtained using equations (3), (5) and (6) with the appropriate boundary conditions, C, will result in actual values of the variables $\bar{\psi}$, $\bar{\varepsilon}$, u, $\bar{v}$, and h (referred to some datum) at every point in the flow. The discussion here is confined to two dimensional flow, although generalisations for the third dimension are straightforward. The boundary conditions, C, are directly determined from the pore geometry and each of the variables calculated will be a function of RN and C.

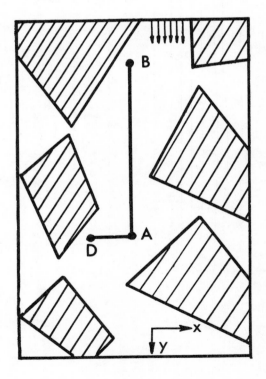

Fig. 4. A porous medium domain

The piezometric head change between any two points can be evaluated using equation (6a) and/or (6b). Thus, if the head loss is required between B and C, equation (6b) must be applied between B and A, whereas equation (6a) applies between A and C. As equations 6a and 6b are essentially similar in format, it is sufficient to consider points along the line BA and thus restrict our attention to equation (6b).

The generalised head loss relation
It follows then from 6b that

$$
(7) \qquad \frac{\gamma L^2}{\mu U} (h_B - h_A) = \alpha + \beta \cdot RN
$$

where

$$
\alpha = - \int_B^A \frac{\partial \bar{\varepsilon}}{\partial \bar{x}} d\bar{y}
$$

$$
\beta = \left(\frac{\bar{v}}{2} \right)_B^A + \int_B^A \bar{u} \bar{\varepsilon} d\bar{y}
$$

and, α and β are both functions of RN and C (the pore geometry). Thus, in general, α and β cannot be considered as constants, even if the pore geometry is fixed. For a particular domain of porous material, the values of $\bar{u}$, $\bar{\varepsilon}$, $\bar{v}$, etc. (and therefore α and β) will be dependent on RN.

Darcy flow linear laminar regime
If creeping flow is assumed, then $RN \to 0$ and equation 7 yields

$$
(8) \qquad \frac{\gamma L^2}{\mu U} (h_B - h_A) = \alpha
$$

where α will be a constant for a particular domain if $RN = 0$.
If the distance $BA = \Delta \bar{y}$ then

$$
(9) \qquad U = \frac{\gamma L^2}{\mu} \frac{\Delta \bar{y}}{\alpha} i
$$

where

$$
i = \frac{(h)_A^B}{\Delta \bar{y}}
$$

is the piezometric head gradient.
If the seepage velocity is chosen as the representative velocity, U, then (9) can be written as
$$
(10) \qquad U = Ki.
$$

This relation is of the same form as Darcy's equation in which K is defined as the permeability. Darcy derived this relation experimentally so that his values of 'i' and 'K' would represent the spatial average of these values.

It is apparent from equation (9) that the permeability can be defined over any length $\Delta \bar{y}$ and that a representative value for a medium would result if sufficient pores were passed in moving from B to A. The microscopic or point value of permeability is obtained by allowing $\Delta \bar{y} \to 0$ and this value would vary from point to point.

Irmay introduced the hydraulic conductivity a_0, which is the reciprocal of K, thus

$$(11) \qquad a_0 = 1/K = \frac{\mu}{\gamma L^2} \frac{\alpha}{\Delta \bar{y}}$$

It is stressed that a_0, K and α are constants for a particular domain only if RN is zero.

The effect of the porosity function on α is not immediately obvious (see Kozeny flow) because, although at zero RN the velocity profiles in the narrowest gaps between particles will approach a parabolic distribution, in the wider spaces and in the wake regions at the rear of particles, reverse flow patterns confuse the prediction of velocity profiles.

The Carman-Kozeny relation

Kozeny (1927) and Carman (1937) introduced a relation which is widely used for creeping flows. This relation attempts to define the value of $(\alpha/\Delta \bar{y})$ as a function of porosity by introducing the hydraulic radius theory. In particular, the Carman-Kozeny relation for creeping flow can be deduced from equation (9) by the following assumptions

(1) U = seepage velocity,

(2) L = hydraulic radius, R, and

(3) $\left(\dfrac{\alpha}{\Delta \bar{y}}\right) = \dfrac{K_1}{n}$ where n is the porosity and K_1 was introduced as a con-

stant for all media and given the value of 5. Thus

$$(12) \qquad i = \frac{\mu}{\gamma} \cdot \frac{U}{n} \cdot \frac{K_1}{R^2}$$

and by introducing the specific surface (S_0) defined by

$$S_0 = \frac{\text{surface area of particles}}{\text{volume of particles}}, \text{ the Carman-Kozeny relation is obtained}$$

$$(13) \qquad i = \frac{\mu}{\gamma} U K_1 S_0^2 \frac{(1-n)^2}{n^3} \ .$$

A great variety of alternative porosity factors have been posed and it is very unlikely that the simple relation required to validate the Carman-Kozeny factor will be universally applicable. Irmay (1958) did obtain the porosity relation in (13) by assuming a parabolic type velocity distribution within the pores, however, as explained above this assumption cannot be universally justified.

The numerical solutions obtained for Figure 1 and a great amount of experimental evidence, e.g. Lord (1955) and Rose (1951) show that the porosity factor in (13) is unsatisfactory. In fact, an alternative porosity factor of $[(1 - n)^{1.5}/n^3]$ would appear to fit all of these results, covering a range of porosity from .36–.99, much better than any other relation known to the author. However, this factor has been obtained empirically and it would be unwise, in view of the great variety of possible pore shapes, to suggest that it is universally applicable.

FORCHHEIMER FLOW

If the creeping flow assumption of Darcy's Law is relaxed, then equation (7) is pertinent in which α and β are functions of RN and the pore geometry. However, if a limited range of flows (RN's) is considered such that the values of $\bar{\varepsilon}, \bar{u}, \bar{V}$, etc. remain approximately constant between A and B, then α and β will also remain constant. This assumption implies that the velocity profiles between A and B remain constant over the range of flows considered and is obviously untrue; however, in the same way that Darcy's law can be used for non zero creeping velocities, this assumption validates the Forchheimer relation (14) over a limited range of flows.

Thus, $i = \mu U/\gamma L^2 \cdot (1/\Delta \bar{y})[\alpha + \beta . RN]$, i.e.

$$(14) \qquad\qquad i = a_0 U + b_0 U^2$$

where a_0 and b_0 are constants in the Forchheimer equation, a_0 is the hydraulic conductivity of Irmay (11)

$$b_0 = \frac{\beta}{Lg\Delta \bar{y}}$$

U is the seepage velocity and L a representative length.

The numerical results presented indicate typical flow ranges over which the Forchheimer equation is valid with an acceptable accuracy.

Tek (1957) adopted a relation which is essentially a Forchheimer equation and his experimental data fits the relation with good agreement for RN from 0 to 5 but with decreasing accuracy for RN from 5 to 10. A number of other empirical relations are also of the Forchheimer type, e.g. Lindquist (1931), however, many relations have been deduced to fit a particular set (or sets) of experimental data and appear to have no rational basis.

THE NUMERICAL RESULTS

The numerical solutions are summarised in Table I for the various arrangement of Figure 1 that have been considered. The results are given in terms of H_G', which is defined as

$$(15) \qquad H_G = \frac{\alpha + \beta.RN}{\Delta \bar{y}}$$

where $\Delta \bar{y}$ is taken as the length of one bay (e.g. from $I = 3$ to $I = 35$ in Figure 2) and RN is defined as in (5), with U as the seepage velocity and L as the side of the (larger) square block (i.e. 'a' in Figure 1).

For creeping flows—the macroscopic permeability for the porous media as defined in 10, becomes

$$(16) \qquad K = \frac{1}{a_0} = \frac{\gamma L^2}{\mu} \cdot \frac{1}{H_G}$$

the Kozeny constant K_1 (as in (13)) will be given by

$$(17) \qquad K_1 = H_G \frac{n^3}{(1-n)^2} \cdot \frac{1}{S_0^2 L^2}$$

whereas for non-creeping flows the Forchheimer relation, (14), becomes

$$(18) \qquad i = (a_0 + b_0 U)U = \mu/\gamma L^2 \cdot H_G \cdot U.$$

TABLE I.

Numerical solutions for flows through the idealised models of Figure 1

Model — Figure 1a					Figure 1b
$a/b = 1$ $n = 0.75$	$a/b = 2$ $n = 0.56$	$a/b = 2/3$ $n = 0.84$	$a/b = 2/5$ $n = 0.918$	$a/b = 1/4$ $n = 0.96$	$a/b = 1$ $n = 0.56$
RN H_G	RN H_G	RN H_G	RN H_G	RN H_G	RN H_G
0.0 18.78	0.0 123.2	0 6.62	0.0 1.86	0 0.55	0.0 368.0
.05 18.79	6.6 123.9	4 6.75	2.86 1.89	2 0.60	2.5 369.2
.25 18.80	33.3 128.4	20 7.20	14.29 2.06	10 0.67	5. 372.9
.5 18.81	50.0 130.3	40 7.50	28.57 2.20	20 0.77	15. 394.1
2.5 18.86	66.6 131.3	80 7.67	57.14 2.31	40 0.78	25. 417.2
5.0 19.02	133.3 134.3				50. 484.2
10.0 19.42					100. 509.6
15.0 19.80	$a/b = 4$ $n = 0.36$	$a/b = 1/2$ $n = 0.889$	$a/b = 1/3$ $n = 0.937$	$a/b = 1/5$ $n = 0.972$	*Comments*
25.0 20.22					
50.0 20.82	RN H_G	RN H_G	RN H_G	RN H_G	$RN = UL/\nu$
100.0 21.44	0 876.2	0.0 3.21	0. 1.16	0.0 0.30	U = Seepage veloc.
150.0 21.62	8 879.6	3.3 3.30	2.5 1.20	1.6 0.31	L=particle size, a
200.0 21.66	40 898.3	16.6 3.57	12.5 1.32	8.3 0.41	$H_G = (a+\beta.RN)/\bar{y}$
250.0 21.62	50 901.8	33.3 3.72	25.0 1.42	16.60 1.42	
500.0 21.62	80 911.9	66.6 3.84	50.0 1.50	3.3 0.45	(See equation 15)
					n = porosity

Fig. 5
$RN = 0$, $a=b$, $n=0.75$
Streamline plot

Fig. 6
$RN=5$, $a=b$, $n=0.75$
Streamline plot

Fig. 7
$RN=200$, $a=b$, $n=0.75$
Streamline plot

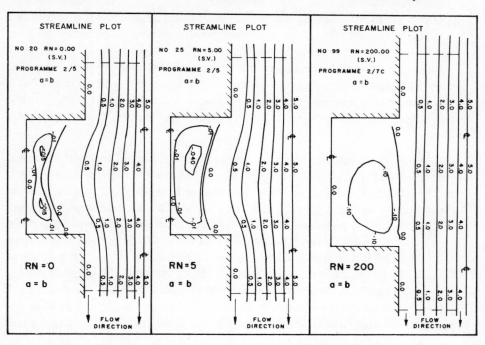

DISCUSSION OF RESULTS

Range of application of Darcy's Law. Darcy's law is applicable over any flow flow range if H_G remains constant (or approximately so). The tabulated nume-- cal results indicate that H_G varies with RN for any given model arrangement. However, as indicated in the analyttical treatment above, the error involved in the application of Darcy's law over low flow ranges is small. Thus, if the model of Figure 1a is considered, the permeability at zero RN can be applied for RN below 5, with an error of less than 1%, providing the porosity (n) is less than 0.75. If the porosity is greater than 0.75 or the model of Figure 1b is considered then the same accuracy for Darcy's law requires a smaller range of flows.

Forchheimer's relation. As RN is increased, the effect of the inertia terms is observed in the tabulated results of H_G. It is apparent that the value of β is not constant. A study of the flow patterns with increasing RN illustrates the variation in the velocity profiles. At very low RN the flow penetrates the wake region behind the particles, but as RN increases, the line of separation between the wake vortex bubble and the main flow moves towards the main flow channel with the main streamlines becoming virtually parallel. Under these circumstances β approaches zero and, in fact, a Darcy law could be applied over the particular range of flows involved with, of course, a different value of α or K from that which pertains at creeping flows. The tendency for this approach to parallel flow is reduced by introducing additional particles or aligning the particles at some other angle to the flow. It is doubtful whether the constant values of H_G indicated at RN values from 200 to 500 for $a/b = 1$ would ever be attained in practice because the onset of turbulence would introduce additional nonlinear terms.

In the analytical discussion of Forchheimer's relation, it was shown that this law is applicable over limited ranges of flow. The numerical results support this. Thus, consider the model of Figure 1.1a with $a = b$ (i.e. $n = 0.75$) then at $RN = 0$, $H_G = 18.78$. If this value is assumed constant, then $\beta/\Delta\bar{y}$ decreases from .06 at $RN = 0.5$ to .0056 at $RN = 500$. If now the value of $\beta/\Delta\bar{y}$ at $RN = 5$ (i.e. .042) is selected as a constant value, it can be used over the range of flows from $RN = 0$ to 50 with a maximum error of 2% which occurs at $RN = 25$. When considering experimental results, the value of $\alpha/\Delta\bar{y}$ at zero RN is not always accurately known. Thus, the experimental results could be described in terms of a Forchheimer law even more accurately than indicated above by selecting an appropriate value of α. Finally, it should be noted that if the values of α and β, used above, are assumed for $RN = 100$, the error involved is 7%. It should be stressed then that experimental determinations of the Forchheimer coefficients should never be extrapolated outside the experimental range.

Porosity relations. The Carman-Kozeny porosity relation can be studied by considering the creeping flow solutions in Table I. Table II gives the value of the Kozeny constant — K_1 in equation (17) for these flows.

TABLE II

Kozeny Constant — K_1 Values (Figure 1a — Model)

$a/b =$	4	2	1	.67	.50	.40	.33	.25	.20
$n =$	.36	.56	.75	.84	.889	.918	.937	.96	.972
$K_1 =$	6.2	7	7.9	9.6	11.4	13.5	15.4	19.6	22.14

The following points should be noted — the value of 'K' is obviously not constant with respect to porosity and therefore the $C - K$ porosity function is inadequate, particularly at high porosities. The variation of K_1 with porosity indicated in Table II compares favourably with the variations given by Happel and Brenner for flow perpendicular to an array of circular cylinders. If the 'tortuosity factor' suggested by Fowler and Hertel (1940) is adopted to compare flow through random and regular arrangements of particles, the modified values of K_1 are coincident with the experimental values of K_1 determined by Lord (1955) for flow through wool fibres with a porosity range of .7 to .99. To obtain this modified K_1, Fowler and Hertel suggest that the calculated value of K_1 should be divided by 1.33. For $n < 0.6$, these modified values compare favourably with the relation deduced experimentally by Rose (1951).

Finally, if the alternative porosity factor suggested in the discussion of the macroscopic flow equations above, viz. $[(1-n)^{1.5}/n^3]$, is used in place of the $C - K$ factor, it is found that the variation of K_1 with porosity is confined within the range 2.5–3.7 for all the results of Happel and Brenner, Rose, Lord (wool fibres) and the numerical results presented herein. It would appear then that this alternative porosity factor is more reliable than the conventional one.

THE FLOW PATTERNS

A number of interesting features of flow within porous materials can be distinguished by studying the flow patterns produced by the computer. Some of these patterns are illustrated in Figures 5 to 11.

The development of the wake bubble varies for the different arrangements; however, a summary of the patterns obtained, following a study of all the solutions, can be given as follows: For $a = b$ at zero RN, the vortex bubble appears to have a double barrelled shape and as a/b becomes less than 1, i.e

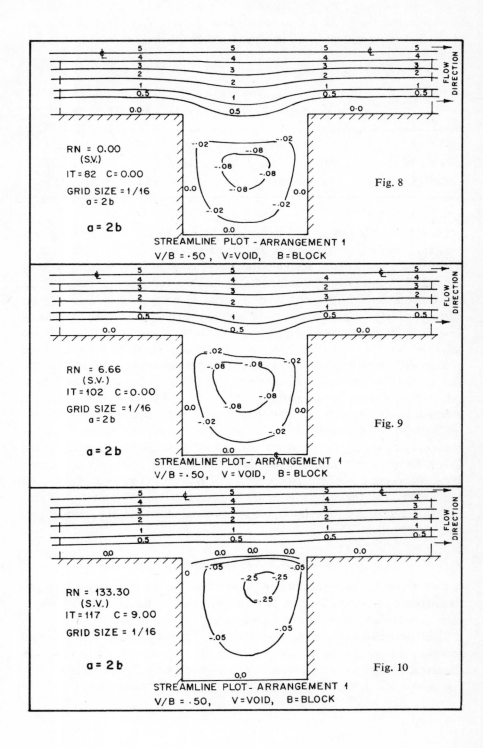

RN = 0.00
(S.V.)

IT = 82   C = 0.00

GRID SIZE = 1/16
a = 2b

a = 2b

Fig. 8

STREAMLINE PLOT - ARRANGEMENT 1
V/B = ·50,   V = VOID,   B = BLOCK

RN = 6.66
(S.V.)

IT = 102   C = 0.00

GRID SIZE = 1/16
a = 2b

a = 2b

Fig. 9

STREAMLINE PLOT - ARRANGEMENT 1
V/B = ·50,   V = VOID,   B = BLOCK

RN = 133.30
(S.V.)

IT = 117   C = 9.00

GRID SIZE = 1/16

a = 2b

Fig. 10

STREAMLINE PLOT - ARRANGEMENT 1
V/B = ·50,   V = VOID,   B = BLOCK

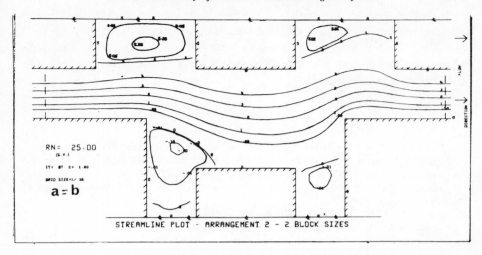

Fig. 11

higher porosities, this shape disappears and the dividing streamline penetrates
to the base of the cavity giving two vortex bubbles, one in each corner of the
cavity. On the other hand, as a/b is increased, i. e. $n < .75$, the bubble tends to
grow in the other direction (at right angles to the flow direction) until a second
very weak vortex, rotating in the opposite direction to the dominant vortex,
becomes evident with $a/b = 4.$ The vortex bubble tends to increase in size
with RN as the separating streamline — separating the main flow and the
vortex flow — becomes parallel with the base of the cavity and joining the two
corners of the particles.

The symmetry of the flow at zero RN is evident and the similarity of flow
patterns, in the main flow patterns, in the main flow channel, at different
Reynolds numbers is also obvious.

CONCLUSIONS

The enormous calculational potential of digital computers has made it possible
to consider the analysis of the complete fluid dynamics equations within
porous materials. An extremely simplified model has been considered in this
paper; however, the problem has already been extended to a three-dimensional
model involving an idealised arrangement of cubes. The three-dimensional
problem presents additional logistic difficulties in handling and interpreting
the increased output; however, the possibilities of computer graphics and
movies will undoubtedly reduce these difficulties. A more efficient analytical
method introduced by Chorin (1967) is recommended for three-dimensional
problems.

The simplified model analysed has shown the sufficiency of Darcy's and
Forchheimer's laws within specified ranges of flow. The solutions obtained

indicate that a revised porosity function should be incorporated in the Carman-Kozeny relation and that a detailed description of the flow characteristics is possible. Further work is required to consider variations in particle sizes, shapes and orientation and to extend the analysis by statistical methods into complex pore geometries.

REFERENCES

CARMAN, P. C. (1937), *Trans. Inst. Chem. Engrs.*, London, *15*, 150–166.

CHORIN, A. . (1967), *A.E.G. Res. & Dev. Rep. No. NYO*-1480–82, New York Univ.

FOWLER, S. L. and HERTEL, K. L. (1940), *J. Appl. Phys.*, *11*, 496.

HAPPEL, J. AND BRENNER, H. (1965), *Low Reynolds Number Hydro-dynamics*, Prentice Hall, Inc., Englewood Clifls, *N. J.*

IRMAY, S. (1956), La Houille Blanche, No. 3, 419.

IRMAY, S. (1958), *Trans Am.. Geophys. Un.*, *39*, (4):702–707.

KOZENY, J. (1927), Wassers in Boden, *Setz-Ber. Weiner Akad.*, *Abt. IIa, 136*, 271–306.

LINDQUIST, E. (1933), 1st Congress Large Dams, Stockholm, *ö*, 81–101.

LORD, E. (1955), *J. Text. Inst.*, *46*, T191-213.

ROSE, H. E. (1951), *Some Aspects of Fluid Flow*, Edward Arnold & Co., London, pp. 136–162.

SCHEIDEGGER, A. E. (1953), *Producers monthly 17*, 10):17–23.

STARK, K. P. (1958), Proc. IASH/AIHS — UNESCO Conf., *Tucson, Pub.* No. 80 AIHS, *2*, 635–649.

STARK, K. P. AND VOLKER, R. E. (1967), *Res. Bull. 1, Dept. Eng., Univ. College, Townsville, Australia.*

TEK, M. R. (1957), *Trans. AIMME, 210*, 376–378.

THOM, A. AND APELT, C. J. A. (1961), *Field Computations in Engineering and Physics*, D. van Nostrand Co., Ltd.

WATSON, K. K. (1964), *Proc. 4th Aust. N. S. Conf. on Soil Mech. and Found. Eng.*, 37–40.

WRIGHT, D. E. (1968), *J. ASCE, Hyd. Div.*, July *11–14*, 851–871.

UNIVERSITY COLLEGE OF TOWNSVILLE, NOW, JAMES COOK UNIVERSITY
OF NORTH QUEENSLAND, TOWNSVILLE, QUEENSLAND, (4810), AUSTRALIA

PROBLEMS CONCERNING SOLUTION
OF STEADY AND UNSTEADY GROUNDWATER
FLOW BY STATISTICAL METHODS

V. Halek and M. Novak

This paper deals with the groundwater flow in a pervious medium consisting of $J + 1$ layers. Limiting lines between adjacent layers are parallel. Stratification may be inclined in respect to the horizontal line. But theoretically it is admissible that this case may be changed into a case where stratification is considered to be horizontal. The existence of a uniform flow regime in the strata is considered as an important condition. Water flows mainly horizontally and no influence of vertical resistance component is predominant in the layers. We then tacitly assume a linear pressure increase on the vertical line. We also admit the occurrence of a free water level with its positional changes depending on time. The corresponding vertical velocity drops linearly from the maximum on the free water level to the zero on the surface of the relatively impervious bed.

In general for a horizontal plane x, y it is possible to assume a perpendicular, and along it to calculate the arbitrary position of the free water level, where pressure is considered to be equal to zero. Thus according to known laws the value of the potential for a point on the free water level is determined. If these values are integrated along the z-axis with respect to all possible positions of the water level, then the function Φ is obtained and may be expressed as

$$(1) \qquad \Phi = - h \left[k_{J+1} \frac{h}{2} + \sum_{J=1}^{J} T_J(k_J + k_{J+1}) \right] + \frac{1}{2} \sum_{J=1}^{J} T_J^2(k_J - k_{J+1})$$

where h = distance of point on free waters level in limits of $J + 1$ layer from surface of relatively impervious bed (Fig. 1).

k_j = permeability coefficient of jth layer

T_j = distance of jth layer upper boundary from surface of relatively impervious bed.

The function Φ is obviously a formal variant of Girinsky's potential known from the theory of surface filtration [6]. It is advantageous that the derivatives of the function Φ along the coordinate axes offer directly discharge components in the direction of these axes.

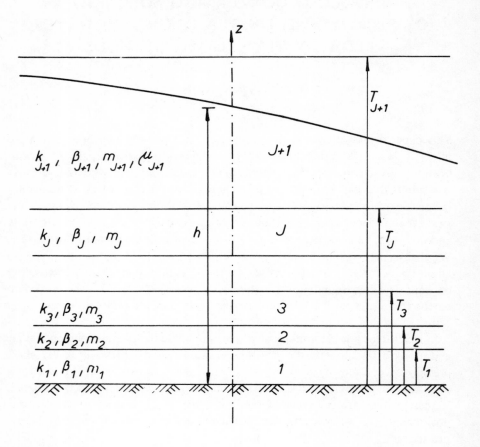

Fig. 1

Fig. 1. Problems concerning solution of steady and unsteady groundwater flow by statistical
        methods

Further suppose that each layer is elasticly compressible and consequently
its material has a certain modulus of compressibility β_J. Under the free water
level along any perpendicular the average value β is

(2)
$$\beta = \frac{\beta_{J+1}h + \sum\limits_{J=1}^{J} T_J(k_J - k_{J+1})}{h}.$$

Similarly the porosity m_J of each layer is practically considered a constant. Along the given perpendicular the average porosity m is

$$(3) \qquad m = \frac{m_{J+1}h + \sum\limits_{J=1}^{J} T_J(m_J - m_{J+1})}{h} .$$

Suppose the unsteady state flow in time t and the free water level moving in the limits of $(J + 1)$ layer. Then it may be easily proved that under the given conditions the following equation results

$$(4) \qquad \frac{\partial \Phi}{\partial t} = a \left(\frac{\partial^2 \Phi}{\partial x^2} + \frac{\partial^2 \Phi}{\partial y^2} \right)$$

where

$$(5) \quad a = \frac{k_{J+1} + \dfrac{1}{h} \sum\limits_{J+1}^{J} T_J(k_J - k_{J+1})}{\left[\beta_{J+1} + \dfrac{1}{h} \sum\limits_{J=1}^{J} T_J(\beta_J - \beta_{J+1}) + \dfrac{m_{J+1} + \dfrac{1}{h} \sum\limits_{J=1}^{J} T_J(m_J - m_{J+1})}{E_v} \right] + \dfrac{\mu_{J+1}}{h}}$$

γ_v — specific volume weight of water

μ_{J+1} — active porosity of $(J + 1)$ layer

In general the Eq. (4) is nonlinear. However, it may be linearized if h in Eq. (5) equals a suitably chosen average value h_s.

Then the solution of Eq. (4) is more convenient under certain boundary and initial conditions. Suppose that the process is investigated in an area closed by a curvature S along which is valid

$$(6) \qquad \Phi = \Phi(x_s, y_s, t) .$$

The initial condition (time $t = 0$) may then express the steady state as

$$(7) \qquad \Phi = \Phi(x, y, 0) .$$

It is obvious that from the value $\Phi(x, y, t)$ we may derive the ordinate of the point on the free water level according to Eq. (8), which follows from Eq. (1)

$$(8) \qquad h = \frac{- \sum\limits_{J=1}^{J} T_J(k_J - k_{J+1}) \pm \sqrt{\left[\sum\limits_{J=1}^{J}(T_J(k_J - k_{J+1}) \right]^2 - 2k_{J+1} \left[\Phi - \frac{1}{2} \sum\limits_{J=1}^{J} T_J^2(k_{J+1}) \right]}}{k_{J+1}}$$

By finding the value Φ the task is solved. It is helpful to substitute

$$(9) \qquad \Delta\Phi(x, y, t) = \Phi(x, y, t) - \Phi(x, y, 0) .$$

Instead of Eq. (4) there results

(10)
$$\frac{\partial(\Delta\Phi)}{\partial t} = a\left[\frac{\partial^2(\Delta\Phi)}{\partial x^2} + \frac{\partial^2(\Delta\Phi)}{\partial y^2}\right]$$

where

(11)
$$\Delta\Phi(x, y, 0) = 0$$

and on the boundary S

(12)
$$\Delta\Phi = \Delta\Phi(x_s, y_s, t).$$

Eq. (10) is parabolic and its theory has been worked out quite sufficiently both in mathematics and mathematical physics. The presented references (e.g. [1], [2], [3], [7], [8]) may be applied and the well-known methods may be adapted to serve our formulations and peculiarities of defining the problem of the ground water movement.

First one should notice that the function $\Delta\Phi$ characterizes the difference Δh, showing how the position of the free water level changes during the non-stationary process. Besides other factors, the movement is caused by elastic deformation and accumulation. Over a unit of plan surface the following volume of liquid is accumulated

(13)
$$dV = -\frac{\Delta\Phi}{a}.$$

Imagine now that the process is followed retrospectively in time $\tau < t$, and Eqs. (10), (11), (12) are valid. There is one difference, namely τ instead of t, to be considered.

Further consider inside S the function $\Delta\Phi^*(x, y, t)$ which has continuous derivatives of the first and second order and which equals zero on the boundary S. The function $\Delta\Phi^*$ is given by Eq. (10). Since it is operated on in the interval $(t - \tau)$ it may be expressed as

(14)
$$-\frac{\partial(\Delta\Phi^*)}{\partial\tau} = a\left[\frac{\partial^2(\Delta\Phi^*)}{\partial x^2} + \frac{\partial^2(\Delta\Phi^*)}{\partial y^2}\right].$$

For the interior of the region

(15)
$$\Delta\Phi^*(x, y, 0) = 0.$$

On the boundary

(16)
$$\Delta\Phi^*(x_s, y_s, t) = 0.$$

These conditions may be satisfied physically by a source placed at the point $A(x, y)$ whose function $\Delta\Phi^*$ is not continuous. But in all the remaining region the continuity is preserved. Visualize first the action of the source so that at time τ it suddenly appears and immediately vanishes. In the interior, due to elasticity and accumulation, the liquid volume is increased by the value V. After some time the liquid has to flow out across the boundary S.

The functions $\Delta\Phi$, $\Delta\Phi^*$ are scalar and it is possible to consider the derivative of their product with respect to τ

$$(17) \qquad \frac{\partial[\Delta\Phi\Delta\Phi^*]}{\partial\tau} = a\left\{\Delta\Phi^*\left[\frac{\partial^2(\Delta\Phi)}{\partial x^2} + \frac{\partial^2(\Delta\Phi)}{\partial y^2}\right] - \Delta\Phi\left[\frac{\partial^2(\Delta\Phi^*)}{\partial x^2} + \frac{\partial^2(\Delta\Phi^*)}{\partial y^2}\right]\right\}.$$

Both sides of the equation are integrated, and the values of the integral from zero to t is calculated. Eq. (15) is valid when $t = 0$. Then

$$(18)$$

$$[\Delta\Phi\Delta\Phi^*]_t = a \int_0^t \left\{\Delta\Phi^*\left[\frac{\partial^2(\Delta\Phi)}{\partial x^2} + \frac{\partial^2(\Delta\Phi)}{\partial y^2}\right] - \Delta\Phi\left[\frac{\partial^2(\Delta\Phi^*)}{\partial x^2} + \frac{\partial^2(\Delta\Phi^*)}{\partial y^2}\right]\right\} dt.$$

A spatial integration can be performed over S, while changing the sequence of integration

$$(19) \qquad \int\int [\Delta\Phi\Delta\Phi^*] dx dy$$

$$= a \int_0^t \int\int\left\{\left\{\int\int\Delta\Phi^*\left[\frac{\partial^2(\Delta\Phi)}{\partial x^2} + \frac{\partial^2(\Delta\Phi)}{\partial y^2}\right] - \Delta\Phi\left[\frac{\partial^2(\Delta\Phi)}{\partial x^2} + \frac{\partial^2\Delta\Phi}{\partial y^2}\right] dx dy\right\}\right\} d\tau.$$

On the left side the function $\Delta\Phi$ is considered as the value of the unknown function of the point $A(x, y)$, and is time-dependent. It should be noted that $\Delta\Phi^*$ characterizes the effect of a source which suddenly appears and immediately vanishes. And the surface integral gives the liquid volume by which the original volume is increased. From the function $\Delta\Phi^*$ is follows that

$$[\Delta\Phi\Delta\Phi^*]_t dx dy = \Delta\Phi(x, y, t) \int\int \Delta\Phi_t^* dx dy$$

$$= -\Delta\Phi(x, y, t) \int\int a d V dx dy = -\Delta\Phi(x, y, t) a V.$$

Green's formula is used on the right side of Eq. (19) and Eqs. (16), (12) are considered to be valid. After rearrangement it becomes

$$(20) \qquad \Delta\Phi(x, y, t) = \frac{1}{V} \int_0^t \left[\int\int_s \Delta\Phi(x_s, y_s, \tau) \frac{\partial(\Delta\Phi^*)}{\partial n} dS\right]$$

where $\partial(\Delta\Phi^*)/\partial n$ is the derivative along the inward directed normal of the surface s. Physically this expression represents the discharge per unit length ·of the boundary. If $\Delta\Phi(x_s, y_s, \zeta) = $ constant, then the function under the integral epresents the volume of liquid V_s, flowing in time t across the boundary S. This special case may be expressed

(21)
$$\frac{\Delta\Phi(x,y,t)}{\Delta\Phi(x_s,y_s,t)} = \frac{V_s}{V} = f(x,y,t) .$$

It is obvious that the given task will be solved under assumption that we know the mathematical description of the nonsteady state process caused by the suddenly appearing and vanishing source. After insertion of the proper expression into Eq. (20) the value will result from the prescribed operations. It is also possible to use the function $\Delta\Phi$ as in Eq. (21), which may be determined by calculation or by experiment. According to the principle of superposition, it is valid in general for the boundary condition on S that

(22)
$$\Delta\Phi(x,y,t) = \int_0^t f(x,y,t-\tau)\frac{\partial}{\partial\tau}[\Delta\Phi(x_s,y_s,\tau)]d\tau .$$

Using the finite differences $\Delta t = t_i - t_{i-1}$

(23)
$$\Delta\Phi(x,y,t) = \sum_{i=0}^{n} [\Delta\Phi(x_s,y_s,t-t_{i+1}) - \Delta\Phi(x_s,y_s,t-t_i)]f(x,y,t-t_i)$$

or, alternatively

(24)
$$\Delta\Phi(x,y,t) = \int_0^t \Delta\Phi(x_s,y_s,\tau)\frac{\partial}{\partial\tau}[f(x,y,t-\tau)]d\tau$$

and using the same finite differences

(25)
$$\Delta\Phi(x,y,t) = \sum_{i=1}^{n} \Delta\Phi(x_s,y_s,t)[f(x,y,t-t_i) - f(x,y,t-t_{i+1})] .$$

The principle of superposition makes possible further consideration. Assume, for example, that in time t at the point x,y several coincident sources suddenly appear and vanish one after another in the same place. For the short moment of the source existence

$$V = Qd\tau$$

where Q is the average discharge.

We may assume that after a regular repetition of the process a volume W will penetrate into the region from the sources. Then

$$W = \int_0^t Qd\tau = Qt .$$

The total volume W_s will penetrate through the boundary s in time t.

Then instead of Eq. (21) the following expression may be accepted

(26)
$$f(x,y,t) = \frac{W_s}{W} .$$

If the described process lasts a very long time we find that the flow gains a steady character and

$$W_s \to Q_s t$$

where Q_s is the discharge through the boundary s and

(27) $$f(x, y, t) = \frac{Q_s}{Q} = \text{const.}$$

1. FORMAL ANALOGY BETWEEN STEADY GROUND-WATER FLOW AND LAWS OF ACCIDENTAL MOVEMENT OF POINT

It is practical and advantageous to apply Eq. (20) for solving ground-water movement because it makes the solution of the problem easier in the case where no boundary condition is constant along the boundary S.

Imagine, for example, that the boundary s is divided into I parts so that $\Delta\Phi(x_s, y_s, \tau) = \Delta\Phi_i(x_s, y_s, \tau) = \text{const.}$

(28) $$\Delta\Phi(x, y, t) = \frac{1}{V} \int_0^t \left[\sum_{i=1}^I \Delta\Phi_i(x_s, y_s, \tau) \int_{s_i} \frac{\partial(\Delta\Phi^*)}{\partial n} ds \right] d\tau .$$

Equation (28) is simplified to express the steady flow. It becomes

(29) $$\Delta\Phi(x, y) = \sum_{i=1}^I \Delta\Phi_i \frac{Q_i}{Q}, \quad i = 1, 2, 3, \cdots, I$$

(30) $$Q_i = \int_{s_i} \frac{\partial(\Delta\Phi^*)}{\partial n} ds .$$ (30)

In accordance with the usual theory of plane steady filtration one may also use the function $\Delta\psi^*$, which makes possible finding Q_i in ith part,

(31) $$Q_i = \int_{s_i} d(\Delta\psi^*) ds .$$

Now another problem should be considered. This is the problem mentioned by I. Petrowsky [11] who has investigated the laws of accidental movement of a point. The movement originates at $A(x, y)$ inside a region E. The movement is supposed to proceed in steps on elementary straight paths of the same lengths and forming a continuity. The results of the consideration prove the existence of probability $p(x, y)$ that after a finite number of elementary steps a particle will hit some point of the set of points inside E which

(32) $$p(x, y) = \int p(\xi, \eta)\alpha(x, y, dE) .$$

$\alpha(x, y, dE)$ is the probability that the point will hit the given element of the boundary E. It is proved that the probability of hitting the whole boundary equals certainty and $\alpha(x, y, E) = 1$. I. Petrowsky further proves the relation between this law and the general partial differential equation of elliptic type for the function $U(x, y)$. He considers Dirichlet's problem, i.e. the mathematical description of a phenomenon in the field assuming that on the boundaries of the region the values $U(x, y)$ are given.

For the present we may neglect the general conclusions of I. Petrowsky's work [11] and preferably analyze Eq. (32) as being valid for the whole region inside E, except at the point at which the movement starts.

It may be expected that all the points of a small circle circumscribing the starting point inside E may be hit with the same probability. For the circumference of the circle $\alpha = 1$. If the circle is divided into ω number of equal parts, then the probability of hitting the points on one of the parts is

$$\alpha = 1/\omega .$$

Assume now that inside E a circle is drawn around the starting point $A(x, y)$, and on its circumference assume a centre of another circle having the same radius. The coordinates of the second circle are x_1, y_1. When moving from the point x, y the probability of hitting the point x_1, y_1 is the same as the probability of hitting the point x, y when starting from x_1, y_1.

Consider further a ω polygon with the centre A and with circles drawn around its apexes. All of them intersect at the point A (Fig. 2a). Investigate the movement originating at A and consider only ω part of the circle drawn around A and intersecting all the apexes of the ω polygon. The probability of hitting one of the apexes lying on one part of the circle is denoted by

$$\frac{1}{\omega} \rho(x_J, y_J) .$$

On the contrary if we start in succession from all the apexes of the ω polygon, then the probability of hitting the point may be found reciprocally as the sum of partial probabilities

$$(33) \qquad \rho(x, y) = \frac{1}{\omega} \sum_{J=1}^{\omega} \rho(x_J, y_J), \quad J = 1, 2, 3, \cdots \omega .$$

Eq. (23) satisfies the theorem of the average value. The validity of this theorem is one of the specific properties of harmonic functions. When considering infinitely small polygons and assuming $\omega \to \infty$. Then for all the system of neighbouring elements

$$\frac{\partial^2 \rho}{\partial x^2} + \frac{\partial^2 \rho}{\partial y^2} = 0 .$$

Consider a small circle with centre at A. We may be sure that the movement started from A will hit some point of the circle. At this point another

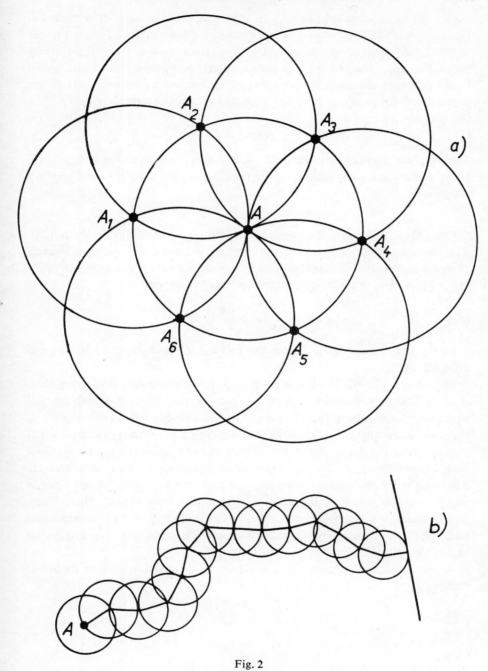

Fig. 2

Fig. 2. Problems concerning solution of steady and unsteady groondwater flow by statistical methods

circle will be constructed, and again, it is sure that the next movement will hit some point of the circle. The line connecting the centres of these successively constructed circles characterizes the track passed from the starting point (Fig. 2b). The whole complex of similarly performed movements results in the fact that the highest probability of hitting occurs at the points in the nearest vicinity of A. It may be deduced that in the nearest vicinity of A the probability is

$$(35) \qquad\qquad\qquad p(x, y) = 1.$$

If we place the starting point on E we find that the assumption is disproved, because there is no possibility for the point to move on E. On the boundary therefore

$$(36) \qquad\qquad\qquad p(x, y) = 0.$$

Eqs. (35) and (36) give the boundary conditions for the solution of Eq. (34). The harmony of a function is conditioned by its continuity and the existence of its derivative. In this case these conditions are not satisfied only by the starting point A. For other points inside E it may be written

$$(37) \qquad\qquad\qquad d\rho = \frac{\partial\rho}{\partial x}dx + \frac{\partial\rho}{\partial y}dy.$$

In Eq. (37) $\partial\rho/\partial x, \partial\rho/\partial y$ denote the increase of probability along the coordinate axes.

Then one path will be chosen from all the collection of paths going from A to E. This path should be the most expectant (in the sense of mathematical expectation) for hitting from the point inside chosen before. Of course, on this most expectant path the probability of hitting the points changes so that it drops from the maximum in the vicinity of A to the minimum on E. Consider the collection of the most expectant paths inside E. You find there are points on the paths whose connecting line gives the geometric place of points which may be hit with the same mathematical expectation. These lines, $\rho = $ constant, are obviously perpendicular to the courses of the most expectant paths which are denoted by symbols $\sigma = $ const. The condition of orthogonality in Eq. (38) result in Eqs. (3y) and (40).

It is possible to determine the spectrum of the most expectant paths according to

$$(38) \qquad\qquad\qquad \frac{\partial\rho}{\partial x}\frac{\partial\sigma}{\partial x} = \frac{\partial\rho}{\partial y}\frac{\partial\sigma}{\partial y} = 0$$

$$(39) \qquad\qquad\qquad \frac{\partial\rho}{\partial x} = \frac{\partial\sigma}{\partial y}$$

$$(40) \qquad\qquad\qquad \frac{\partial\rho}{\partial y} = -\frac{\partial\sigma}{\partial x}.$$

The derivative of the function $\sigma(x, y)$ along the coordinate axes gives unambiguously the components of changes of probabilitiy in the direction of other axes. Eqs. (39), (40) show that, owing to the continuity $\rho(x, y)$ there exists the total differential

$$(41) \qquad d\sigma = \frac{\partial \sigma}{\partial x} dx + \frac{\partial \sigma}{\partial y} dy \, .$$

However, Eq. (41) is not valid for the starting point A.

The total differential indicates the probability of hitting the element of the line inside E which does not pass through the point A. Along the line starting at the point C and ending at the point D.

$$(42) \qquad \Omega_{CD} = \int_{CD} d\sigma = \sigma_D - \sigma_C$$

The integration path, and therefore the shape of the curve between C, D, may be neglected. Along the closed curve $(C \equiv 0)$, $\Omega_{CD} = 0$.

Further, consider a fixed point C inside E. Between this point and some other is

$$(43) \qquad \Omega = \sigma - \sigma_C$$

After insertion in Eqs. (3y), (40)

$$(44) \qquad \frac{\partial \rho}{\partial x} = \frac{\partial \Omega}{\partial y}$$

$$(45) \qquad \frac{\partial \rho}{\partial y} = - \frac{\partial \Omega}{\partial x} \, .$$

After insertion in Eq. (41)

$$(46) \qquad d\sigma = d\Omega = \frac{\partial \Omega}{\partial x} dx + \frac{\partial \Omega}{\partial y} dy = - \frac{\partial \rho}{\partial y} dx + \frac{\partial \rho}{\partial x} dy \, .$$

Along the curve E where $\rho = 0 = $ constant, the lines $\sigma = $ const. are perpendicular to the boundary and therefore

$$(47) \qquad d\Omega = d\sigma = - \frac{\partial \rho}{\partial n}$$

$$(48) \qquad \Omega = \int_E d\sigma dE$$

and also

$$(49) \qquad \Omega = - \int_E \frac{\partial \rho}{\partial n} dE \, .$$

The authors have not found in the literature a suitable term for the function $\Omega(x, y)$ and with respect to the aim of this paper they call it probability flow. Its importance lies in the fact that it makes possible the fundamental consideration of formal relations between Eqs. (30), (31) and Eqs. (48), (49).

$$\Delta\Phi^*(x, y) \leftrightarrow \rho(x, y)$$

(50)
$$\Delta\psi^*(x, y) \leftrightarrow \sigma(x, y)$$

$$Q_i(x, y) \leftrightarrow \Omega_i(x, y).$$

From the foregoing it is obvious that there is a relation between the probability of hitting a point inside E and Green's function which under steady flow conditions describes the process of the flow from the source into a region, whose boundary S is geometrically similar to E (5). It enables also the solution of harmonic problems in an inhomogeneous region and under more complicated boundary conditions, e.g. $\partial(\Delta\Phi)/\partial n$ along some parts of the boundary S (4). The solution of the problems consists then only of the correct choice of the right method by which the mathematical analogy may be realized.

2. FORMAL ANALOGY BETWEEN GROUNDWATER UNSTEADY FLOW AND TIME DEPENDENT ACCIDENTAL POINT MOVEMENT

Consider that a point movement on an elementary path ds requires a time $d\tau$ (time quantum). Then the velocity of movement is

$$v = \frac{ds}{d\tau}.$$

From this point of view the probability of hitting a point inside E is time-dependent. The nearest vicinity of the starting point $A(x, y)$ will be hit with certainty. If the point is not placed on the boundary, then E may be hit after a certain length of time, depending on the geometric configuration of the region and on the position of the point A. There obviously exists a spectrum of the most expectant paths until the moment of hitting the point. The paths have finite lengths and they end on an abstract boundary which in this circle is a circle centered at the starting point. In the following phases a certain elementary part of the boundary E will be hit with a certain probability in a certain time. The length of the hit part of E increases proportionally to time. We may expect with certainty that all the boundary E will be hit after the elapse of finite time. In substance we follow up the gradual changes of the stationary states analyzed in the foregoing paragraph. After the elapse of a certain time we may determine the value of the probability flow $\Omega(x, y, t)$ on E for each t and we may integrate

(51)
$$\theta_s = \int_0^t \Omega d\zeta = -\int_0^t \left[\frac{\partial\rho}{\partial n} ds \right] d\tau.$$

There exists a formal analogy to Eq. (20), or better to Eq. (26) if $\Delta\Phi(x_s, y_s, t)$ = constant.

$$(52) \qquad f(x, y, t) = \frac{\theta_s}{\theta}$$

where θ is calculated by Eq. (51) for the circle situated in the nearest vicinity of the starting point $A(x, y)$.

The analysis of the situation along the ith part of the boundary S and its geometrically corresponding ith part of the boundary E has a practical significance. It is obviously valid that

$$(53) \qquad \Delta\Phi(x, y, t) = \sum_{i=1}^{I} \Delta\Phi(x_s, y_s)\frac{\theta_i}{\theta} = \sum_{i=1}^{I} \Delta\Phi(x_s, y_s)\frac{W_i}{W}$$

$$(54) \qquad W_i \leftrightarrow Q_i .$$

Of course in practice the use of the analogy requires the analysis of time relations in the both principally different processes.

We are aware of the fact that the formal analogy is not quite exact. For instance Eqs. (10) and (14) indicate that there exists a continuous value change of the dependent variable immediately after the start of the nonstationary process. On the contrary Eq. (52) tacitly supposes that during an accidental movement the boundary E may be hit only after the elapse of a certain time (the principle of quantization of time). Then there may exist a time interval during which the boundary E will not be hit at all. This is more satisfactory for the real hydraulic character of the nonstationary process, however, it does not fully satisfy the linearized equations constructed for the purpose of investigating this phenomenon.

3. REALIZATION OF IMAGINARY ACCIDENTAL MOVEMENT BY MEANS OF COMPUTER

Under the summarizing term "Monte Carlo Method" we understand the solution of different problems of numerical mathematics. It is performed so that for each problem we construct a corresponding probability process with proper parameters satisfying given conditions. The spectrum of accidental movements agrees with this conception [12]. Accidental movements may be realized by using physical means. Mathematical modelling with automatic digital computers offer other possibilities, too. However, the best results are obtained by word for word interpretation. We know the coordinates of the starting point x, y for one piece of path having a sufficiently small but finite length Δs. Imagine that the elementary path is oriented by directional angle α so that the coordinates ξ, η of the track and are

$$\xi = x + \Delta s \cos\alpha, \quad \eta = y + \Delta s \sin\alpha, \quad 0 \leq \alpha < 2\pi .$$

The angle α is a random quantity which may be found by means of the physical generator of random numbers or by means of the numerical generator of pseudo-random numbers. We assume that time Δt is needed for passing through one elementary path. Starting from the point $A(x, y)$ it may be found whether the boundary E is hit after time $n\Delta t$. The boundary E is analytically determined as a whole or at least as parts (lines, parts of circle, or other suitable curves). In a positive case, i.e. after hitting the i-part of the boundary the value of 1 is added in the proper counter. In a negative case the value 0 is added. The accidental track from the point $A(x, y)$ is repeated M-times and we may deduce that the relation between the number N_i of hits in the ith of the boundary and the total number of considered tracks may be expressed

$$\frac{N_i}{M} \approx \frac{\theta_i}{\theta} = f(x, y, n\Delta t).$$

We choose in succession the number n i.e. total time and so we obtain the sufficient number of points to construct the unknown function $f(x, y, t)$ which is further worked out by the method given in paragraph 1.

The description shows that the calculation requires the application of subprogrammes for goniometric functions. More rapid but less suitable is the method supposing an interior imaginary square net constructed so that distance $\Delta x = \Delta y = $ constant. The accidental movement is realized from one point of the net to the other, so that at the given point there are four disposable directions. The accidental choice is possible again by means of the generator of random or pseud-orandom numbers. The principle of solution, however, is the same as in the previous example.

To illustrate the exactness of the results the authors have used a computer for solving a region inside a circle with radius $R = 35$ under assumption $\Delta x = \Delta y = 1$, $a = 1$, $\Delta t = 0.25$. The starting point being at the centre of the circle. The result for $M = 100$ has been compared with the analytical solution of function $f(r, t)$ and plotted (Fig. 3). It is found that the statsistical method works well in practice. However, errors increase for $\Delta t / R^2 \to 0$. Similarly, like other methods of approximation, this one requires a sensitive approach and at least the minimum of experience is supposed for solving practical tasks.

4. CONCLUSION

In this paper the authors have tried to formulate the plane flow affected by the elastic deformation of pervious medium and by the water accumulation in the limits of free water level movement. By means of linearized equations it is possible to describe the unsteady flow in a region closed by a curve S along which a boundary condition is given. A special case of stationary process

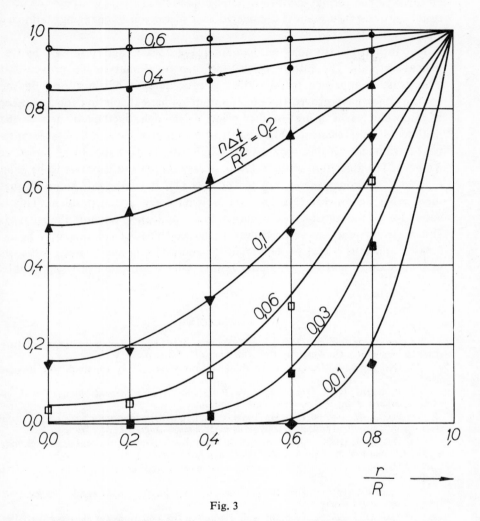

Fig. 3

Fig. 3. Problems concerning solution of steady and unsteady groundwater flow by statistical methods

shows the possibility of using Green's function as a physical interpretation of the flow from the source at point $A(x,y)$ where we try to find the value of the function characterizing the position of free water level. The assumed existence of source may also be used for the solution of nonstationary movement.

Further, it was proved that there exists a formal identity between Green's function and the theory of accidental point movement worked out by I.

Petrowsky. The formal identity is noticeable even in the case where the accidental movement of point is considered as a succession of changing stationary states.

This formal identity may be used for solving water flow problems by statistical methods. The authors think their theoretical analysis has proved that a mistrustful approach to the statistical procedures is baseless. And on the contrary they have proved the possibility of numerical solution for a number of important tasks using another principle than the method of nets or the method of finite elements. The statistical procedures have indisputable practical advantages because they do not require the larger capacity of computer memory. The algorithm of the solution is very simple and it proves to be good even in the inhomogeneous regions of flow [4]. It is stressed, however, that there exists no universal method for solving every actual problem. This is valid also for the statistical procedures whose application requires a preanalysis and a certain practical skill. However, the exactness of solution may be increased and it is useful to determine the optimal relation between the costs of computation and the profitable results which are practically applicable.

REFERENCES

1. BONCH-BRURYEVICH, V. L., AND TYABLIKOV, S. V. Metod funktsyi Greena v statisticheskoy mekhanike. (In Russian), *Gos. Izd. Mat. Lit.*, Moscow.

2. EYDELMAN, S. D. (1964), Parabolicheskiye systemy. (In Russian). *Izd. Nauka*, Moscow.

3. FORSYTHE, G. E. AND WASOW, W. R. *Finite-diference methods for partiel differential equations.* John Wiley, New York–London.

4. HALEK, V. AND NOVAK, M. (1967), Pouziti pravdepodobnostnich metod pri reseni stacionarniho proudeni podzemni vody. *Vodohospodarsky casopis SAV*, c. 3, Bratislava.

5. HALEK, V. (1968), Spektralni metoda reseni stacionarniho proudeni podzemni vody a jeji prakticka realizace pomoci analogu. *Vodni hospodarstvi* c. 1.

6. HALEK, V. (1965), Hydrotechnicky vyzkum 3. Metody analogii v hydraulice. *SNTL* Praha.

7. JOOS, G. (1956), Lehrbuch der theoretischen Physik, *Akademische Verlaggesellschaft Geest und* Leipzig.

8. LEGRAS, J. (1956), Techniques de resolution des equations aux dérivées partielles, Dunod, Psris.

9. METROPOLIS, N. AND ULAM, S. (1949), The Monte Carlo method. *J. Am.rican Statistical Assoc.*, 247.

10. PETROWSKY, I. G. (1961), Lektsii ob uravnyeniyakh s chastnimi proizvodnim (In Russian). *Gos. Izd. Fiz. Mat. Lit*, Moscow.

11. PETROWSKY, I. (1934) Ueber das Irrfahrproblem. *Mathematische Annalen, 109* Springer, Berlin.

12. SHREIDER, YU. A. (1964), *Method of statistical testing.* Elsevier Publ. Co., Amsterdam.

TECHNICAL UNIVERSITY,
BRNO, CZECHOSLOVAKIA

FUNDAMENTALS OF THE CONTINUUM APPROACH FOR CALCULATION OF RATE-INDEPENDENT DEFORMATIONS OF PARTICULATE MEDIA

GERD GUDEHUS

INTRODUCTION

Soil is a particulate medium that generally consists of 3 phases: mineral (e.g. quartz), pore-fluid (e.g. water), and gas (e.g. air). The main aim of soil mechanics is to determine displacements and contact stresses along given surfaces. In many practical cases the displacement field (averaged over single grains) is sufficiently smooth to allow the continuum approach. Generally the deformations are accompanied by a flow of pore-fluid and gas through the mineral skeleton, causing relaxation and creep effects. For many soils these viscous effects could be predicted by calculation of flows, assuming the mineral structure to be rate-independent.

The continuum approach is based on a relation between stresses and displacements, the so-called constitutive equation. Particulate media have the following properties which properly exclude the application of the classical theory of elasticity:

i) the relation between stresses and strains is not only nonlinear, but also hereditary, i.e. it cannot be expressed by ordinary functions;

ii) the material is capable of increasing orientation, i.e. it is not always isotropic;

iii) the deformations often are large enough to exclude the concept of infinitesimal strains.

In this paper first the basic principles of modern nonlinear field theory are laid down and specialized for particulate media. As the results are far too general for any numerical application methods are discussed for getting practicable formulae. The capacity of both modern testing apparatuses and large computers is taken into account.

The desired relations will be derived from *invariance requirements*, denoted by J 1, J 2 etc., and *additional assumptions*, denoted by A 1, A 2 etc. Physical experience is concentrated in these provisions.

119

BASIC PRINCIPLES

J 1: *Coordinate invariance*: no relations should depend on the special choice of a coordinate system.

This requirement is almost a matter of course and generally observed in physics but unfortunately not in soil mechanics. It is satisfied by using tensors (of suitable order) for physical quantities. We mainly use the symbolical notation (for brevity) and sometimes Cartesian components (without loss of generality in this context).

The following development follows the lines of modern classical field theories [11].

The place x of a particle X is a function of time $x(X, t)$ called motion. The instantaneous stress vector t acting upon a surface element with normal n is given through the Cauchy-relation

$$(1) \qquad\qquad\qquad t = Tn.$$

T is the instantaneous stress tensor.

The *principle of determinism* means that the past motion or history of the body B $x(X, t - s)\big|_{\infty}^{s=0}$ determines the present stress $T(X, t)$. Synonymous is the statement

$$(2) \qquad\qquad\qquad T = \overset{s=0}{\underset{\infty}{\mathscr{F}}} [x(X, t-s)] X \in B$$

wherein $\mathscr{F}$ denotes a functional. An important restriction is given by the *principle of local action*: $T(X, t)$ is already determined by the motion of the immediate vicinity of the particle X.

The motion of a neighbouring particle Y can be approximated by

$$(3) \qquad x(Y, t) = x(X, t) + F(X, t)(Y - X) + o(|Y - X|)$$

($X \in C^1$ provided). The tensor F, defined by (3), is the displacement gradient.

A third important principle can be stated as an invariance requirement.

J 2: *constitutive invariance*: constitutive relations should remain unaffected by arbitrary rigid body motions.

An equivalent statement is that the position of the body in space-time is not distinguished in any way (frame-indifference). This principle is tacitly acknowldged by using arbitrary reference systems, e.g. the walls of a laboratory for experimental tests. Unfortunately it is not satisfied by many constitutive equations proposed in soil mechanics.

As a rigid body motion can be described by

$$(4) \qquad\qquad\qquad x^* = a(t) + Q(t)(x - b)$$

(Q is a rotation tensor, satisfying $QQ^T = 1$), the functional relation (2) must be invariant for a rigid body motion of the reference system with arbitrary a, Q, b. From the arbitrariness of a it can be concluded that only relative motions can enter into (2). If (2) is restricted to the approximation of the vicinity motions through E, (2) specializes to

(5)
$$T = \overset{s=0}{\underset{\infty}{\mathscr{F}}} [F(X, t-s)],$$

which is the defining equation of a *simple material*.

To satisfy J 2 one can use tensors for stress and deformation which are invariant against rigid body motions. As any vector v is transformed by (4) into $v^* = Q v$, the Cauchy stress tensor T and the deformation tensor F are according to (1) and (3) transformed into

(6)
$$T^* = QTQ^T; \quad F^* = QF$$

and consequently not frame-indifferent. Suitable frame-indifferent tensors are the Kirchhoff stress tensor

(7)
$$T^K = F^{-1}T(F^{-1})^T \rho/\rho_0$$

(ρ_0, ρ = reference resp. instantaneous density) and the Green deformation tensor

(8)
$$E = \tfrac{1}{2}(F^TF - 1).$$

The frame-indifference of these and other suitable tensors is easily proved [4].

It is possible, as well, to state constitutive relations with frame-dependent tensors. Then J 2 serves as an additional functional equation (as for hypoelastic bodies, e.g. [11]). The above principles were first laid down in full mathematical rigour by Noll [11].

FURTHER RESTRICTION FOR PARTICULATE MEDIA

A 1: Particulate media are simple solids.

This assumption is always tacitly made in soil mechanics but is not a matter of course. There are materials (sometimes called multipolar) for which the local influence upon T is not sufficiently covered by F alone. The physical meaning is that the contact forces between the particles have comparatively wide ranges. This may be true, e.g. for rocks but certainly not for soils. Nevertheless, it would be wrong to say that particulate media are exactly simple solids; A 1 is merely a good approximation.

In [5] deformation is taken as independent variable. This standpoint is adopted here as being useful not only in the theoretical development but also for experiments and numerical applications. There is no physical objection against the dual version that stress is the independent variable. Most generally speaking a constitutive relation is a functional equation

$$(9) \qquad \overset{s=0}{\underset{\infty}{\mathscr{G}}} \, (T, F) = 0$$

containing stress and deformation. We merely assume that (9) can be solved for T but neither T nor F is defined to be the physical "cause" of the process.

The rate-independence of a material can be stated as J 3: *time-scale invariance*: the present stress is not changed by an arbitrary non-decreasing transformation of time $s^* = \phi(s)$.

J 3 simply means: not speed but only order and magnitude of deformations determine the stress. This can only be valid for the mineral constituent. For a 3-phase system like soil one has to apply the theory of mixtures [6] which, however, has not yet been completed for grained constituents. Special difficulties arise for hygroscopic particles as the electrically bound water films cause viscous effects which cannot be explained by pore-water flow. By assuming J 3 we exclude such cases. The physical reason for J 3 is not trivial, it is closely connected with the problem of dry friction.

J 3 is satisfied by taking the arc-length of deformation.

$$(10) \qquad l = \int_0^s \sqrt{\mathrm{tr}(dE\,dE)}$$

(which generally is a Stieltjes-Integral) as a new time variable instead of s [8]. So we can take as constitutive equation, making use of the frame-indifferent tensors (7) and (8).

$$(11) \qquad T^K = \overset{0}{\underset{l=lmax}{\mathscr{F}}} \, E(t - l).$$

Owen has proved that (11) with (10) satisfies J 3 [8]. This general form will be specialized in the following text.

We now have to introduce the concept of *isotropy*. As is well known, particulate media can show a certain orientation. Nevertheless one can assign isotropy to them when applying the definition of Noll [11]: the material may have a reference state which is not changed (and consequently leaves unchanged the constitutive relation) through rotations $Q \in \mathscr{G}$. The so-called isotropy group $\mathscr{G}$ may be very special; e.g. undeformed crystals remain unchanged only for certain special rotations depending on the crystal class. For particulate media one can state

J 4: *isotropy*: one can always find a reference state for which the isotropy group $\mathscr{G}$ is the full orthogonal group o.

It is not necessary for this reference state to be a natural state. In fact, for example a natural soil can have preferred directions immediately after its sedimentation because of gravity resulting in orthotropy. For the validity of J 4 it is only essential that this natural state can be thought of having originated by pure deformation from an isotropic state. If this is true J 4 can be used to get important simplifications.

Direct use of J 4 can be made for certain special motions as was demonstrated by Coleman [1]. These solutions are, however, not accessible to numerical treatment. Functionals have to be replaced, by ordinary functions in any way. This is possible only through an additional smoothness assumption, which we state as

A 2: a small change in deformation causes only a small change in stress.

This assumption is not a matter of course. The so-called "plasticity kink" observed with metals is a counterexample. For particulate media A 2 seems to be reasonable and is backed by experience.

There are different ways of giving A 2 a precise mathematical meaning leading to different possible approximations for (11) subject to J 4 [9].

STATE VARIABLES AND STATE FUNCTIONS

The different series approximations for isotropic functionals still contain so many functions that they are still too complicated for evaluation of experiments and for numerical applications. A further simplification results from

A 3: A set of variables that can be summarized into two symmetric second-order tensors is sufficient to describe the state of the element.

If one makes use of a differential (generally anholonomic) relation between stress and strain components, the coefficients of this relation will depend on the state variables. Evidently the stress tensor is necessary to describe the state. For particulate media it is not sufficient. The structure (array) of the grains can change and influences the future behaviour. It is assumed by A 3 that the structure can be sufficiently described by a symmetric second-order tensor called affinity tensor A. The use of a second order tensor has certain advantages which will become clear in the following text. In this way the memory of the structure is concentrated into A. As a matter of course the structure could be described more precisely by more variables, but A should suffice to allow for typical orientation effects. One should remember that, up to the present, the only geometrical state variable used in soil mechanics is the density. Evidently the permeability is also structure-dependent. Thus in the presence of a pore-fluid A can influence the viscous behavior of soils.

There are different ways of specifying A, e.g.

i) Phenomenological: A is associated to the plastic strain E^p. If one assumes $A = E^p$ as frequently suggested in modern plasticity theory [2] the exper-

Gerd Gudehus

imental determination of A is specially simple. However, E^p generally is not a state variable, especially not for large distortions. For this case the relation between A and E^p has to be modified according to experience. The associate dynamic variable would be T^K. The disadvantage of this specification is that E^p and T^K do not allow a simple physical interpretation.

ii)   Statistical: the quadric $A_{ij}x_i x_j = 1$ gives the probability distribution of the contact normals between the individual particles of an element. There is some evidence that this quadric is an ellipsoid [3]. The probability distribution is of course discrete for a regular array of equal symmetric particles. A_{ij} is only reasonable for irregular particles in a stochastic array. The associate dynamic variable is the instantaneous stress T which can also be interpreted statistically [3]. One can imagine that in a granulate with friction the normals of the contact areas concentrate in the direction of maximum principal stress. On principle both A and T can be measured directly. It is reasonable to associate a change dA with a change dE^p of plastic strain by a linear relation [3]. The coefficients must be chosen in such a way that the relation is anholonomic and suffices to discern between test results.

The appropriate specification of A and its relation to plastic strain is regarded as one of the main open problems in present particulate mechanics.

Use will be made of certain scalar state functions which are assumed to depend on the state variables. They are

   i)   the internal (or elastic) energy U,
   ii)   the plasticity function $f \leq 0$, with $f = 0$ for plastic flow as usually defined and $f < 0$ for an elastic state.

The arguments of U and f can be reduced, because they are isotropic tensor functions. For the choice $A = E^p$ and T^K this follows from J 4 (isotropy): by a rotation Q in the reference state E^p is transformed into QE^pQ^T and T^K into QT^KQ^T. As U and f should be unchanged by arbitrary rotations Q in the reference state, they can depend on E^p and T^K only through the invariants [2]

$$J_1 = \operatorname{tr} E^p; \quad J_2 = \operatorname{tr} E^{p2}; \quad J_3 = \operatorname{tr} E^{p3}; \quad J_4 = \operatorname{tr} T^K; \quad J_5 = \operatorname{tr} T^{K2}; \quad J_6 = \operatorname{tr} T^{K3};$$

(12)

$$J_7 = \operatorname{tr} E^p T^K; \quad J_8 = \operatorname{tr} E^p T^{K2}; \quad J_9 = \operatorname{tr} E^{p2} T^K; \quad J_{10} = \operatorname{tr} E^{p2} T^{K2}.$$

(The proof, resting on group theory, is involved.)

A similar result is obtained for the choice A, T that can be interpreted statistically. A rigid-body rotation transforms A into QAQ^T and T into QTQ^T. By J 2 U and f must be unaffected for any Q and therefore can depend only on 10 invariants (obtained by setting A for E^0 and T for T^K in (12)). The fact that frame-indifference causes isotropic relations is also observed with fluids, see e.g. [6]. We, therefore, have to face calculations with functions of 10 scalar invariants. By omitting the time variable J 3 is automatically satisfied.

RESTRICTIONS FOR DIFFERENTIAL CONSTITUTIVE RELATIONS

We need a differential relation connecting increments of stress and strain. It turns out that one needs further assumptions. These can be stated as thermo-dynamic principles. It is necessary to take conjugate pairs of stress T and strain E in such a way that the increment of external work is $dW = \mathrm{tr}\,T\,dE$. Thermo-dynamic assumptions are associated with dW and have differential or integral form. As pointed out by Hill the choice of each special conjugate pair leads to a special thermodynamic assumption [5]. Evidently, there is a certain ar-bitrariness in such assumptions. Truesdell and Noll [11] regard ther-modynamics as the main open problem in modern nonlinear field theory. In fact the principles and assumptions stated up to here in this paper do not suffice to determine the constitutive relations entirely.

We briefly regard three different continuations dE of a given strain history:

i) *Elastic* continuation. From the energetic definition of the elastic range it is easily shown [4] that

$$(13) \qquad dT_{ij}^{K} = dE_{kl}\partial^2 U/\partial E_{ij}^{e}\partial E_{kl}^{e}$$

is the desired relation for the increment of stress (E_{ij}^{e} is the elastic strain, uniquely associated with the stress [4]).

ii) *Plastic* continuation. The increment dE_{ij} is split into an elastic and a plastic part:

$$(14) \qquad dE_{ij}^{e} + dE_{ij}^{p} = dE_{ij}$$

The plasticity condition is permanently valid:

$$(15) \qquad df = \frac{\partial f}{\partial T_{ij}^{K}}dT_{ij}^{K} + \frac{\partial f}{\partial A_{ij}}dA_{ij} = 0.$$

If we assume a special incremental law for the affinity tensor, e.g. $dA_{ij} = dE_{ij}^{p}$, (13), (14), and (15) furnish 13 equations for 18 unknowns. The missing 5 equations must fix the direction of dE_{ij}^{p} in E_{ij}^{p}-space. It can be shown [4] that the orthogonality of dE_{ij}^{p} to the $f = 0$-surface, usually accepted in plas-ticity theory, is not necessary but only sufficient to satisfy the postulate of irreversibility

$$(16) \qquad \oint T_{ij}^{w}dE_{ij} \geqq 0.$$

Consequently the correct specification of the dE_{ij}^{p}-direction is still an open problem.

iii) *Critical* continuation. Under special stress condition soils are capable of unrestricted isochoric plastic flow. Such states are called critical [10]. A volume-preserving dE_{ij} (if it is not elastic) leaves the structure unchanged, i.e. $dA_{ij} = 0$.

APPROPRIATE TESTS

The constitutive equation of a simple material can be determined experimentally by applying a homogeneous deformation history (F = const. throughout
the sample) and measuring the stress [11]. This implies

J 5: *apparatus invariance*: the measured constitutive relation must not depend on the special testing equipment.

Although J 5 should be a matter of course, experimental technique has reached a sufficient standard, only in the last years. It is now possible to build an
apparatus in which three principal strains in a cubical sample can be varied
arbitrarly, causing stresses that can be measured (technical details of such an
apparatus are omitted in this paper).

Strain-controlled tests have some advantages as compared with the usual
stress-controlled ones. The homogeneity of strains — necessary to satisfy J 5 —
can be strictly maintained through a suitable boundary lubrication. Consequently stress is also homogeneous and can be determined by measuring
resultant forces. Samples with applied surface stresses often become inhomogeneous in critical states. This effect, which excludes the continuum approach,
is avoided in fully strain-controlled tests. It is easier, by the way, to produce a
desired displacement than a desired load. Large deformations yield no additional difficulties.

By applying different rates of deformation the assumed rate-independence
can be easily checked. Of course, the rate of volume change must be slow enough
to avoid non-negligible pore-fluid-pressure. Relaxation effects with properly
viscous media can be studied as well, e.g. with bentonite.

Isotropy has to be observed on filling in the material. The test results must
be invariant with respect to the three possible 90°-rotations. The assumptions
of continuity (A 2) and irreversibility can be easily checked.

By sequences of deformation cycles, one can produce all relevant states of the
sample. The evaluation [4] gives a lot of different scalar invariants according
to (12). One can apply and test different formulae for the affinity tensor and
finally choose the one that best characterized the state. On principle it is also
possible to freeze the sample in any state (or fix it by an appropriate glue) and
make statistical measurements of the structure. Finally, assumptions for the
direction of the plastic strain increment can be improved by direct observation.
It may turn aut that some of the invariants (12) have no essential influence.
No such result can be expected in advance, however. For the approximation
the following statement should be observed.:

J 6: dimensional invariance: the constitutive relations must not depend on
special units.

J 6 is satisfied by introducing a characteristic modulus of elasticity E_0. On
principle, the choice of E_0 is arbitrary but it may be useful to take a modulus

of quartz or felspar, respectively. We then have a set of dimensionless variables and functions $\bar{T} = T/E_0$, $\bar{U} = U/E_0$. E and A are dimensionless by definition, and f can be taken as a dimensionless function without loss of generality. Consequently all material constants emerging in the approximation should be dimensionless. The *Pi*-theorem insures that J 6 allows no further reduction.

REMARKS CONCERNING THE NUMERICAL APPLICATION

If the differential constitutive equations are fixed by experimental data, boundary value problems can be solved (at least on principle) through finitization.

Any deformation process is broken up into sufficiently small steps each of which is governed now by linear constitutive equations. The body is dissected into small elements of simple forms (e.g. tetrahedrons), and x is interpolated by polynomials between the nodal points. Each element is deformed already and suffers a new incremental deformation which is calculated by the linear difference equations of equilibrium (finite element method). A special difficulty arises for plastic continuations: the computer does not "know" in advance whether an incremental deformation is an elastic continuation or not. In a first step one alternative must be assumed, controlled afterwards, corrected etc. in an iterative procedure.

For large deformations, cases of instability may occur. Of course individual slip surfaces cannot be predicted by the finite element method. One has to expect zones of large distortions whereas natural soil may prefer to develop several nearly parallel discrete slip surfaces that are not accessible to the continuum approach.

If the pores are saturated with a fluid, the method indicated above does not generally yield correct solutions. Plastic volume changes cause time-dependent flows that have to be taken into account. The coefficients in the relevant diffusion equation depend on the instantaneous stress and affinity tensors and consequently vary during the (three-dimensional) consolidation process. As yet no effort has been made to solve this problem.

(This paper is a shortened and partly revised version of [4]).

REFERENCES

1. COLEMAN, B. D. (1968), On the use of symmetry to simplify the constitutive equations of isotropic materials, *Proc. Roy. Soc. A. 306*, 449–476.

2. GREEN, A. AND NAGHIDI, P. (1965), A general theory of an elastic-plastic continuum, *Arch. Rat. Mech. Anal.*, 18.

3. GUDEHUS, G. (1968) Gedanken zur statistischen Bodenmechanik, *Der Bauingenieur*, 43, 9, S. 320–326.

4. GUDEHUS, G. (1968), A continuum theory for calculation of large deformations in soils. *Veröff. Inst. f. Bodenmech. u. Felsmech.* Univ. Karlsruhe, Heft 36.

5. HILL, R. (1968), On constitutive inequalities for simple materials, *Journ. Mech. Phys. Sol.*, 229–242.

6. MÜLLER, I. (1968), A thermodynamic theory of mixtures of fluids, *Arch. Rat. Mech. Anal.*

7. OWEN, D. (1968), Thermodynamics of materials with elastic range, *Publ. Carnegie Inst. of Techn., Dept. of Math.*

8. OWEN, D. AND WILLIAMS, W. (1967), On the concept of rate-independence, *Publ. Carnegie Inst. of Techn., Dept. of Math.*.

9. PIPKIN, A. AND RIVLIN, R. (1965), Mechanics of rate-independent materials, *Z. Angew. Math. u. Phys.* 16.

10. SCHOFIELD, A. AND WROTH, P. (1969), *Critical State Soil Mechanics.* McGraw Hill, London.

11. TRUESDELL, C. AND NOLL, W. (1966), The nonlinear field theories of mechanics, *Handbook Phys.*, Vol. III/3. Springer, Berlin.

INSTITUT FÜR BODENMECHANIK UND FELSMECHANIK,
UNIVERSITÄT KARLSRUHE, RICHARD-WILLSTÄTTER-ALLEE,
75 KARLSRUHE, GERMANY

DETERMINISTIC AND STATISTICAL CHARACTERIZATION OF POROUS MEDIA AND COMPUTATIONAL METHODS OF ANALYSIS

A. E. Scheidegger

1. INTRODUCTION

The afternoon session occupied itself with two distinctly different topics, inasmuch as the papers submitted fell into two categories: papers on the structure of porous media, and papers on the actual solution of the equations referring to particular flow problems. We shall therefore review these two subjects in their turn.

2. STRUCTURE OF POROUS MEDIA

(a) *General*

Porous media are very complicated objects. Generally, they can be regarded as a "solid" body with irregular small holes in it. An altervative view is to regard porous media as an array of obstacles in an otherwise empty volume. Usually, one is only interested in those "holes" which are interconnected, inasmuch as one wishes primarily to investigate the patterns of flow of fluids through a porous medium.

Thus it is the pore "channels" that are of primary interest. Evidently, one can approach this subject in two fashions: one can look at nature, and investigate what *particular* porous media really look like, or one can try to fathom out theoretically what porous media *could* look like. By looking *only* at nature, one will necessarily investigate merely a series of particular media, whereas, in making theoretical studies, one could in principle think of all possible porous media: some of the envisaged structures, however, are so outlandish, that no counterparts exist in nature. It is thus, as always, necessary to collate theory and experiment.

(b) *Experimental approach*

1. *General remarks on techniques.* It is difficult to look simply at a porous medium. First of all, porous media are three-dimensional objects, but the eye (i.e. the retina) sees things only in two dimensions. Thus, what is seen, must be somehow transferred from 2 to 3 dimensions. Here, one generally assumes that, at least statistically, any plane section of a particular porous medium is

equivalent to any other such section. If the porous medium is anisotriopc, one must restrict the above generality to the hypothesis that any plane section normal to a given direction is equivalent to any other plane section normal to that given direction.

With these hypotheses, it is in principle possible to make general inferences from the examination of sections of porous media. One cannot, however, easily make a section of a porous medium, without first filling the pores with something. Techniques to do this are well known: injections of Wood's metal above the melting temperature and letting it cool have been used for a long time, also polymerizable materials that become solid after injection.

2. *Progressive pressure technique*. In this connection, the work of Dullien contributed to this session aims at getting a good experimental knowledge of porous media by injecting Wood's metal in liquid form at progressively higher pressure. Wood's metal, for rocks, is a non-wetting fluid, so that the technique is analogous to the well-known mercury injection technique to determine "pore size distribution" by measuring the volume injected at a particular pressure. However, by Dullien's new technique, the process can be stopped at any pressure, the medium can be cut apart and an examination can be made as to where the injected material actually went. The injection pressure applicable on Wood's metal, being a non-wetting fluid, gives a measure of the largest constriction which leads from the outside to a particular point in the porous medium.

Similarly, if a wetting fluid (some polymerizable substance) is used for injection one obtains from the injection pressure a measure of the largest opening on the narrowest path leading from the outside to a particular point in the porous medium. In this fashion, a rather intimate knowledge of the structure of porous media is obtained.

3. *Application of this knowledge*. The detailed knowledge of the structure of a porous medium gives a means to predict the hysteresis in capillary pressure, as well as flow properties, such as the permeability.

(c) *Theory*

1. *Models*. The theory of the configuration of the pore channels in a porous medium is based on the idea of constructing theoretical *models*. There are basically two different classes of such models that can be envisaged: deterministic models and probabilistic models.

2. *Deterministic models*. The deterministic models are based upon the solution of the basic differential flow equations for viscous fluids (viz. the Navier-Stokes equations) for configurations of channels. Because the Navier-Stokes equations are difficult to solve, the channels (representing the boundary condition) must be chosen of a relatively simple form so the solution can be effected.

Two approaches have been tried: to take some sort of a bundle of capillaries and study the flow through each of them, or to regard the fluid as a bulk mass into which obstacles (representing the solid part of the porous medium) are put. The capillaric models have been reviewed in the paper by Scheidegger and Liao contributed to another session of this Symposium; the theory of "obstacle models" has been discussed at length in a book by Happel and Brenner (1965) and is touched upon the paper of Stark reviewed in this session.

Accordingly, while most solutions of the Navier-Stokes equations neglect the inertia terms, i.e. they discuss only "creeping" flow, Stark keeps the inertia terms and considers the plane flow around regular arrays of obstacles. Two cases are discussed: (i) all obstacles are of the same size., (ii) there are two sizes of obstacles. Stark obtains a first integral of the Navier-Stokes equations analytically and then solves the resulting equation by a new numerical technique.

Stark obtains very encouraging results: the Forchheimer relation, the Kozeny equation and, for low flow velocities, the Darcy formula can be duly deduced. This shows that, with a sufficient sophistication, many of the observed facts can be explained by deterministic models. Of course, it is difficult to expand this approach to 3-dimensions, owing to difficulties in solving the Navier-Stokes equations under all but the most simple boundary conditions.

3. *Probabilistic models.* In contrast to the deterministic models, in which the porous medium is simplified to the point where the Navier-Stokes equations can be solved, the probabilistic models attack the problem of flow through a porous medium from the opposite extreme: it is assumed that our knowledge of such media will never suffice to follow the flow of fluid particles in detail through them. Thus, although the flow path of a particle is, in effect, completely determined, it can be regarded in view of our ignorance, as a random walk. For this type of approach, it is the statistical structure of porous media that is importance.

With regard to the present symposium, this type of approach has been touched upon by the paper of Scheidegger and Liao, which is being refereed in another session. Let it just be stated that the randomness can be put not only into the paths of the particles, but also into the flow configurations that result. In this case, these configurations are represented by mathematical "graphs" of which average properties, taken over ensembles, can be calculated. The results agree reasonably well with the facts gleaned from nature.

(d) *State of the art*

Surveying the present state of the art of characterizing the structure of porous media. we note that first of all experimental techniques have now been carried to a high degree of refinement. Generally, one must inject some substance that solidifies thereafter into the medium, so that, when the medium is cut apart, it is clear which parts have been invaded by the fluid and which are solid. Microscopy will then provide a means to get specific information on the porous medium. The pertinent literature is touched upon in the papers contributed to this session.

Turning to the theoretical part, we note that two approaches have been proposed: deterministic models and stochastic models of porous media. Again, there have been two types of deterministic models: models based on representing a porous medium essentially as a bundle of capillaries, which have been briefly summarized in the paper of Scheidegger and Liao in this symposium, and models based essentially on representing a porous medium as a series of obstacles in a fluid stream. The general state of knowledge regarding the latter was summarized in the book by Happel and Brenner (1965) cited above ; new developments were presented in the paper by Stark in the present session.

With regard to probabilistic models, apart from the work refereed in this report, a comprehensive study by Matheron (1967) must be mentioned. Matheron builds up a theory of porous media based on set theory and describes a possible characterization of "random media." It is difficult to see, however, at this stage how the theory can be collated with experiments. A practical result of Matheron's theory is that the bulk permeability in a medium randomly composed of blocks of various permeabilities, is in the two-dimensional case half-way and in the three-dimensional case two-third of the way between the harmonic and arithmetic mean of the (weighted) component permeabilities.

3. ANALYTICAL-NUMERICAL METHODS

(a) *General*

The second subject of the afternoon's session referred to the problem of finding actual solutions for given flow problems in porous media. Thus, let us assume that the basic flow equation (Darcy's law, Forchheimer equation etc.) has been chosen. The boundary conditions, the permeability distribution in the porous medium, etc. all are given. The problem, then, is to find the flow pattern that will evolve in time.

Mathematically, the above problem reduces to the solution of a system of nonlinear partial differential equations. This is a difficult subject, but solutions for many particular cases are available. Several new solutions and techniques have been proposed which were discussed in the afternoon's session. These are, as usually, split into two distinct classes: analytical techniques, and numerical (computer) techniques.

(b)   *Analytical   techniques*

1. *General remarks.* The system of differential equations describing the flow through a particular porous medium with given boundary, initial conditions, etc. is so complicated that no entirely general solutions can be found. Thus, one usually has to make simplifications: perhaps steady state conditions have to be assumed, gravity is neglected, or a linearization of the flow equations is somehow effected. In this fashion, a whole host of solutions can be obtained.

2. *Permeability inhomogeneity.* The work reported by Gheorgitza, and discussed in this session, is specifically concerned with an inconstant permeability k; i.e. k is a function of the space-coordinates x, y, z. Generally, Gheorgitza considers steady state Darcy flow with gravity.

The cases considered fall into three classes: (I) barriers are present, (II) k is block-wise constant, (III) k satisfies an equation of the type

$$\text{lap}(k^{1/2}) + \alpha^2 k^{1/2} = 0$$

where α is some constant ("Helmholtz-inhomogenity") which may be zero ("Laplace-inhomogeneity") and (IV) the medium contains a cavity.

Gheorgitza discusses the possibilities of solutions for the above cases. In the case of cavities ("feeding" contours, impermeable contours), analytical solutions can be obtained in the plane case by conformal mapping techniques. Matching conditions have to be formulated at the boundaries in a manner similar to the analogous problems in elasticity theory (rigid inclusions, voids in a stressed elastic medium).

It is clear that analytical techniques can be carried further to more and more complicated cases almost without end.

(c)   *Numerical techniques*

1. *General remarks.* Attempts can be made to solve the system of differential equations describing the flow of a fluid through a particular porous medium with given boundary, initial conditions etc. by numerical means.

The available techniques are many and varied, and have been made possible by modern computer technology. Generally, they consist in approximating differential quotients by difference quotients and then solving the resulting system of difference equations step by step.

2. *A new random technique.* Halek and Novak, in the work refereed here approached the numerical solutions of the system of differential equations describing flow through porous media in an entirely different fashion. They

noted that in many problems of groundwater flow, the flow equation can be reduced to the following

$$\frac{\partial \Phi}{dt} = a\left(\frac{\partial^2 \Phi}{\partial x^2} + \frac{\partial^2 \Phi}{\partial y^2}\right)$$

where Φ is some flow potential and a is a complicated function which depends on x and y.

It is well known that the above equation also arises in the theory of diffusion (it is, in fact, called "diffusivity equation"). Instead of solving now this diffusivity equation, the opposite approach is taken to find its solution by representing it by a random-walk process (characteristic of diffusion) that can be modeled on a computer. Thus, the superposition of suitably chosen random processes (effected numerically on a computer) duly produces an approximate solution of the diffusivity equation characteristic of groundwater flow.

The method appears to give good results in groundwater investigations as is evidenced by the graphs presented by Halek and Novak.

(d) *State of the art*

Surveying the present state of the art of solving particular flow problems in porous media systems, we note that one is basically faced with the necessity of solving a system of nonlinear partial differential equations with appropriate boundary, initial etc. conditions.

Thus, the methods available in the theory of flow through porous media are essentially those available in the theory of partial differential equations: analytical and numerical methods. In special cases, the equations can be linearized, which effects a great simplification of the problem.

Specific applications of analytical methods to porous media problems are scattered widely over the literature. Summaries in book form have been produced with specific regard to groundwater flow by Polubarinova-Kochina (1952), and with regard to flow through nonhomogeneous porous media by Oroveanu (1963). The work reported in this session represents a refinement thereon.

With regard to numerical methods of solving partial differential equations, the literature is vast. Collatz (1960), some time ago, has set down the basic requirements involved, but techniques are constantly improving and becoming more involved, in concordance with the development of computer technique. The limit has obviously not yet been reached.

4. QUESTIONS FOR DISCUSSION

1. In what direction should further research into the geometrical characteri-

zation of porous media (experimental and theoretical) go? Which are the most important parameters ?

2. What type of flow patterns in porous media are the most important ? For what cases should the solution of the basic flow equation be pushed?

3. Which methods of solution of the basic flow equations appear the most promising? What is the relative virtue of analytical *versus* numerical methods? Which should be emphasized?

REFERENCES

COLLATZ, L. (1960) *The Numerical Treatment of Differential Equations.* 3rd ed. Berlin, Springer.

HAPPEL, J. AND BRENNER, H. (1965) *Low Reynolds-number Hydrodynamics.* Englewood Cliffs, Prentice-Hall, N. J..

MATHERON, G. (1967) Élement pour une théorie des milieux poreux. Masson, Paris.

Oroveanu, T. (1963) Scurgerea fluidelor prin meddii poroase neomogene. Academ. Rep. Pop. Romine, Bucharest.

POLUBARINOVA-KOCHINA, P, YA. (1952) Teoriya dvizheniya gruntovykh vod. Gosizdat tekh-teoret. lit/ Moscow.

DEPARTMENT OF MINING, METALLURGY,
   AND PETROLEUM ENGINEERING,
   UNIVERSITY OF ILLINOIS,
    URBANA, ILLINOIS, 61801, U.S.A.

THE RECIPROCITY PRINCIPLE IN FLOW THROUGH HETEROGENEOUS POROUS MEDIA

G. A. Bruggeman

Geohydrological investigations of groundformations in the southwest deltaic area of the Netherlands by the author have shown that, in a horizontal layered formation, consisting of two leaky aquifers, separated and confined by semi-pervious strata, the drawdown distribution of the piezometric head in the upper aquifer as a result of draining the lower aquifer by means of a steady well, equals the drawdown distribution in the lower aquifer, caused by pumping the upper aquifer with the same discharge. In this particular case I could prove this phenomenon mathematically. However, the question arose, whether this reciprocity of drawdown and discharge would be valid as a general principle. This question can be answered in the affirmative as the following proof, given by Dr. Josselin de Jong in a session of the Hydrological Colloquium in Holland, shows.

1. GENERAL PROOF OF THE RECIPROCITY PRINCIPLE

Theorem (Fig. 1): *The drawdown of the piezometric head at an arbitrary*

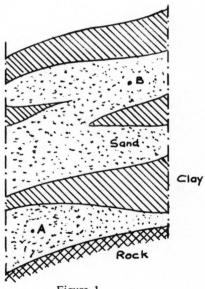

Figure 1

point *A* in a heterogeneous anisotropic porous medium, caused by pumping with a discharge *Q* in another arbitrary point *B* at the time *t* after the beginning of the pumping is equal to the drawdown in *B* as a result of pumping in *A* with the same discharge *Q* and after the time *t* (Reciprocity principle).

Proof. At first this theorem will be proved for *steady flow*. Let the points *A* and *B* belong to a closed groundwater body with volume *V*, whose boundary is a piecewise smooth orientable surface *S* (Fig. 2). Assume that pumping in *A* with a discharge Q_A gives a steady drawdown distribution ϕ and a bulk velocity $\bar{q}$ (vector notation) and pumping in *B* with a discharge Q_B causes a steady drawdown distribution ψ and a bulk velocity $\bar{p}$ in that groundwater body:

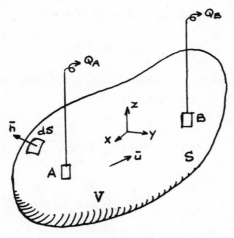

Figure 2

ϕ, ψ and the components of $\bar{p}$ and $\bar{q}$ (p_x, p_y, p_z, q_x, q_y, and q_z) be functions of *x*, *y* and *z*.

In this case the reprocity principle holds that ϕ_B (the drawdown at *B*, caused by pumping in *A*) = ψ_A (the drawdown at *A*, caused by pumping in *B*) if $Q_A = Q_B$.

This can be proved by means of Green's theorem that can be derived from the divergence theorem of Gauss.

For the groundwater body in Fig. 2 the divergence theorem can be written:

$$(1) \qquad \iint_S \bar{u} \cdot \bar{n} \, dS = \iiint_V \operatorname{div} \bar{u} \, dV$$

where $\bar{u}$ (x, y, z) is a vector function, regular in *V* and $\bar{n}$ is the outer unit normal vector of the surface *S*.

For $\bar{u}$ we choose the vector which in every point (except in A and B) of the groundwater body is defined as

$$\bar{u} = \phi\bar{p} - \psi\bar{q}$$

Consequently formula (1) takes the form

$$\iint_S (\phi\bar{p} - \psi\bar{q}) \cdot \bar{n}\, dS = \iiint_V \{\text{div}(\phi\bar{p}) - \text{div}(\psi\bar{q})\}dV$$

The integrand of the double integral may be written as

(2) $$\phi\bar{p} \cdot \bar{n} - \psi\bar{q} \cdot n = \phi p_n - \psi q_n$$

in which p_n and q_n are the components of $\bar{p}$ and $\bar{q}$ in the direction of the outer normal of S. So the left hand side of Eq. (2) becomes

$$\iint_S (\phi p_n - \psi q_n)dS$$

From the definition of the divergence of a vectorfuntion the integrand of the triple integral in Eq. (2) becomes

$$\text{div}(\phi\bar{p}) - \text{div}(\psi\bar{q}) =$$

$$= \frac{\partial(\phi p_x)}{\partial x} + \frac{\partial(\phi p_y)}{\partial y} + \frac{\partial(\phi p_z)}{\partial z} - \frac{\partial(\psi q_x)}{\partial x} - \frac{\partial(\psi q_y)}{\partial y} - \frac{\partial(\psi q_z)}{\partial z}$$

$$= p_x \frac{\partial\phi}{\partial x} + p_y \frac{\partial\phi}{\partial y} + p_z \frac{\partial\phi}{\partial z} + \phi\,\text{div}\,\bar{p} - q_x \frac{\partial\psi}{\partial x} - q_y \frac{\partial\psi}{\partial y} - q_z \frac{\partial\psi}{\partial z} - \psi\,\text{div}\,\bar{q} = 0.$$

This last expression is equal to zero as $\text{div}\,\bar{p} = \text{div}\,\bar{q} = 0$ is the condition of incompressibility of the ground (steady flow). Furthermore we know that

$$p_x = -K_x \frac{\partial\psi}{\partial x}, \quad q_x = -K_x \frac{\partial\phi}{\partial x}$$

$$p_y = -K_y \frac{\partial\psi}{\partial y} \quad q_y = -K_y \frac{\partial\phi}{\partial y}$$

$$p_z = -K_z \frac{\partial\psi}{\partial z} \quad q_z = -K_z \frac{\partial\phi}{\partial z} \quad \text{(Darcy's law)}$$

Consequently:

$$p_x \frac{\partial\phi}{\partial x} - q_x \frac{\partial\psi}{\partial x} = -K_x \frac{\partial\psi}{\partial x}\frac{\partial\phi}{\partial x} + K_x \frac{\partial\phi}{\partial x}\frac{\partial\psi}{\partial x} = 0 \text{ and so on.}$$

From this, Green's theorem follows:

(3) $$\iint_S (\phi p_n - \psi q_n)\, dS = 0$$

A condition for Eq. (1) is, that no sources or sinks exist within the body, enclosed by S. So we have to choose for S the surface, indicated by Fig. 3 where $S = S^1 + S_A + S_B$ and S^1 = the outer boundary of the groundwater body. S_A and S_B are the surfaces of the filters used by pumping in A and B.

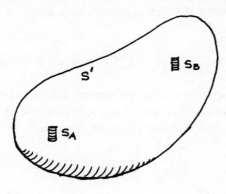

Figure 3

In practice somewhere in the groundwater body a boundary S^1 (finite or infinite) where the conditions are such that the pumping has no influence on the piezometric head or the bulk velocity normal to the surface, exists. The change of head or the change of velocity at such a surface will be zero:

$$\phi(S^1) = \psi(S^1) = 0 \quad \text{or} \quad p_n(S^1) = q_n(S^1) = 0$$

and Eq. (3) becomes:

$$(4) \qquad \iint_{S_A} (\phi p_n - \psi q_n)\, dS + \iint_{S_B} (\phi p_n - \psi q_n)\, dS = 0$$

When pumping takes place in A the drawdown equals ϕ_A resp. ϕ_B everywhere at the filter surfaces S_A resp. S_B and also pumping in B gives ψ_A and ψ_B for the drawdown at S_A resp. S_B. Therefore we can write (4) as:

$$(5) \quad \phi_A \iint_{S_A} p_n dS - \psi_A \iint_{S_A} q_n dS + \phi_B \iint_{S_B} p_n dS - \psi_B \iint_{S_B} q_n dS = 0$$

$\iint_{S_A} q_n dS$ represents the bulk velocity normal to the filter A, caused by pumping in A and integrated along the filter surface S_A and is equal to the discharge Q_A. Also $\iint_{S_B} p_n dS = Q_B$.

$\iint_{S_A} p_n dS$ represents the discharge in A during the pumping in B and will consequently be zero. In a similar way $\iint_{S_B} q_n dS$ equals zero. These values substituted in (5) gives

$$-\psi_A Q_A + \phi_B Q_B = 0$$

or

$$\psi_A = \phi_B \text{ if } Q_A = Q_B$$

Secondly the theorem will be proved for *unsteady flow*, ϕ, ψ, $\bar{p}$, $\bar{q}$ and the components of $\bar{p}$ and $\bar{q}$ are now also functions of the time t and the divergence of the bulk velocity is unequal to zero as a result of the compressibility of the ground:

$$(6) \qquad \operatorname{div} \bar{p} = - S_s \frac{\partial \psi}{\partial t} \text{ and } \operatorname{div} \bar{q} = - S_s \frac{\partial \phi}{\partial t}$$

where S_s is the specific storage (in general a function of x, y and z).

The volume integral in the divergence theorem of Gauss applied to the vector $\bar{u} = \phi\bar{p} - \psi\bar{q}$ (see Eq. (2)) now becomes:

$$\iiint_V (\phi \operatorname{div} \bar{p} - \psi \operatorname{div} \bar{q})\, dV$$

and so

$$(7) \qquad \iint_S (\phi p_n - \psi q_n)\, dS = \iiint_V S_s \left(\psi \frac{\partial \phi}{\partial t} - \phi \frac{\partial \psi}{\partial t} \right) dV$$

The dependence of time can be eliminated by applying the Laplace integral transformation to the functions that depend on t, which means for example the function $\phi(x, y, z, t)$ that is multiplied by e^{-st} and integrated with respect to t from zero to infinity, thus obtaining a new function $\widehat{\phi}(x, y, z, s)$ written shortly as $\widehat{\phi}$ in which the variable t has vanished and has been replaced by s; as in the formula

$$\widehat{\phi}(s) = \int_0^\infty e^{-st}\, \phi(t)dt$$

Also $\widehat{\bar{q}}(s) = \int_0 e^{-st} \bar{q}(t)dt$ etc.

Instead of the vector $\bar{u} = \phi\bar{p} - \psi\bar{q}$ which depends on time and the coordinates x, y and z we now want to start from the vector: $\bar{V}(s) = \widehat{\phi}\widehat{\bar{p}} - \widehat{\psi}\widehat{\bar{q}}$ which depends on s and x, y and z.

Applying the divergence theorem to this vector, we find:

$$(8) \qquad \iint_S (\widehat{\phi}\widehat{\bar{p}} - \widehat{\psi}\widehat{\bar{q}}) \cdot n\, dS = \iiint_V \{\operatorname{div}(\widehat{\phi}\widehat{\bar{p}}) - \operatorname{div}(\widehat{\psi}\widehat{\bar{q}})\} dV$$

Now

$$\operatorname{div}(\widehat{\phi}\widehat{\bar{p}}) = \frac{\partial(\widehat{\phi}\widehat{p}_x)}{\partial x} + \frac{\partial(\widehat{\phi}\widehat{p}_y)}{\partial y} + \frac{\partial(\widehat{\phi}\widehat{p}_z)}{\partial z} = \widehat{p}_x \frac{\partial \widehat{\phi}}{\partial x} + \widehat{p}_y \frac{\partial \widehat{\phi}}{\partial y} + \widehat{p}_z \frac{\partial \widehat{\phi}}{\partial z} + \widehat{\phi} \operatorname{div} \widehat{\bar{p}}$$

and

$$\operatorname{div}(\widehat{\psi}\widehat{\bar{q}}) = \widehat{q}_x \frac{\partial \widehat{\psi}}{\partial x} + \widehat{q}_y \frac{\partial \widehat{\psi}}{\partial y} + \widehat{q}_z \frac{\partial \widehat{\psi}}{\partial z} + \widehat{\psi} \operatorname{div} \widehat{\bar{q}}$$

According to Darcy's law $p_x = - K_x(\partial \psi/\partial x)$ etc.

Taking the Laplace transform of this equation we obtain:

$$\widehat{p}_x = -K_x\left(\widehat{\frac{\partial\psi}{\partial x}}\right) = -K_x\frac{\partial\widehat{\psi}}{\partial x} \text{ etc.}$$

(K_x etc. assumed independent of t) as

$$\left(\widehat{\frac{\partial\psi}{\partial x}}\right) = \int_0^\infty e^{-st}\frac{\partial\psi}{\partial x}dt = \frac{\partial}{\partial x}\int_0^\infty e^{-st}\psi dt = \frac{\partial\widehat{\psi}}{\partial x}.$$

$\operatorname{div}\bar{p} = -S_s(d\psi/\partial t)$ and $\operatorname{div}\widehat{p} = -S_s\left(\widehat{\frac{\partial\psi}{\partial t}}\right)/$

As $\left(\widehat{\frac{\partial\psi}{\partial t}}\right) = \int_0^\infty e^{-st}\frac{\partial\psi}{\partial t}dt = \int_0^\infty e^{-st}\partial\psi = [\psi e^{-st}]_0^\infty + s\int_0^\infty \psi e^{-st}dt$

$$= -\psi(0) + s\widehat{\phi}$$

while the drawdown ψ at the time $t = 0$ is equal to zero we find:

$$\left(\widehat{\frac{\partial\psi}{\partial t}}\right) = s\widehat{\psi} \text{ and } \operatorname{div}\widehat{p} = -S_s s\widehat{\psi}$$

In the same way we find: $\operatorname{div}\widehat{q} = -S_s s\widehat{\phi}$

Thus, evaluating Eq. (8) we obtain at last:

$$\iint_S (\widehat{\phi}\widehat{p}_n - \widehat{\psi}\widehat{q}_n)dS = \iiint_V S_s(-s\widehat{\phi}\widehat{\psi} + s\widehat{\psi}\widehat{\phi})dV.$$

If we choose the same s for the transformation of ψ and ϕ, which is permitted as s is an arbitrary parameter, the integrand of the triple integral equals zero and consequently also in the case of unsteady flow Green's theorem holds:

$$\iint_S (\widehat{\phi}\widehat{p}_n - \widehat{\psi}\widehat{q}_n)dS = 0$$

Mathematical analysis of this expression in a similar way as has been done for steady flow shows that

$$\widehat{\phi}_B\widehat{Q}_B = \widehat{\psi}_A\widehat{Q}_A$$

for every value of the parameter s.

If $\widehat{Q}_A$ is the same function of s as $\widehat{Q}_B$, $\widehat{\psi}_A$ will be the same function of s as $\widehat{\phi}_B$ and inasmuch the inverse Laplace transform of a function is essentially unique, we can conclude that ϕ_B is the same function of t as ψ_A if Q_B is the same function of t as Q_A.

2. APPLICATION TO A HORIZONTAL "DOUBLE AQUIFER"

Fig. 4 is a diagrammatic representation of a horizontal artesian system, consisting of two elastic leaky aquifers, separated and confined by semipervious strata. Both aquifers will be drained in succession by means of a steadily discharging pumping well that is screened only throughout the aquifer in question.

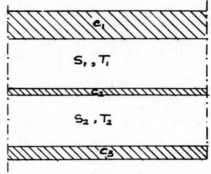

Figure 4

The following assumptions are made:

1.   The aquifers are individually elastic, homogeneous, isotropic, uniform in thickness and infinite in areal extent. The formation constants are for the upper aquifer $T_1 = K_1D_1$ and S_1 and for the lower aquifer $T_2 = K_2D_2$ and S_2 where

T = coefficient of transmissivity (dimension length²/time)
K = coefficient of permeability (length/time)
D = the thickness of the aquifer (length)
S = storage coefficient (dimensionless) = DS_s where
S_s = specific storage (length^{-1})

2.   There is no storage in the semipervious layers; the resistance to ground-waterflow of the three semipervious layers is respectively c_1, c_2 and c_3 where

$$c = \frac{D^1}{K^1} \text{ (dimension of time)}$$

D^1 = thickness of the layer (length) and
K^1 = coefficient of permeability of the semipervious layer (length/time).

3.   The formation constants T, S and c remain constant with time and constant in the space of the layer they characterize.

4.   Groundwater flow as a result of pumping occurs vertically in the semipervious layers and horizontally in the aquifers.

5.   The system is overlain and underlain by bodies of groundwater in which the piezometric heads will not be influenced by pumping in the aquifers.

6.   Before the beginning of the pumping ($t = 0$) the groundwater is in rest and the piezometric head is the same everywhere.

7.   The steady discharge of the pumping well draining the upper aquifer be Q_1 and the drawdown distributions in the upper and lower aquifers as a

result of this pumping be ϕ_1 and ϕ_2 respectively (see Fig. 5a) whereas pumping the lower aquifer with a steady discharge Q_2 causes drawdown distributions ψ_1 and ψ_2 (see Fig. 5b). ϕ_1, ϕ_2, ψ_1 and ψ_2 are functions of the horizontal distance r to the vertical axis of the pumping well and of time:

$$\phi = \phi(r,t) \text{ and } \psi = \psi(r,t).$$

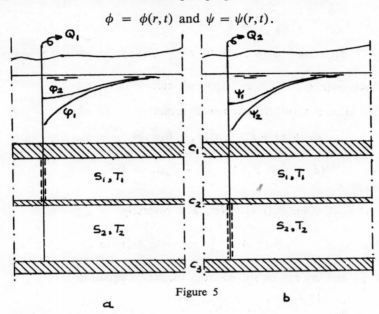

Figure 5

a   b

a. Drawdown distributions by pumping the upper aquifer (see Fig. 5a).

It can easily be shown that the boundary value problem for this system in polar coordinates can be written down as:

$$\frac{\partial^2 \phi_1}{\partial r^2} = \frac{1}{r}\frac{\partial \phi_1}{\partial r} - (a_1 + b_1)\phi_1 + b_1\phi_2 - \beta_1 \frac{\partial \phi_1}{\partial t} = 0$$

$$\frac{\partial^2 \phi_2}{\partial r^2} + \frac{1}{r}\frac{\partial \phi_2}{\partial r} - (a_2 + b_2)\phi_2 + a_2\phi_1 - \beta_2 \frac{\partial \phi_2}{\partial t} = 0$$

$$\phi_1(r,0) = \phi_2(r,0) = 0 \qquad\qquad r\frac{\partial \phi_1}{\partial r_2}(0,t) = -P_1$$

$$\phi_1(\infty,t) = \phi_2(\infty,t) = 0 \qquad\qquad r\frac{\partial \phi_2}{\partial r}(0,t) = 0$$

in which

$$a_1 = \frac{1}{T_1 c_1} \qquad b_1 = \frac{1}{T_1 c_2} \qquad \beta_1 = \frac{S_1}{T_1} \qquad P_1 = \frac{Q_1}{2\pi T_1}$$

$$a_2 = \frac{1}{T_2 c_2} \qquad b_2 = \frac{1}{T_2 c_3} \qquad \beta_2 = \frac{S_2}{T_2} \ .$$

Laplace transformation with respect to t gives:

$$\frac{\partial^2 \bar{\phi}_1}{\partial r^2} + \frac{1}{r} \frac{\partial \bar{\phi}_1}{\partial r} - (a_1 + b_1 + \beta_1 s)\bar{\phi}_1 + b_1\bar{\phi}_2 = 0$$

$$\frac{\partial^2 \bar{\phi}_2}{\partial r^2} + \frac{1}{r} \frac{\partial \bar{\phi}_2}{\partial r} - (a_2 + b_2 + \beta_2 s)\bar{\phi}_2 + a_2\bar{\phi}_1 = 0$$

$$\bar{\phi}_1(\infty,s) = \bar{\phi}_2(\infty,s) = 0; \quad r\frac{\partial \bar{\phi}_1}{\partial r}(0,s) = -\frac{P_1}{s} \quad r\frac{\partial \bar{\phi}_2}{\partial r}(0,s) = 0$$

in which $\bar{\phi}_1(r,s) = \int_0^\infty e^{-st}\phi_1(r,t)dt$ etc.

After Hankel transformation with respect to r we obtain:

$$(a_1 + b_1 + \beta_1 s + \alpha^2)\breve{\bar{\phi}}_1 - b_1\breve{\bar{\phi}}_2 = \frac{P_1}{s}$$

$$(a_2 + b_2 + \beta_2 s + \alpha^2)\breve{\bar{\phi}}_2 - a_2\breve{\bar{\phi}}_1 = 0$$

in which

$$\breve{\bar{\phi}}_1(\alpha,s) = \int_0^\infty rJ_0(\alpha r)\bar{\phi}_1(r,s)\,dr \text{ etc. and reversely}$$

$$\bar{\phi}_1(r,s) = \int_0^\infty \alpha J_0(\alpha r)\breve{\bar{\phi}}_1(\alpha,s)\,d\alpha$$

Now $\breve{\bar{\phi}}_1$ and $\breve{\bar{\phi}}_2$ can be solved from these two equations:

$$\breve{\bar{\phi}}_1 = \frac{P_1(a_2 + b_2 + \beta_2 s + \alpha^2)}{s} \cdot \frac{1}{cs^2 + (a\alpha^2 + d)s + \alpha^4 + b\alpha^2 + e}$$

and

(9)
$$\breve{\bar{\phi}}_2 = \frac{P_1 a_2}{s} \cdot \frac{1}{cs^2 + (a\alpha^2 + d)s + \alpha^4 + b\alpha^2 + e}$$

in which $a = \beta_1 + \beta_2$, $b = a_1 + b_1 + a_2 + b_2$, $c = \beta_1\beta_2$, $d=(a_1+b_1)\beta_2 + (a_2 + b_2)\beta_1$ and $e = a_1a_2 + b_1b_2 + a_1b_2$. Instead of

$$cs^2 + (a\alpha^2 + d)s + \alpha^4 + b\alpha^2 + e$$

we write $c(s + x_1) \cdot (s + x_2)$ in which

$$x_1 + x_2 = \frac{a\alpha^2 + d}{c} \quad \text{and} \quad x_1 x_2 = \frac{\alpha^4 + b\alpha^2 + e}{c} .$$

Obviously x_1 and x_2 are the roots of the quadratic equation:

$$cx^2 - (a\alpha^2 + d)x + \alpha^4 + b\alpha^2 + e = 0 \quad (x_1 \text{ and } x_2 \text{ both positive}).$$

The equations (9) now become:

$$\breve{\bar{\phi}}_1(\alpha,s) = \frac{P_1(a_2 + b_2 + \beta_2 s + \alpha^2)}{s \cdot c(s + x_1)(s + x_2)}$$

(10)
$$\breve{\phi}_2(\alpha, s) = \frac{P_1 a_2}{s} \frac{1}{c(s + x_1)(s + x_2)}.$$

The inverse Laplace transform of the function

$$\breve{F}(\alpha, s) = \frac{1}{c(s + x_1)(s + x_2)} = \frac{1}{c(x_1 - x_2)}\left(\frac{1}{s + x_2} - \frac{1}{s + x_1}\right)$$

can easily be found as:

$$\breve{F}(\alpha, t) = \frac{1}{c(x_1 - x_2)}(e^{-x_2 t} - e^{-x_1 t})$$

and so the inverse Laplace transform of $1/s \breve{F}(\alpha, s)$ becomes:

$$c \frac{1}{(x_1 - x_2)} \int_0^t (e^{-x_2 \tau} - e^{-x_1 \tau}) d\tau = \frac{1}{c x_1 x_2}\left(1 + \frac{x_2}{x_1 - x_2}e^{-x_1 t} - \frac{x_1}{x_1 - x_2}e^{-x_2 t}\right).$$

Consequently we obtain for the inverse Laplace transform of (10):

$$\breve{\phi}_1(\alpha, t) = \frac{P_1(a_2 + b_2 + \alpha^2)}{c x_1 x_2}\left(1 + \frac{x_2}{x_1 - x_2}e^{-x_1 t} - \frac{x_1}{x_1 - x_2}e^{-x_2 t}\right)$$

$$+ \frac{P_1 \beta_2}{c(x_1 - x_2)}(e^{-x_2 t} - e^{-x_1 t})$$

(11)

$$\breve{\phi}_2(\alpha, t) = \frac{P_1 a_2}{c x_1 x_2}\left(1 + \frac{x_2}{x_1 - x_2}e^{-x_1 t} - \frac{x_1}{x_1 - x_2}e^{-x_2 t}\right).$$

Inverse Hankel transformation gives the final solution:

$\phi_1(r, t)$

$$= \frac{P_1}{c} \int_0^\infty \frac{\alpha(a_2 + b_2 + \alpha^2)}{x_1 x_2} J_0(\alpha r)\left(1 + \frac{x_2}{x_1 - x_2}e^{-x_1 t} - \frac{x_1}{x_1 - x_2}e^{-x_2 t}\right)d\alpha$$

(12)
$$- \frac{P_1 \beta_2}{c} \int_0^\infty \frac{\alpha}{x_1 - x_2} J_0(r\alpha)(e^{-x_1 t} - e^{-x_2 t})d\alpha$$

$\phi_2(r, t)$

$$= \frac{P_1 a_2}{c} \int_0^\infty \frac{\alpha}{x_1 x_2} J_0(\alpha r)\left(1 + \frac{x_2}{x_1 - x_2}e^{-x_1 t} - \frac{x_1}{x_1 - x_2}e^{-x_2 t}\right)d\alpha$$

in which x_1 and x_2 are functions of α which can be solved from the quadratic equation:

$$cx^2 - (a\alpha^2 + d)x + \alpha^4 + b\alpha^2 + e = 0$$

while

$$a = \beta_1 + \beta_2 \qquad\qquad a_1 = \frac{1}{T_1 c_1} \qquad a_2 = \frac{1}{T_2 c_2}$$

$b = a_1 + b_1 + a_2 + b_2$

$c = \beta_1 \cdot \beta_2$ $b_1 = \dfrac{1}{T_1 c_2}$ $b_2 = \dfrac{1}{T_2 c_3}$

$d = (a_1 + b_1)\beta_2 + (a_2 + b_2)\beta_1$

$e = a_1 a_2 + b_1 b_2 + a_1 b_2$ $\beta_1 = \dfrac{S_1}{T_1}$ $\beta_2 = \dfrac{S_2}{T_2}$ $P_1 = \dfrac{Q_1}{2\pi T_1}$

By direct substitution of these final solutions of ϕ_1 and ϕ_2 in the original differential equation and by applying the boundary conditions, the correctness of the solutions has been verified.

b Drawdown distributions by pumping the lower aquifer (see Fig. 5b)

The differential equations in this case are the same as the case of pumping the upper aquifer, but there is a difference with respect to the boundary conditions:

$$\frac{\partial^2 \psi_1}{\partial r^2} + \frac{1}{r}\frac{\partial \psi_1}{\partial r} - (a_1 + b_1)\psi_1 + b_1\psi_2 - \beta_1 \frac{\partial \psi_1}{\partial t} = 0$$

$$\frac{\partial^2 \psi_2}{\partial r^2} + \frac{1}{r}\frac{\partial \psi_2}{\partial r} - (a_2 + b_2)\psi_2 + a_2\psi_1 - \beta_2 \frac{\partial \psi_2}{\partial t} = 0$$

$$\psi_1(r,0) = \psi_2(r,0) = 0 \qquad r\frac{\partial \psi_1}{\partial r} 0,t) = 0$$

$$\psi_1(\infty,t) = \psi_2(\infty,t) = 0 \qquad -r\frac{\partial \psi_2}{\partial r}(0,t) = P_2$$

in which $P_2 = Q_2/2\pi T_2$ and the other constants are the same as in a.

The solutions of this boundary value problem are derived in a similar fashion as in the foregoing problem, running as follows:

$$\psi_1(r,t) = \frac{P_2 b_1}{c} \int_0^\infty \frac{\alpha}{x_1 x_2} J_0(\alpha r)\left(1 + \frac{x_2}{x_1 - x_2}e^{-x_1 t} - \frac{x_1}{x_1 - x_2}e^{-x_2 t}\right)d\alpha$$

$$\psi_2(r,t) = \frac{P_2}{c} \int_0^\infty \frac{\alpha(a_1 + b_1 + \alpha^2)}{x_1 x_2} J_0(\alpha r)\left(1 + \frac{x_2}{x_1 - x_2}e^{-x_1 t} - \frac{x_1}{x_1 - x_2}e^{-x_2 t}\right)d\alpha$$

(13)

$$- \frac{P_2\beta_1}{c} \int_0^\infty \frac{\alpha}{x_1 x_2} J_0(\alpha r)(e^{-x_1 t} - e^{-x_2 t})d\alpha$$

in which $P_2 = Q_2/2\pi T_2$ and the other parameters and constants have the same values as in Eq. (12).

According to the reciprocity principle ϕ_2 in Eq. (12) should be equal to ψ_1 in Eq. (13) if $Q_1 = Q_2$ and indeed the right hand sides of both equations have the same integral form and differ only with respect to the constants before the integral signs. But if we assume that $Q_1 = Q_2 = Q$ we easily find:

$$\frac{P_2 b_1}{c} = \frac{Q}{2\pi T_2} \cdot \frac{1}{T_1 c_2} \cdot \frac{1}{c} = \frac{Q}{2\pi T_1} \cdot \frac{1}{T_2 c_2} \cdot \frac{1}{c} = \frac{P_1 a_2}{c}$$

and consequently $\phi_2(r,t) = \psi_1(r,t)$ if $Q_1 = Q_2$ for all values of r and t.

c. Solutions for the steady state

As time becomes effectively large the storage in the aquifers will give out and the pumping well will be sustained almost entirely by leakage passing through the semipervious layers; thus, a steady state of flow will be essentially realized. Consequently the steady drawdown equations in which the storage coefficients must have disappeared can be derived from the unsteady ones by letting t approach to infinity.

Putting $t = \infty$ in the Eqs. (12) and (13) we obtain:

$$\phi_1(r) = P_1 \int_0^\infty \frac{\alpha(a_1 + b_2 + \alpha^2)}{\alpha^4 + b\alpha^2 + e} \cdot J_0(\alpha r) d\alpha$$

$$\phi_2(r) = P_1 a_2 \int_0^\infty \frac{\alpha}{\alpha^4 + b\alpha^2 + e} J_0(\alpha r) d\alpha$$

(12a)

$$\psi_1(r) = P_2 b_1 \int_0^\infty \frac{\alpha}{\alpha^4 + b\alpha^2 + e} J_0(\alpha r) d\alpha$$

(13a)

as

$$\psi_2(r) = P_2 \int_0^\infty \frac{\alpha(a_1 + b_1 + \alpha^2)}{\alpha^4 + b\alpha^2 + e} J_0(\alpha r) d\alpha$$

$$x_1 x_2 = \frac{\alpha^4 + b\alpha^2 + e}{c}$$

Putting $\alpha^4 + b\alpha^2 + e = (\alpha^2 + y_1)(\alpha^2 + y_2)$ where $y_1 + y_2 = b$ and $y_1 y_2 = e$ we find

$$\frac{1}{\alpha^4 + b\alpha^2 + e} = -\frac{1}{y_1 - y_2}\left(\frac{1}{\alpha^2 + y_1} - \frac{1}{\alpha^2 + y_2}\right)$$

and

$$\frac{a_2 + b_2 + \alpha^2}{\alpha^4 + b\alpha^2 + e} = -\frac{1}{y_1 - y_2}\left(\frac{a_2 + b_2 - y_1}{\alpha^2 + y_1} - \frac{a_2 + b_2 - y_2}{\alpha^2 + y_2}\right).$$

Substituting this in Eqs. (12a) and (13a), integrals of the form

$$\int_0^\infty \frac{\alpha}{\alpha^2 + k^2} J_0(\alpha r) d\alpha$$

occur, which may be replaced by $K_0(kr)$.

Thus 12a and 13a become:

$$\phi_1 = -\frac{P_1}{y_1 - y_2}\{(a_2 + b_2 - y_1)K_0(r\sqrt{y_1}) - (a_2 + b_2 - y_2)K_0(r\sqrt{y_2})\}$$

(12b)

$$\phi_2 = -\frac{P_1 a_2}{y_1 - y_2} \cdot \{K_0(r\sqrt{y_1}) - K_0(r\sqrt{y_2})\}$$

$$\psi_1 = -\frac{P_2 b_1}{y_1 - y_2}\{K_0(r\sqrt{y_1}) - K_0(r\sqrt{y_2})\}$$

(13b)

$$\psi_2 = -\frac{P_2}{y_1 - y_2}\{a_1 + b_1 - y_1)K_0(r\sqrt{y_1}) - (a_1 + b_1 - y_2)K_0(r\sqrt{y_2})\}$$

By means of a skilful substitution and some mathematical analysis we can obtain

$$\phi_1 = -\frac{P_1}{z_1 - z_2}\{z_2 K_0(r\sqrt{y_1}) - z_1 K_0(r\sqrt{y_2})\}$$

(12c)

$$\phi_2 = \frac{P_1}{z_1 - z_2}\{K_0(r\sqrt{y_1}) - K_0(r\sqrt{y_2})\}$$

$$\psi_1 = -\frac{z_1 z_2 P_2}{z_1 - z_2}\{K_0(r\sqrt{y_1}) - K_0(r\sqrt{y_2})\}$$

(13c)

$$\psi_2 = \frac{P_2}{z_1 - z_2}\{z_1 K_0(r\sqrt{y_1}) - z_2 K_0(r\sqrt{y_2})\}$$

and consequently:

(14)
$$\phi_1 + z_1 \phi_2 = P_1 K_0(r\sqrt{y_1})$$
$$\phi_1 + z_2 \phi_2 = P_1 K_0(r\sqrt{y_2})$$

and

(15)
$$\psi_1 + z_1 \psi_2 = P_2 z_1 K_0(r\sqrt{y_1})$$
$$\psi_1 + z_2 \psi_2 = P_2 z_2 K_0(r\sqrt{y_2})$$

in which z_1 and z_2 are the roots of the quadratic equatior.

$$a_2 z^2 + (a_2 - a_1 + b_2 - b_1)z - b_1 = 0$$

and

$$y_{1,2} = a_1 + b_1 - a_2 z_{1,2}$$

d. Determination of the formation constants

With the help of two pumping tests it is possible, if a steady state has been attained, to determine the five formation constants T_1, T_2, c_1, c_2 and c_3 (see Fig. 4) starting from the Eqs. (12c), (13c), (14) and (15).

Successfully the next steps have to be taken:

1. The pumping test in the upper aquifer yields values for ϕ_1, ϕ_2 and Q_1 whereas the pumping test in the lower aquifer gives data concerning ψ_1, ψ_2 and Q_2.

2. Verify whether $\phi_2 = \psi_1$ if $Q_1 = Q_2$.

This follows from Eq. (13c) as

$$z_1 z_2 = - \frac{b_1}{a_2}$$

(product of the roots of a quadratic equation)

or

(16)
$$z_1 z_2 = - \frac{T_2 c_2}{T_1 c_2} = - \frac{T_2}{T_1}$$

and thus

$$z_1 z_2 P_2 = - \frac{T_2}{T_1} \cdot \frac{Q_2}{2\pi T_2} = - \frac{Q_2}{2\pi T_1} = -P_1 \quad \text{if } Q_1 = Q_2$$

3. Plot ϕ_1 and ψ_2 semilogarithmically against the distance r; determine T_1 and T_2 approximatively by means of Jacob's straight-line method. Calculate the product $z_1 z_2$ according to Eq. (16).

4. The first equation of Eq. (15) multiplied by z_2 gives if we put $Q_1 = Q_2$:

$$z_2 \psi_1 + z_1 z_2 \psi_2 = z_1 z_2 P_2 K_0(r\sqrt{y_1}) = - P_1 K_0(r\sqrt{y_1})$$

and from Eq. (14):

$$z_2 \psi_1 + z_1 z_2 \psi_2 = - \phi_1 - z_1 \phi_2 .$$

Inasmuch $\phi_2 = \psi_1$ we obtain:

$$(z_1 + z_2)\psi_1 = -z_1 z_2 \psi_2 - \phi_1$$

or:

(17)
$$z_1 + z_2 = \frac{(T_2/T_1)\psi_2 - \phi_1}{\psi_1} .$$

By substituting for several values of r the accessory values of ψ_2, ϕ_1 and ψ_1 in Eq. (17) we can determine the sum $z_1 + z_2$. Together with $z_1 z_2$ the values of z_1 and z_2 can be calculated.

5. Plot the curves $\gamma_1 = \phi_1 + z_1 \phi_2$ and $\gamma_2 = \phi_2 + z_2 \phi_2$ of Eq. (14) on a logarithmic paper and determine by means of Theis' type-curve method the values of T_1, y_1 and y_2. In this case the type-curve is formed by a logarithmic plot of $K_0(u)$ against u.

Check these results by doing the same for the curves

$$\delta_1 = \frac{1}{z_2}\psi_i + \psi_2 \quad \text{and} \quad \delta_2 = \frac{1}{z_2}\psi_1 + \psi_2 \quad \text{of Eq. (15)}$$

which yields values for T_2 y_1 and y_2.

6. As y_1, y_2, z_1 and z_2 are given now, the values of a_1, a_2, b_1 and b_2 can be solved and consequently the values of c_1, c_2 and c_3 with help of the already determined values of T_1 and T_2.

3

Theory of coupled processes in porous media, including heat and mass transfer and polyphase flow phenomena

ÉTUDE EXPERIMENTALE DE LA CONVECTION NATURELLE EN MILIEU POREUX

C. Thirriot et S. Bories

1. INTRODUCTION

Les instabilités qui se développent dans un fluide soumis à des différences de température, peuvent donner naissance à des mouvements de convection naturelle qui dépendent des conditions extérieures et des propriétés du fluide.

L'étude de ces mouvements est très difficile dans le cas général, aussi est-on conduit à n'aborder l'analyse de ces phénomènes que dans le cas de configurations particulièrement simples.

Ainsi, l'étude de la convection naturelle dans des couches fluides horizontales chauffées par dessous, retient encore, après plus de 60 ans l'intérêt des chercheurs.

Depuis les expériences effectuées par Bénard en 1900 [1] un nombre considérable de travaux ont été réalisés sur ce thème, mais ce n'est que depuis une vingtaine d'années seulement que l'on a entrepris quelques études sur les courants thermoconvectifs dans des couches poreuses.

La recherche d'explications, concernant certains phénomènes naturels et les considérations économiques liées à la détermination des coefficients de transfert de chaleur, nécessitent cependant, de plus en plus, une connaissance approfondie du phénomène de la convection naturelle en milieu poreux.

Après avoir évoqué la situation des recherches dans ce domaine, nous nous proposons de présenter, dans ce qui suit, les premiers résultats que nous avons obtenus en ce qui concerne l'organisation des mouvements de convection dans une couche poreuse saturée horizontale chauffée par dessous et présentant une surface libre. Nous serons ainsi amené à donner un aperçu de l'installation expérimentale et à présenter le procédé de visualisation utilisé.

2. INSTABILITÉ D'UNE COUCHE FLUIDE HORIZONTALE DONT LA DENSITÉ VARIE D'UNE FACON CONTINUE SUIVANT LA VERTICALE

Si l'on considère une couche liquide maintenue entre deux plans isothermes horizontaux portés à des températures différentes T_1 et T_2, avec $T_2 > T_1$, on peut montrer simplement, à partir du calcul de l'énergie potentielle, que suivant l'orientation du gradient de température (vers le haut ouvers le bas) les systèmes respectifs correspondants seront instable ou stable.

Dans le premier cas, une perturbation accidentelle est alors susceptible d'amorcer des mouvements qui tendront à faire évoluer le système en assurant l'échange des places entre les masses chaudes et froides.

Si les parois sont bonnes conductrices et uniformément chauffées, le mouvement va s'entretenir par suite du réchauffement de la masse froide descendante et du refroidissement de la masse chaude ascendante.

Mais cette analyse sommaire ne doit pas nous amener à conclure pour autant qu'une différence de température si faible soit-elle, doit nécessairement donner naissance à des mouvements de convection naturelle.

En effet, l'action simultanée des forces de frottement dues à la viscosité et de la vitesse de refroidissement des particules qui en s'élevant cèdent de la chaleur au fluide environnant (ce qui peut se traduire par une diminution de la force motrice d'origine archimédienne) exige, pour qu'il puisse y avoir développement des instabilités, une différence de température minimum, en deça de laquelle toute perturbation se trouve amortie.

Le régime de convection établi est donc précédé par un régime préconvectif stable (pas de mouvement) caractérisé par une distribution de température linéaire qui correspond à la conduction pure $T = T_1 - \beta z$, et le passage, de l'état préconvectif à l'amorce de mouvements, ne peut s'effectuer qu'à partir d'une certaine différence de température critique.

3. ANALYSE BIBLIOGRAPHIQUE SOMMAIRE

1°. *Recherches théoriques*

Par analogie avec les travaux sur la ·onvection naturelle dans les couches fluides horizontales chauffées par dessous, l'analyse de ce phénomène en milieu poreux a été réalisée en deux parties :

1. Détermination du critère d'amorce de la convection
2. Étude de la configuration de l'écoulement

Les équations qui régissent la convection naturelle en milieu poreux résultent des équations générales de continuité, du mouvement, de l'énergie et de l'équation d'état du fluide.

Un certain nombre d'hypothèses simplificatrices, dont la plus importante consiste à admettre ρ_f densité du fluide constante (sauf dans le terme qui traduit l'action de la pesanteur), permettent de résoudre d'un point de vue théorique, le système d'équations auquel obéit le phénomène.

Parmi les hypothèses admises dans le cas des couches poreuses nous citerons encore :

— écoulement lent — perméabilité uniforme — viscosité du fluide constante — effets de rayonnement négligeables — pas de réaction chimique entre le milieu poreux et le fluide.

Dès lors et compte tenu des hypothèses précédentes, le système d'équation s'écrit:

(1) $$\frac{\rho}{\varepsilon} \frac{\partial \vec{V}}{\partial t} = -\vec{\text{grad}}\, P - \frac{\mu}{K} \vec{V} - \rho \vec{g}$$

(2) $$\frac{DT}{Dt} = k\nabla^2 T$$

[4]

(3) $$\text{div}(\rho \vec{V}) = 0$$

(4) $$\rho = \rho_0(1 - \alpha\Delta T).$$

Ce système a été résolu par Lapwood [2] en admettant tout d'abord que les perturbations de pression de température et de vitesse résultant du mouvement sont faibles. Il est alors possible de linéariser les équations et, par suite, après élimination de la pression, on se ramène à la résolution de l'équation aux dérivées partielles.

(5) $$\left(\frac{1}{\varepsilon} \frac{\partial}{\partial t} + \frac{\mu}{K\rho}\right) \left(\frac{\partial}{\partial t} - k\nabla^2\right) \nabla^2 w = g\alpha\beta\nabla_1^2 w$$

La condition de développement des instabilités est déterminée en faisant

$$\frac{\partial}{\partial t} = 0 \quad [11], \text{ ce qui conduit à l'équation}$$

(6) $$\nabla^4 w = \frac{K\rho g\alpha\beta}{\mu k} \nabla_1^2 w \quad \text{avec} \quad \nabla^2 = \frac{\partial^2}{\partial x^2} + \frac{\partial^2}{\partial y^2} + \frac{\partial^2}{\partial z^2}$$

$$\nabla_1^2 = \frac{\partial^2}{\partial x^2} + \frac{\partial^2}{\partial y^2}$$

équation qui est résolue en cherchant une solution particulière de la forme $w = w(z) \sin ln \sin my$ (1 et m nombres associés aux directions x et y).

On obtient alors le gradient de température critique:

$$\beta_c = \frac{\Delta T_c}{H} = \frac{Ra_c \mu k}{K\rho g\alpha H^2} \qquad Ra_c = \text{cte qui dépend des conditions aux limites}$$

qui traduit la différence de température minimum à assurer entre les deux faces de la couche poreuse pour qu'une perturbation de longueur d'onde donnée ait des chances d'être entretenue. Le nombre adimensionnel:

$$\frac{K\rho g\alpha\beta_c H^2}{\mu k} = Ra_c$$

qui rend compte du rapport des forces de pesanteur aux forces de frottement et que l'on peut également déterminer par l'analyse dimensionnelle est appelé nombre de Rayleigh critique.

Pour de petites perturbations, l'équation (6) fournit également la répartition des températures et la distribution des vitesses dans la couche poreuse. Les solutions sont alors du même type que celles rencontrées dans le cas des couches fluides.

Il faut signaler cependant, à propos des conditions aux limites, que leur nombre étant plus important que le nombre de constantes à déterminer, il n'est pas possible de les considérer toutes. Les conditions relatives aux contraintes tangentielles qui s'explicitent en fonction des dérivées de la vitesse sont écartées et par suite la théorie précédente n'est plus valable au voisinage immédiat des frontières.

Dans une étude récente, Katto et Masuoka [3] ont montré toutefois, en ajoutant à l'équation du mouvement $(1)_2$ un terme supplémentaire qui rend compte de la résistances visqueuse: $\mu \nabla V$, et en prenant en considération toutes les conditions aux limites, que le gradient de température critique, déterminé par la théorie précédente, restait valable tant que le rapport d/H (d diamètre des grains) n'excédait pas 0,15.

Parmi les autres recherches théoriques, nous citerons les travaux de Rogers [4], tout à fait semblables à ceux de Lapwood, les travaux de Elder [5] qui par voie numérique a déterminé la répartition des températures et la distribution des vitesses dans le cas bi-dimensionnel du modèle Hele-Shaw et enfin les travaux de Wooding [6].

En ce qui concerne les recherches théoriques portant principalement sur le critère de convection, nous citerons les travaux de Giambelli [7].

2°. *Études expérimentales*

Les quelques travaux de caractère expérimental, réalisés sur la convection naturelle dans des couches poreuses horizontales, ont eu essentiellement pour objet la détermination du nombre de Rayleigh critique et la recherche d'une loi de variation du coefficient de conductivité effective en fonction du nombre de Rayleigh.

Seul, Elder s'est intéressé à l'aspect mécanique des fluides du phénomène et a vérifié, dans le cas bi-dimensionnel du modèle Hele-Shaw, le bon accord d'un point de vue qualitatif des mouvements de convection observés avec les résultats de ses calculs.

En ce qui concerne les valeurs du nombre de Rayleigh critique pour une couche poreuse comprise entre deux plans imperméables et conducteurs, nous avons rassemblé dans le tableau ci-dessous les résultats qui sont actuellement connus.

Auteurs	Horton 1954	Lapwood 1948	Giambelli 1955	Scshneider 1953	Dirksen 1967	Elder 1966	Katto 1967
Théorie	39,5	39,5	12				39,5
Expérience	6 à 12,5		12	39,5	2,5 à 7,6	39,5	39,5

Si les valeurs obtenues par Horton et Giambelli peuvent être contestées par le fait que leur modèle expérimental ne satisfait point aux conditions admises dans le développement de la théorie (milieu poreux de grandes dimensions horizontales par rapport à l'épaisseur) il en va différemment des résultats obtenus par Dirksen.

Nous ne discuterons pas ici ces résultats (nous attendrons pour cella d'avoir terminé nos expériences portant sur le critère de stabilité) mais nous ferons remarquer cependant que la dispersion des valeurs ne permet pas à l'heure actuelle d'apporter une conclusion définitive à ce problème d'amorce de la convection.

De l'analyse bibliographique que nous avons faite, il ressort cependant qu'aucune étude expérimentale des courants de convection dans une couche poreuse horizontale n'a été effectuée.

Si les solutions, fournies par la théorie de la stabilité linéaire, laissent prévoir une organisation cellulaire du type de Bénard encore faut-il le vérifier expérimentalement. Tel est l'objet et l'originalité de notre étude.

4. INSTALLATION EXPÉRIMENTALE. PROCÉDÉ DE VISUALISATION

L'installation expérimentale est constituée des éléments suivants:

1°. Une cuve en matiére plastique de 70 cm de long, 50 cm de large et 8 cm de hauteur, dont l'isolation thermique latérale est assurée outre par la matière plastique constituant les parois, par un matelas de mousse polyuréthane de 3 cm d'épaisseur.

La partie inférieure de la cuve est une plaque d'AU G4 chauffée par des résistances thermocoax placées dans des rainures.

2°. Une chaîne de régulation de température maintenant la plaque chauffante à température constante.

3°. Un thermocouple pouvant être déplacé verticalement et monté sur un chariot qui peut se déplacer dans un plan horizontal parallèle à la couche poreuse.

4°. Un bac thermostaté permettant de régler la température de référence T_0 des thermocouples à la valeur désirée afin d'accroître la précision des mesures.

5°. Un système d'enregistrement et de mesure des températures

6°. Un système d'éclairage et de photographie.

La figure 1 donne une vue d'ensemble de l'installation.

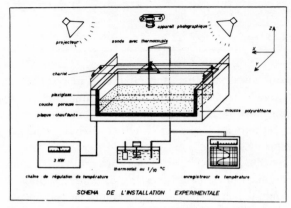

Figure 1. Schéma de l'installation experimentale

La couche poreuse est constituée par des billes de verre (dans nos expérien-ces, nous avons utilisé des billes de 3 et 5 mm). L'épaisseur de cette couche est réglée avant d'introduire les billes dans la cellule au moyen d'un couvercle emboîtable dont la distance au font peut être réglée à volonté,

Le modèle ayant été vibré et mis en place, le fluide de saturation (huile de silicone S1 510) est alors introduit par des orifices disposés latéralement. Le couvercle est ensuite enlevé et la chaîne de régulation mise en route pour une température permettant d'assurer un nombre de Rayleigh correspondant à un régime franchement convectif: $Ra > 27,5$ [2]

MÉTHODE DE VISUALISATION

A cause des difficultés que présente la visualisation des écoulements en milieu poreux et à plus forte raison dans des conditions non isothermes, nous avons utilisé un procédé fondé sur l'observations causées par les mouvements de convection sur la surface libre du fluide de saturation.

En effet, en régime de convection stationnaire, les courants viennent se fer-mer à la surface. En pulvérisant sur celle-ci de fines particules d'aluminium, il doit par conséquent être possible d'observer les mouvements superficiels qui résultent des mouvements à l'intérieur de la couche.

Avant d'appliquer ce procédé aux milieux poreux, nous avons eu cependant le souci de contrôler la validité de notre hypothèse en réalisant quelques expé-riences a vec descouches fluides.

Au cours de ces expériences, nous avons tout d'abord observé les tour-billons thermoconvectifs au moyen de particules *en suspension*. Ensuite, en saupoudrant la surface, nous avons observé effectivement que les particules venaient se placer sur le contour des cellules.

En conclusion de ces expériences, il apparaissait donc possible à partir d'observations superficielles de metre en évidence une organisation cellulaire résultant de courants de convection.

Dans le cas d'une couche poreuse présentant une surface libre, on peut cependant faire remarquer que le résultat d'une altération peut tout aussi bien provenir de mouvenents ayant leur origine dans la pellicule fluide superficielle. En pratique, ceci peut être évité si l'on choisit, compte tenu des expressions donnant la valeur des gradients critiques en couche fluide et en couche poreuse, des épaisseurs telles que la convection ne puisse pas apparaître dans la pellicule fluide.

Ainsi, pour une couche fluide présentant une surface libre:

$$\beta_f = 1100 \ \frac{vk_f}{g\alpha h^4}$$

et pour une couche poreuse, en adoptant la valeur fournie par la théorie linéaire: $Ra = 27,5$

$$\beta_m = 27,5 \ \frac{vk_m}{g\alpha K H^2} \qquad \beta_m \ll \beta_f \qquad H^2 \gg \frac{27,5}{1100} \ \frac{k_m}{k_f} \ \frac{h^4}{K}.$$

Si l'on prend $hn = 0,1$ cm avec des billes de verre pour milieu poreux et de l'huile de silicone pour fluide de saturation, il vient $H \gg 0,5$ cm.

5. RÉSULTATS DES VISUALISATIONS. CONTROLE DES RÉSULTATS DÉDUITS DES OBSERVATIONS PAR LA MESURE DE LA RÉPARTITION DES TEMPERATURES. LONGUEURS D'ONDE

1°. *Visualisation*

Les photographies 2, 3, 4, 5 et 6 représentent les visualisations superficielles des courants de convection dans des couches poreuses de 2,5–3,4,5 et 6 cm d'épaisseur.

Les particules d'aluminium saupoudrées en surface ont été chassées du centre des cellules et sont venues s'accumuler sur les contours des polygones délimitant ainsi les cellules de convection. Les centres des polygones correspondent aux jets ascendants des courants chauds, tandis que les contours délimitent la zone de descente des courants froids. Il en résulte au sein de la couche poreuse une organisation cellulaire tout à fait analogue au phénomène qui apparaît dans les couches fluides et que nous avons représentée sur la figure (7).

Les photographies précédentes sont relatives à des couches poreuses constituées par des billes de verre de 3 mm de diamètre, mais des expériences analogues effectuées avec des billes de 5 mm n'ont pas fait apparaître de modification dans l'organisation, la grandeur et la forme des cellules.

VISUALISATION DES COURANTS SUPERFICIELS

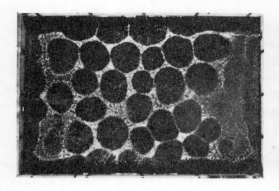 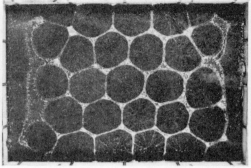

H=2,5 cm
Photo n° 2.

H=3 cm
Photo n° 3

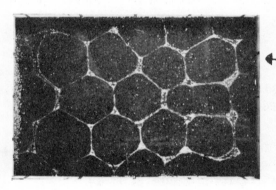

Relevé des témperatures

H=4 cm
Photo n° 4.

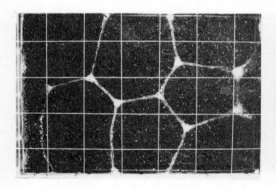 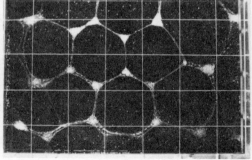

H=5 cm
Photo n° 5.

H=6 cm
Photo n° 6.

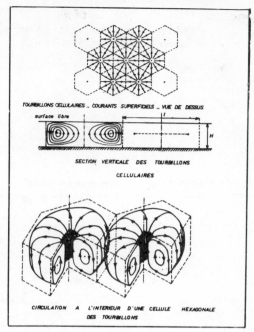

TOURBILLONS CELLULAIRES _ COURANTS SUPERFICIELS _ VUE DE DESSUS

SECTION VERTICALE DES TOURBILLONS
CELLULAIRES

CIRCULATION A L'INTERIEUR D'UNE CELLULE HEXAGONALE
DES TOURBILLONS

Figure 7

2°. *Contrôle des observations par la mesure des températures*

Si les observations superficielles sont bien significatives d'une distribution périodique des vitesses, il doit en résulter pour la température un phénomène semblable.

Ainsi, des mesures de température effectuées le long d'horizontales coupant les cellules doivent permettre d'observer des répartitions périodiques en accord avec l'organisation superficielle. Les photos (8 et 9) qui traduisent les résultats des mesures de température effectuées le long d'une horizontale, dans la pellicule fluide superficielle, et le long d'une horizontale située dans le même plan vertical que la précédente mais au milieu de la couche poreuse, confirment, d'une manière évidente, la validité des résultats fournis par la visualisation.

En outre, des mesures réalisées dans une cellule unitaire, le long de trois verticales situées: au centre (zone de courants chauds ascendants), sur les bords (zone de courants froids descendants) et dans la zone intermédiataire (conduction) confirme une fois de plus la validité de nos observations (photo 10.)

A titre d'information, nous citerons encore les essai de vérification réalisés en utilisant de l'huile de silicone colorée. Des injections effectuées à la partie inférieure de la couche poreuse ont permis d'observer effectivement la remontée de l'huile colorée vers le centre des cellules.

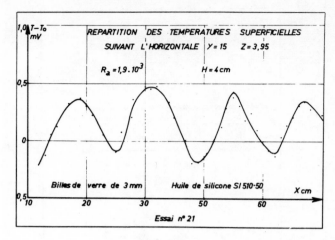

Figure 8

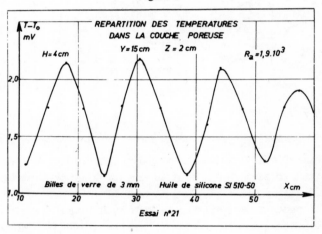

Figure 9

3°. *Longueur d'ondes des cellules*

Le dépouillement d'une vingtaine d'essais nous a permis de déterminer la valeur approchée de la longueur d'onde des cellules. Sur la figure 11, nous avons porté en fonction de l'épaisseur de la couche poreuse le rapport λ/H ou λ représente la longueur d'onde. Chaque segment porté sur cette figure englobe la plage des différents résultats obtenus pous une valeur de H donée, ainsi que 'intervalle de confiance des mesures.

Nous remarquerons que la valeur moyenne 3,2, est à raprocher des résultats obtenus dans le cas des couches fluides par:

Rayleigh: valeur théorique: $\lambda/H = 3,29$

Benard: valeur expérimentale: $3,34 < \lambda/H > 3,27$

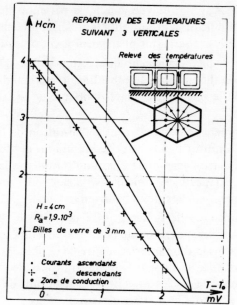

Figure 10

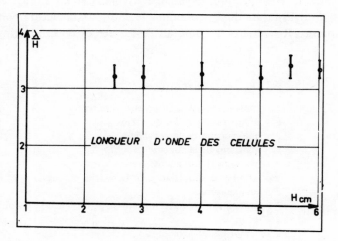

Figure 11

6. CONCLUSION

Si la seule intuition pouvait laisser prévoir l'analogie des phénomènes de convection naturelle en chonche fluide et en couche poreuse, il était cependant fondamental de rechercher autant que possible une conclusion scientifique.

Nous pensons, de ce point de vue, que les expériences de visualisation présentées dans cette publication et qui ont été incontestablement confirmées par les relevés de température satisfont pleinement à cet objectif.

Nous ferons remarquer également qu'il est possible d'étudier désormais, dans quelle mesure la théorie de la stabilité linéaire permet d'appréhender d'une manière convenable le phénomène macroscopique de la convection naturelle en couche poreuse.

Enfin, pour terminer, nous signalerons qu'en raison de la complexité de ce phénomène, le Laboratoire de l'Institut de Mécanique des Fluides de Toulouse, a déjà abordé sous le titre général de Thermofiltration, un large programme de recherche dont les perspectives immédiates, faisant suite à ce premier travail, portent principalement sur les points suivants:

— Étude expérimentale de la répartition des vitesses — Vérification des schémas théoriques fournis par l'étude de la stabilité linéare — Étude en modèle Hele-Shaw au moyen d'un interféromètre de la phase transitoire d'amorce des instabilités — Étude des coefficients de transfert de chaleur globaux — Étude des phénomènes de microconvection et coefficients de transfert locaux — Influences de contact dans la matière poreuse.

Cette étude a été effectuée sous le patronage de la Division Forage et Production de l'Institut Français du Pétrole dans le cadre d'une action concertée de l'Association des Recherches Techniques de Forage et de Production.

NOTATIONS

Cp chaleur spécifique a pression constante
d diamétre des billes de verre
g accélération de la pesanteur
H épaisseur de la couche
K perméabilité, ε porosité
Ra nombre de Rayleigh pour la couche poreuse $= g\ a\beta\ H2\ K/\mu\ k$
T température
ΔT différence de température etre la plaque chaude et froide
t temps
uvw composantes de la vitesse
xyz coordonnées
a coef. de dilatation cubique
β gradient de température
k diffusivité thermique
λ conductivité thermique
μ viscosité
v viscosité cinématique
ρ densité
$(\ )f$ valeurs relatives au fluide
$(\ )m$ valeurs raeltives au milieu poreux saturé

Huille de silicone

ρ à 25° C=0,99

$a=0.0096$

$Cp = 0,372$ cal/g°C

$\lambda = 3,5 \ 10^{-4}$cal/s/cm/°C

Bille de verre

$\rho = 2,5$ g/cm^4

$\lambda = 21,5 \ 10^{-4}$ cal/s/cm °C

$C= 0,18$ cal/g °C

BIBLIOGRAPHIĖ

1. BENARD (1900), Tourbillons cellulaires dans une nappe liquide. *Rev. Géné. Sci.* 11, 1261–1271; 1309–1328.

2. LAPWOOD (1948) Convection of a fluid in a porous medium. *Proc. Camb. Phil. Soc.* 44, 508–521 .

3. KATTO AND MASUOKA (1967) Criterion for the ouset of convective flow in a fluid in a porous medium. *J. Heat and Mass. Transfer 10*, 297–309.

4. HORTON AND ROGERS (1945), Convection currents in porous media, *J. Appl. Phys.* 16, 367.

5. MORRISSON, ROGERS ET HORTON (1949), Convective currents in porous media. II: Observation of conditions at ouset of convection. *J. Appl. Phys. 20*, 1027–1029.

6. ELDER (1967) Steady free convection in a porous medium heated from below. *J. Fluid Mech. 27* août *1*, 29–48.

7. WOODING (1957), *J. Fluid. Mech. 2*, 273 et six autres articles.

8. GIAMBELLI (1956), Fenomena convectivi di fluidi in letti porosi. *La Termotecnica.*

9. SCHNEIDER 11ème Int. Cong. of Refrigeration paper 11–4, Munich.

10. DIRKSEN (1966), Journal of Society of Petroleum Engineers (September).

11. CHANDRASEKAR, Hydrodynamic and hydromagnetic stability.

12. ROGERS, SCHLIBERY ET MORRISSON (1951), Convection currents in porous media IV: Remark on the theory. *J. Appl. Phys. 22*, 1471–1479.

13. ROGERS ET MORRISSON (1950), Convection currents in porous media III: Extended theory of critical gradient. *J. Appl. Phys. 21*, 1177–1180.

INSTITUT DE MÉCANIQUE DES FLUIDES DE TOULOUSE,
    2, RUE CHARLES-CAMICHEL,
        31, TOULOUSE (FRANCE)

GROWTH OF A VAPOUR BUBBLE
IN A POROUS MEDIUM

S. G. B<small>ANKOFF</small>

INTRODUCTION

We consider the growth of a spherical gas bubble in a statistically uniform porous medium initially filled with liquid. A bubble is here defined to be the region from which liquid has, at least in part, been displaced by gas. The envelope of this region is taken to be a spherical surface, of radius $R(t)$, where t is the time after growth initiates. Evidently, this definition is adequate only for times such that $r_c/R \ll 1$, where r_c is the mean pore radius, and we limit ourselves to consideration of these later, or asymptotic, stages of bubble growth.

The bubble may grow either by isothermal precipitation of gas from a super-saturated solution, as may occur upon release of external pressure, or by diffusion of heat from a superheated liquid. The analysis of the second case is somewhat more complicated, but it is of considerable practical importance. In the present work the asymptotic growth of a vapour bubble in a porous medium initially filled with uniformly superheated liquid is considered. It is shown that the bubble growth proceeds as the square root of time, analogously to the "exact" solution for bubble growth in a liquid of uniform superheat, to the first order of approximation.

CONSERVATION EQUATIONS

Consider a vapour bubble whose centre is at rest with respect to the solid, with growth controlled by diffusion of heat to the bounding surface. We assume that the porous solid is locally at the same temperature as the fluid within the pores, equivalent to the assumption of rapid local equilibration over a distance scale of r_c. The significance of this assumption will be treated in more detail below. To make the formulation more general, we also consider a volumetric heat source to be present in the solid. Focussing our attention on the liquid-solid region first, the continuity equation may be written:

$$(1) \qquad r^2 u(r,t) = R^2 u(R^+,t) \qquad r > R$$

where $u(r,t)$ is the pore velocity averaged over a spherical surface of radius r at time t, and $u(R^+,t)$ is the average liquid velocity just outside the bubble wall.

166

The pressure equation is assumed to be Darcy's Law:

$$u = -K_2 \frac{\partial p}{\partial r}$$

(2)

where K_2 is the Darcy coefficient of the liquid-solid region (region 2), assumed to be a constant, and $p = p(r, t)$ is the pressure. Combining (1) and (2), we find that

$$\frac{\partial}{\partial r} \left(r^2 \frac{\partial p}{\partial r} \right) = 0$$

(3)

so that, upon integrating twice,

$$\frac{p(r, t) - p_\infty}{p(R) - p_\infty} = \frac{R}{r} \qquad r > R(t)$$

(4)

where the external pressure, $p_\infty \equiv p(\infty, t)$, is taken to be a constant.

The heat equation for the liquid is

$$\varepsilon \rho_f c_f \left[\frac{\partial T_f}{\partial t} + u \frac{\partial T_f}{\partial t} \right] = \varepsilon \frac{k_f}{r^2} \frac{\partial}{\partial r} \left(r^2 \frac{\partial T_f}{\partial r} \right) - h(T_f - T_s)$$

(5)

$$r \geqq R, \quad t > 0$$

where ε = porosity, or pore volume fraction
  ρ_f = liquid density
  c_f = liquid specific heat
  k_f = liquid thermal conductivity, assumed to be independent of temperature
  T_s = temperature of solid
  T_f = temperature of liquid
  h = volumetric heat transfer coefficient based on total volume

For the solid the heat flow is similarly described by

$$(1 - \varepsilon) \rho_s c_s \frac{\partial T_s}{\partial t} = \Gamma + (1 - \varepsilon) \frac{k_s}{r^2} \frac{\partial}{\partial r} \left(r^2 \frac{\partial T_s}{\partial r} \right) + h(T_f - T_s)$$

(6)

where $\Gamma = \Gamma(t)$ is the volumetric heat source strength per unit of total volume, assumed to be a function of time only.

Upon subtracting (5) from (6), the result is

$$\frac{\partial}{\partial t}(T_s - T_f) - \frac{1}{r^2} \frac{\partial}{\partial r} \left(r^2 \frac{\partial}{\partial r}(D_s T_s - D_f T_f) \right) - u \frac{\partial T_f}{\partial r} -$$

(7)

$$\frac{\Gamma}{(1 - \varepsilon) \rho_s c_s} - h(T_f - T_s) \left[\frac{1}{(1 - \varepsilon) \rho_s c_s} + \frac{1}{\varepsilon \rho_f c_f} \right] = 0$$

where $D_s = k_s / \rho_s c_s$ is the thermal diffusivity of the solid, and similarly D_f for the liquid. If $D_s \cong D_f$, the first two terms on the left-hand side represent the accumulation and diffusion rates of the local temperature difference, $(T_s - T_f)$. With small pore sizes, we expect that $|\partial T_s/\partial t| \gg |(\partial T_s - T_f))/\partial t|$ and similarly $|\nabla^2 T_s| \gg |\nabla^2(T_s - T_f)|$, since for a bounded heat release rate and liquid velocity, $(T_s - T_f) \rightarrow 0$ everywhere and for all time as $r_c \rightarrow 0$. Hence, from (5) and (6), these terms can be neglected in an order-of-magnitude estimate. Note that the last term in (7) is not negligible, even for small pore radii, even though $(T_f - T_s) \rightarrow 0$ as $r_c \rightarrow 0$, since h then becomes large. To see this, one can write $h = (Nu)_{av}k_f/r_c a$, where the average Nusselt number is nearly two, for sufficiently small pore sizes, and a, the surface area per unit volume, is $O(r_c^2)$. Hence $h = O(r_c^{-1})$. One thus finds that

$$
(8) \qquad \frac{T_s - T_f}{\Delta T_\infty} \cong \frac{\dfrac{\Gamma}{(1-\varepsilon)\rho_s c_s} + u\,\dfrac{\partial T_f}{\partial r}}{h\Delta T_\infty \left[\dfrac{1}{(1-\varepsilon)\rho_s c_s} + \dfrac{1}{\varepsilon \rho_s c_s} \right]}
$$

where $\Delta T_\infty \equiv T(\infty, t) - T_{sat}$, the superheat of the liquid at infinity, is a characteristic temperature difference. Thus, for small pore sizes or small temperature gradients and source strengths, the solid and liquid temperatures can be taken to be very nearly equal. This will often be true nearly everywhere except in a thin, thermal boundary layer next to the bubble wall, where temperature gradients are quite large. A combined energy equation obtained by adding (5) and (6) can then be employed:

$$
(9) \qquad c_2 \left[\frac{\partial T_2}{\partial t} + \gamma_2 u\,\frac{\partial T_2}{\partial r} \right] = \Gamma + \frac{k_2}{r^2}\frac{\partial}{\partial r}\left(r^2 \frac{\partial T_2}{\partial r} \right) \qquad r \geq R,\ t > 0
$$

where $c_2 = \varepsilon \rho_f c_f + (1-\varepsilon)\rho_s c_s$; $k_2 = \varepsilon k_f + (1-\varepsilon)k_s$; $\gamma_2 = \varepsilon \rho_s c_s/c_2$; and $T_2 \equiv T_f = T_s$ nearly everywhere. This approximation may be in error in $R < r < R + \delta_T$, where δ_T is the thickness of the thin thermal boundary layer surrounding the bubble, so that the error in the boundary conditions at the bubble surface is $O(\delta_T)$. This completes the statement of the conservation equations for the solid-liquid region.

Similar equations can be deduced for the gas-solid region. Assuming for the moment that this region contains no liquid (this restriction will be later removed), the continuity equation states

$$
(10) \qquad \frac{\partial \rho_g}{\partial t} + \frac{1}{r^2}\frac{\partial}{\partial r}(\rho_g u r^2) = 0
$$

where the gas density is given by the perfect gas law:

$$
(11) \qquad \rho_g = \frac{p}{BT_g}
$$

Here B is the gas constant in mass units. Assuming that the gas and solid temperatures are locally identical, we have

$$T_g(r,t) = T_s(r,t) \equiv T(r,t); \quad r < R(t).$$

The flow is again considered to follow Darcy's Law:

$$(12) \qquad u = -K_1 \frac{\partial p}{\partial r}; \quad r < R(t)$$

so that from (10)–(12) we have

$$(13) \qquad \frac{\partial}{\partial t}\left(\frac{p}{T_1}\right) - \frac{K_1}{r^2}\frac{\partial}{\partial r}\left(\frac{r^2 p}{T_1}\frac{\partial p}{\partial r}\right) = 0; \quad r < R$$

The heat equation in region 1 is similar to that in region 2:

$$(14) \qquad c_1\left[\frac{\partial T_1}{\partial t} + y_1 u\frac{\partial T_1}{\partial r}\right] = \Gamma + \frac{k_1}{r^2}\frac{\partial}{\partial r}\left(r^2\frac{\partial T_1}{\partial r}\right); \quad r < R(t), \ t > 0$$

where the symbols have similar meanings to those in equation (7), mutatis mutandis.

INITIAL AND BOUNDARY CONDITIONS

It is assumed that initially the liquid and solid are everywhere at the same temperature, T_i, and that the liquid is at rest. Furthermore, the bubble radius is considered to be negligible at zero time. These are expressed by :

$$(15) \qquad T_2(r,0) = T_i; \quad r > 0$$

$$(16) \qquad R(0) = 0; \quad u(r,0) = 0.$$

At infinity $u \to 0$ and $\partial T/\partial r \to 0$, so that equation (9) becomes

$$(17) \qquad \frac{\partial T_\infty}{\partial t} \equiv \frac{\partial T_2(\infty,t)}{\partial t} = \frac{\Gamma(t)}{c_2}$$

Hence if there are no volume heat sources

$$(18) \qquad T_\infty = T_i = \text{constant}.$$

The conditions at the origin are

$$(19) \qquad \lim_{r \to 0}\left[r^2\frac{\partial T_1}{\partial r}(r,t)\right] = 0$$

obtained by multiplying (14) by r^2, integrating from 0 to r, and letting $r \to 0$, and

$$(20) \qquad \frac{\partial p}{\partial r}(0,t) = 0$$

from (12) by symmetry.

The matching conditions at the interface can be derived from integral balances. A mass balance on the bubble requires that

$$\frac{d}{dt}\int_0^R \rho_g r^2 dr = \left(\frac{dR}{dt} - u(R^+,t)\right)\rho_f R^2$$

(21)
$$= \frac{1}{B}\int_0^R \frac{\partial}{\partial t}\left(\frac{p}{T_1}\right)r^2 dr + \frac{R^2}{B}\frac{dR}{dt}\left(\frac{p}{T_1}\right)_{r=R}$$

upon employing (11).

An enthalpy balance on the bubble yields

$$\frac{d}{dt}\int_0^R (c_1\Delta T_1 + \varepsilon\lambda\rho_g)r^2 dr = \frac{R^3\Gamma}{3} + k_2 R^2 \left.\frac{\partial T_2}{\partial r}\right|_{r=R^+}$$

$$+ \left(\left(\frac{dR}{dt}\right) - u(R^+,t)\right)R^2\varepsilon\rho_f c_f\Delta T_2(R^+,t)$$

where $\Delta T_1 = T_1 - T_{sat}$; $\Delta T_2 = T_2 - T_{sat}$, and $T_{sat} = g(p_\infty)$. Here $T = g(p)$ represents the saturation temperature as a function of pressure, and surface tension effects on the bubble interior pressure have been considered negligible.

Finally, the equilibrium temperature condition at the bubble surface implies that:

(23) $T_1(R^-,t) = T_2(R^+,t) = g(p_R)$

where $p_R \equiv p(R(t),t)$.

Some simplification of equations (21) and (22) is possible. A material balance at the bubble wall requires that

(24) $(\rho_f - \rho_g)\frac{dR}{dt} = [\rho u]_{r=R}$

where $[\rho u]_{r=R} \equiv \rho_f u(R^+,t) - \rho_g u(R^-,t)$, is the jump in mass velocity at the bubble wall. In general, the mass velocity of the gas at $r = R$ is small compared to that of the liquid, and, in fact, vanishes, upon integrating equation (10) radially from 0 to R, if ρ_g is a constant. Hence,

(25) $\frac{dR}{dt} - u(R^+,t) \cong \frac{\rho_g}{\rho_f}\frac{dR}{dt}$

which agrees with (21) if $\partial p_g/\partial t = 0$ within the bubble. We shall show, that to the zeroth-order, and essentially to the first-order, p and T are constant within the bubble. Hence (25) is an excellent approximation, and

(26) $u(r,t) = u(R^+,t)\frac{R^2}{r^2} = \left(1 - \frac{\rho_g}{\rho_f}\right)\frac{R^2}{r^2}\frac{dR}{dt}$

from (1) and (25).

UNIFORM INITIAL SUPERHEAT; NO VOLUME SOURCES

Consider now the growth of a bubble, beginning at $t = 0$ with zero radius, in a porous medium filled with a uniformly superheated liquid at a temperature $T_i > T_{sat}$. If the growth rate is sufficiently small, the pressure at the bubble wall will be essentially that at infinity, and the temperature will be the saturation temperature at this pressure [1, 2]. Hence we expand T_1, T_2, R, and p in perturbation series as follows:

(27)
$$
\begin{aligned}
T_1 &= T_{10} + T_{11} + \cdots\cdots \\
T_2 &= T_{20} + T_{21} + \cdots\cdots \\
R &= R_0 + R_1 + \cdots\cdots \\
p &= p_0 + p_1 + \cdots\cdots
\end{aligned}
$$

where it is assumed that $T_{10}(R_0, t) = g(p_\infty)$.

Substituting (26) and (27) into (14) and (9), and collecting zeroth-order terms, one obtains

(28)
$$
c_1 \frac{\partial T_{10}}{\partial t} = \Gamma + \frac{k_1}{r^2} \frac{\partial}{\partial r}\left(r^2 \frac{\partial T_{10}}{\partial r}\right); \quad r < R_0
$$

(29)
$$
c_2 \frac{\partial T_{20}}{\partial t} + \frac{b_2}{r^2} R_0^2 \frac{dR_0}{dt} \frac{\partial T_{20}}{\partial r} = \Gamma + \frac{k_2}{r^2} \frac{\partial}{\partial r}\left(r^2 \frac{\partial T_{20}}{\partial r}\right); \quad r > R_0
$$

where $b_2 \equiv \gamma_2(1 - \rho_g/\rho_f)$, and the convective term in (14) has been considered to be negligible compared to the conduction term.

The initial and boundary conditions are:

$$
T_{20}(r, 0) = T_i = T_{20}(\infty, t) - \frac{1}{c_2}\int_0^t \Gamma(t')dt'
$$

(30)
$$
T_{20}(Z_0, t) = T_{10}(R_0, t) = T_{sat}
$$

$$
R_0(0) = 0 .
$$

Assume now $\Gamma(t) = 0$, $t \geq 0$, corresponding to the absence of heat sources. Then a solution of (28) which satisfies (30) is

(31)
$$
T_{10}(r, t) = T_{sat} \quad r < R_0 .
$$

Equation (29) has a similarity solution, as noticed first by Kirkaldy [3], and also by Birkhoff et al. [4] and Scriven [5]. Define

(32)
$$
\eta = \frac{r}{\sqrt{D_2 t}} ; \quad \beta_0 = \frac{R_0(t)}{\sqrt{D_2 t}}
$$

$$
D_2 = \frac{k_2}{C_2}; \quad \theta_{20} = \frac{T_{20} - T_i}{T_{sat} - T_i}
$$

Then (29) and (30) become

$$(33) \qquad \left(\frac{b_2\beta_0^3}{2\eta^2} - \frac{2}{\eta} - \frac{\eta}{2}\right)\frac{d\theta_{20}}{d\eta} = \frac{d^2\theta_{20}}{d\eta^2}$$

subject to $\theta_{20}(\beta_0) = 1$; $\theta_{20}(\infty) = 0$, with the solution

$$(34) \qquad \theta_{20} = \frac{\Phi(\eta, \beta_0)}{\Phi(\beta_0, \beta_0)}$$

where

$$35) \qquad \Phi(\eta, \beta_0) = \int_\eta^\infty \frac{1}{z^2} \exp\left[-\frac{\beta_0^3 b_2}{2z} - \frac{z^2}{4}\right] dz \ .$$

To determine the growth constant, β_0, equation (22) is expanded to the zeroth-order:

$$(36) \qquad \frac{1}{R_0^2}\int_0^{R_0} \frac{\partial}{\partial t}(c_1 T_{10} + \varepsilon\lambda\rho_{g0})r^2 dr + \varepsilon\lambda\rho_{g0}\frac{dR_0}{dt} = \frac{R_0\Gamma}{3} + k_2\frac{\partial T_{20}}{\partial r}\bigg|_{r=R_0^+}$$

since $\Delta T_{10}(R_0, t) = \Delta T_{20}(R_0, t) = 0$.

To the same order of approximation (13) can be written:

$$(37) \qquad \frac{\partial p_0}{\partial t} + \frac{k_1}{2r^2}\frac{\partial}{\partial r}\left(r^2\frac{\partial p_0}{\partial r}\right) = 0; \quad r < R_0$$

subject to

$$p_0(R_0, t) = p_\infty$$

$$(38) \qquad \frac{\partial p_0(0, t)}{\partial r} = 0$$

$$R_0(0) = 0$$

which is satisfied by

$$(39) \qquad p_0(r, t) = p_\infty; \qquad r < R_0 \ .$$

Hence,

$$(40) \qquad \rho_{g0} = \frac{p_p}{BT_{10}} = \frac{p_\infty}{Bg(p_\infty)}$$

a constant, and the integral in (36) vanishes. Thus, for $\Gamma = 0$, (36) reduces to

$$(41) \qquad k_2\frac{\partial T_{20}}{\partial r}\bigg|_{r=R_0} = \varepsilon\lambda\rho_{g0}\frac{dR_0}{dt}$$

Making use of (32) and (34) (41) may be written as

$$(42) \qquad \frac{\partial T_{20}}{\partial r}\bigg|_{r=R_0} = \frac{-\Delta T_i}{\sqrt{D_2 t}} \frac{\Phi'(\beta_0, \beta_0)}{\Phi(\beta_0, \beta_0)} = \frac{\varepsilon \lambda \rho_{g0}}{k_2} \frac{\beta_0}{2} \sqrt{\frac{D_2}{t}}$$

where $\Phi'(\beta_0, \beta_0) \equiv \dfrac{\partial \Phi(\beta_0, \beta_0)}{\partial \eta}$, and where $\Delta T_i \equiv T_i - T_{sat}$ is the initial super-heat. Hence, β_0 is the real positive root of

$$(43) \qquad \beta_0 + 2J \frac{\Phi'(\beta_0, \beta_0)}{\Phi(\beta_0, \beta_0)} = 0$$

where $J \equiv (c_2 \Delta T_i)/(\varepsilon \lambda \rho_{g0})$ is a modified Jakob number.*

FIRST-ORDER CORRECTION

A corrected estimate of the pressure at the bubble wall can be obtained from (2) and (26):

$$(44) \qquad -K_2 \frac{\partial}{\partial r}(p_0 + p_1) = \left(1 - \frac{\rho_g}{\rho_f}\right) \frac{1}{3r^2} \frac{d}{dt}(R_0 + R_1)^3.$$

Noting that $p_0 = p_\infty$, and assuming that R_1 represents a small correction on R_0, this equation yields

$$(45) \qquad \frac{\partial p_1}{\partial r} = \left(\frac{\rho_g}{\rho_f} - 1\right) \frac{1}{3K_2 r^2} \left\{\frac{dR_0^3}{dt} + O\left(\frac{d}{dt}(R_0{}^2 R_1)\right)\right\}$$

subject to $p_1(\infty, t) = 0$.

Upon integrating from R_0 to ∞, and neglecting higher-order terms, one obtains

$$(46) \qquad p_1(R_0, t) = \left(1 - \frac{\rho_g}{\rho_f}\right) \frac{R_0}{K_2} \frac{dR_0}{dt} = \frac{\beta_0^2 D_2}{2K_2} \left(1 - \frac{\rho_g}{\rho_f}\right)$$

upon using (32). The first-order temperature correction at the bubble wall can now be obtained by expanding (23);

$$(47) \qquad T_{21}(R_0{}^+, t) = T_{11}(R_0{}^-, t) = g'(p_\infty)p_1(R_0, t)$$

Hence a measure can be obtained off the dimensionless temperature correction from (46), (47) and (30):

$$(48) \qquad \psi_1 \equiv \frac{T_{21}(R_0, t)}{\Delta T_i} = \frac{\beta_0^2 D_2}{2K_2} \left(1 - \frac{\rho_g}{\rho_f}\right) \frac{g'(p_\infty)}{T_i - g(p_\infty)} \ .$$

---

* Among other helpful comments, Dr. P. G. Kosky has brought to the attention of the author the work of Labuntsov, et al. [6], who give the following three-term expansion for the root of (43):

$$\beta_0 \cong \left(\frac{12}{\pi}\right)^{\frac{1}{2}} J\left\{1 + \frac{1}{2}\left(\frac{\pi}{6J}\right)^{\frac{1}{2}} + \frac{\pi}{6J}\right\}^{\frac{1}{2}}$$

If $\psi_1 \ll 1$, $R_0(t)$ is a good approximation of the bubble radius. Note that ψ_1 is time-independent.

The first-order corrections to the bubble pressure and temperature may now be calculated. Upon substituting (12) into (14) and expanding, the first-order terms are:

$$
(49) \qquad
c_1 \left[\frac{\partial T_{11}}{\partial t} - \gamma_1 K_1 \left\{ \frac{\partial p_1}{\partial r} \frac{\partial T_{10}}{\partial r} + \frac{\partial p_0}{\partial r} \frac{\partial T_{11}}{\partial r} \right\} \right]
$$

$$
= \frac{k_1}{r^2} \frac{\partial}{\partial r} \left(r^2 \frac{\partial T_{11}}{\partial r} \right); \quad r < R_0 + R_1
$$

where the terms in the braces vanish, by virtue of (31) and (39).

The continuity equation within the bubble to the same order yields, from (13)

$$
(50) \qquad
\frac{\partial}{\partial t} \left(\frac{p_1}{p_0} - \frac{T_{11}}{T_{10}} \right) - \frac{K_1}{r^2} \frac{\partial}{\partial r} \left(r^2 \frac{\partial p_1}{\partial r} \right) = 0; \quad r < R_0 + R_1
$$

From (45) the second term vanishes. The boundary conditions are given by (46) and, from (20) and (23):

$$
(51) \qquad \frac{\partial p_1(0,t)}{\partial r} = p_1(\infty,t) = 0
$$

$$
(52) \qquad T_{11} = T_{21} \text{ at } r = R_0 + R_1; \qquad t > 0
$$

$$
(53) \qquad T_{21}(\infty,t) = T_{21}(r,0) = 0
$$

Then (49) and (50), subject to (46), (51) and (52), are satisfied by

$$
(54) \qquad
\begin{aligned}
T_{11}(r,t) &= p_1(R_0 + R_1, t) g'(p_\infty) \\
&= T_{11}(R_0 + R_1, t); \qquad r \leq R_0 + R_1
\end{aligned}
$$

where

$$
(55) \qquad
p_1(r,t) = \frac{\beta_0^2 D_2}{2K_2} \left(1 - \frac{\rho_g}{\rho_f} \right); \qquad r \leq R_0 + R_1,
$$

a constant, from (46) and (47). Thus, both to the zeroth and first-order, the temperature and pressure within the bubble are uniform and constant, with the magnitude of the first-order correction depending upon the zeroth-order growth constant and the properties of the liquid-filled region.

The first-order terms in the expansion of the enthalpy equation (9) become, upon using (26).

$$
(56) \qquad
c_2 \frac{\partial T_{21}}{\partial t} + \frac{b_2}{r^2} \left[R_0^2 \dot{R}_0 \frac{\partial T_{21}}{\partial r} + R_0^2 \dot{R}_1 \frac{\partial T_{20}}{\partial r} + 2R_1 R_0 \dot{R}_0 \frac{\partial T_{20}}{\partial r} \right]
$$

$$
= \frac{k_2}{r^2} \frac{\partial}{\partial r} \left(r^2 \frac{\partial T_{21}}{\partial r} \right); \quad r \geq R_0 + R_1
$$

where $\dot{R}_0 \equiv \dfrac{dR_0}{dt}$.

In this equation the first term in the square brackets represents the correction to the convective heat flux due to the temperature correction, and the second and third terms that due to the correction to the local velocity. Were it not for these latter terms (56) would be exactly analogous to the zeroth-order equation (29) with no heat sources, with a solution of the same form.

The temperature correction must vanish initially and at infinity:

$$\text{(57)} \qquad T_{21}(r,0) = T_{21}(\infty,t) = 0 \ .$$

The matching condition at the bubble wall comes from the first-order expansion of (22), taking into consideration (54) and (55):

$$R_0^2 \dot{R}_0 (c_1 T_{11} + \varepsilon\lambda\rho_{g1}) + \varepsilon\lambda\rho_{g0}(R_0^2 \dot{R}_1 + 2R_0 R_1 \dot{R}_0) =$$

$$\text{(58)} \qquad 2k_2 R_0 R_1 \frac{\partial T_{20}}{\partial r}\bigg|_{r=R} + k_2 R_0^2 \frac{\partial T_{21}}{\partial r}\bigg|_{r=R} + R_0{}^2 R_1 \Gamma$$

Taking $\Gamma = 0$ and noting that the coefficient of R_1 in this equation vanishes, by virtue of (41), one obtains, after integrating (58),

$$\text{(59)} \qquad R_1 + R_0\delta = \frac{k_2}{\varepsilon\lambda\rho_{g0}} \int_0^t \frac{\partial T_{21}(R,t')}{\partial r} dt'$$

where

$$\delta \equiv \frac{c_1 T_{11} + \varepsilon\lambda\rho_{g1}}{\varepsilon\lambda\rho_{g0}} \ .$$

An estimate of the integral term can be obtained by neglecting the local velocity correction terms in (56). Defining

$$\text{(60)} \qquad \beta_1 = \frac{R_0 + R_1}{\sqrt{D_2 t}}; \qquad \theta_{2i} = \frac{T_{21} - T_i}{T_{sat} + T_{11} - T_i}$$

we have the solution

$$\text{(61)} \qquad \theta_{21} = \frac{\Phi(\eta, \beta_1)}{\Phi(\beta_1, \beta_1)} \ .$$

As in (42)

$$\text{(62)} \qquad \frac{\partial T_{21}}{\partial r}\bigg|_{r=R0+R1} = \frac{-\Delta T_i + T_{11}}{\sqrt{D_2 t}} \frac{\Phi'(\beta_1, \beta_1)}{\Phi(\beta_1, \beta_1)}$$

whence (59), (60) and (62) give

$$\text{(63)} \qquad \beta_1 - \beta_0(1-\delta) = \frac{2k_2(-\Delta T_i + T_{11})}{D_2 \varepsilon\lambda\rho_{g0}} \frac{\Phi'(\beta_1, \beta_1)}{\Phi(\beta_1, \beta_1)}$$

from which the corrected growth constant, β_1, can be determined.

The bubble, to the first order, thus grows nearly as the square root of time, with an error determined by the magnitude of the neglected velocity correction terms in (56).

RELATED PROBLEMS

Suppose (as is usually the case) that the bubble does not displace all the liquid from the porous medium, but instead leaves a constant fraction $\varepsilon' < \varepsilon$. If there is no source term ($\Gamma = 0$), ε' will be constant in space and time within the bubble. The zeroth-order expansions of the energy equation (28) and (29) then remain unchanged, except for the substitutions

$$(64) \qquad c_1' = (1 - \varepsilon)c_s\rho_s + \varepsilon'c_f\rho_f + (\varepsilon - \varepsilon')c_g\rho_g$$

$$(65) \qquad k_1' = (1 - \varepsilon)k_s + \varepsilon'k_f + (\varepsilon - \varepsilon')k_g$$

and now he material balance at the bubble wall requires that

$$(66) \quad (\varepsilon - \varepsilon')(\rho_f - \rho_g)\dot{R} = \varepsilon\rho_f u_f(R^+, t) - \varepsilon'\rho_f u_f(R^-, t) - (\varepsilon - \varepsilon')\rho_g u_g(R^-, t)$$

Again neglecting velocities within the bubble, (46) thus now becomes

$$(67) \qquad u(r, t) = \left(1 - \frac{\varepsilon'}{\varepsilon}\right)\left(1 - \frac{\rho_g}{\rho_f}\right)\frac{R^2}{r^2}\dot{R}$$

Similarly, (41) and (43) are applicable, with the substitution $\varepsilon \to \varepsilon - \varepsilon'$. The first-order corrections follow similarly.

ACKNOWLEDGMENT

I wish to acknowledge with gratitude helpful discussions with G. F. Hewitt and P. G. Kosky, of A. E. R. E., Harwell, England. This work was conducted under the auspices of the United Kingdom Atomic Energy Authority, and appeared as Report AERE–R5772 (1968).

REFERENCES

1. PLESSET, M. S. AND S. ZWICK (1952), *J. Appl. Phys.*, *23*, 95.
2. BANKOFF, S. G. (1963), *Appl. Sci. Res.*, A*12*, 567.
3. KIRKALDY, J. J. (1958), *Can. J. Phys.*, *36*, 446.
4. BIRKHOFF, G., R. S. MARGULIES AND W. A. HORNING (1958), *Phys. Fluids*, *1*, 201.
5. SCIVEN, L. E. (1959), *Chem. Eng. Sci.*, *10*, 1.
6. LABUNTSOV, D. A., B. A. KOL'CHUGIN, V. S. GOLOVIN, E. A. ZAKHAROVA AND L. N. VLADIMIROVA (1964), *High Temperature*, *2*, No. 3, 404–409.

CHEMICAL ENGINEERING DEPARTMENT, NORTHWESTERN UNIVERSITY,
EVANSTON, ILLINOIS 60201

ON THE FLOW OF TWO IMMISCIBLE FLUIDS IN FRACTURED POROUS MEDIA

JACOB BEAR AND CAROL BRAESTER

INTRODUCTION

There may be three orders of inhomogeneity in a porous medium (Fig. 1).

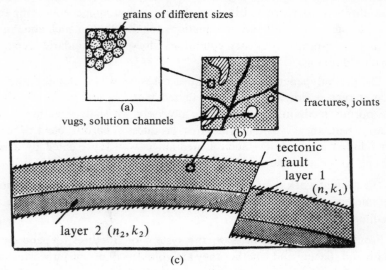

grains of different sizes

(a)

vugs, solution channels

(b)

fractures, joints

tectonic
fault
layer 1
(n, k_1)

layer 2 (n_2, k_2)

(c)

Fig. 1. Inhomogeneity in a porous medium: (a) of the first order (b) of the second order (c) of the third order.

(i)   The first is due to inhomogeneity in the granular structure of the rock; it results from the pore- or grain-size distribution, the variability in grain-shapes etc. The length characterizing this inhomogeneity is the mean (or median) grain (or pore) size.

(ii)   The second is due to the presence of macrojoints, small tectonic fissures, vugs and solution channels which form a network of interconnected channels. The fractures may be randomly oriented or they may have preferential directions. The system of channels may be such that the entire porous domain is made up of a large number of porous blocks, each completely surrounded by the void-pace of the channel and fracture system, or the channel and fracture system may form an interconnected network of narrow passages imbeded

177

in the porous matrix. The length characterizing this type of inhomogeneity is the size of the porous block surrounded by fractures, or the mean length of a channel between adjacent junctions.

(iii)   Inhomogeneity resulting from the presence of distinct layers or other forms of zoning of different media (e.g. different permeability).

By applying the continuum approach to the inhomogeneity of the first type, the actual medium is replaced by a continuum which is *locally* homogeneous, although continuous variations in permeability may be present. In the latter case we may have $k = k \ (x, y, z)$.

The third type of inhomogeneity, characterized by abrupt changes in permeability is usually treated by considering each homogeneous zone separately (e.g. stating mass conservation equations for the fluid in such zone) and introducing appropriate boundary conditions along the boundaries separating the various domains.

The present paper deals with those cases in which inhomogeneity of the second type is present and is distributed more or less uniformly throughout the porous medium domain. More precisely we shall treat here only those cases where the system of fractures produces "porous blocks" completely surrounded by fractures. Such a medium is well represented by a rock containing joints at a more or less uniform spacing.

The present work is an attempt to state the equations and analyze the medium parameters which describe immiscible displacement in a fractured porous medium.

From the standpoint of flow, the system of fractures, depending on their width and spacing and whether they are void or full of fine impervious material, may introduce a permeability which is much higher than that of the porous medium without the fractures. The relative contribution to the overall permeability of the fractured rock depends, however, on the fracture system as well as on the permeability of the porous blocks. In some cases because of the size of the blocks, although of a rather low permeability, the contribution of the blocks to the overall permeability may equal or exceed that of the fracture system. On the other hand, the main contribution to the porosity, or to the storage capacity of the rock, comes from the porous blocks and not from the fractures. Also, in two phase flow, other phenomena occur at the interfaces between the blocks and the fractures (e.g. various capillary phenomena).

Descriptions of fractures (oil) reservoirs are given by Gibson (1948), Pirson (1953) Wilkinson (1953), Daniel (1954), Freeman and Natanson (1959), Cheuca (1959), Rats (1963) and Pîrvu (1965). Rats' studies of the sandstone of Ordovician flysh indicate an average spacing of 4 to 16 cm between neighbouring fractures, and an average fracture width of 0.3 to 1.4 mm. Daniel describes

Errata: The names of the Authors in the running heads of pages 180–202 should be *J. Bear and C. Braester*.

joints of 0.1–0.2 mm wide free of any minerals in the Ain Zalah Field in Iraq. The same width is indicated also in the Kirkuk oil fiield, with a joint spacing of 0.15 to 0.9 m. Pîrvu describes an orthogonal system of 0.05 to 0.1 mm wide fissures in the calcareous Moesic Platform of Rumania.

THE REPRESENTATIVE ELEMENTARY VOLUME

The only feasible way of analytically treating flow of fluids through porous media is by the *continuum approach*. According to this approach, the actual porous matrix filled with a flowing homogeneous fluid is replaced by a fictitious continuum, to each mathematical point of which (whether in the solid or in the void space) medium parameters such as porosity or permeability are assigned. The values of these parameters at a point are obtained by taking the average of the local values of the considered parameters over a small porous medium volume (= *physical point*) for which the considered point is a centroid. In the continuum obtained by repeating this procedure for all points of the considered domain, all medium parameters vary continuously from point to point. In the same way continuously varying averaged fluid and flow parameters (e.g. saturation, pressure, velocity) may also be obtained. The continuum approach is applicable also to polyphase flow by defining a separate eontinuum for each of the fluids (at its respective saturation) with each continuum entirely filling the flow domain. We thus obtain overlapping (or interpenetrating) continua.

Essential in the continuum approach is the definition of a *representative elementary* volume (REV), or *physical point*, over which averages of medium, fluid and flow properties are performed. The common procedure, following Prandtl and Tietjens (1934), is to enclose the considered point within the porous medium domain by a sequence of ever decreasing volumes ΔU_i $(i = 1, 2, 3 \cdots)$. For each of these volumes the average value of the considered property is determined.

The procedure is illustrated for the case porosity (Bear 1969). Consider a mathematical point M inside the domain occupied by the porous medium, and a volume ΔU_i (say having the shape of a sphere) for which M is its centroid. For a sequence of volumes ΔU_i $(\Delta U_1 > \Delta U_2 > \Delta U_3 \cdots)$ for all of which M remains the centroid, we may determine the corresponding porosity n_i:

$$n_i = (\Delta U_v)_i / \Delta U_i$$

where $(\Delta U_v)_i$ is the volume of voids within ΔU_i.

For large values ΔU_i, the ratio n_i may undergo gradual changes as $(\Delta U)_i$ is reduced, especially when the considered domain is inhomogeneous (e.g. layers of soil). Below a certain value of ΔU_i, depending on the distance of M from such boundaries of inhomogeneity, these changes or fluctuations tend to decay, leaving only small-amplitude fluctuations which are due to the random distribution of pore sizes in the neighborhood of M. However, below a certain value ΔU_0 we suddenly observe large fluctuations in the ratio n_i. This happens as the dimensions of ΔU_i approach those of a single pore. Finally, as $\Delta U_i \to 0$, converging on the mathematical point M, n_i will become either one or zero, depending on whether M is inside a pore or inside the solid matrix of the medium. Figure 2 shows the relationship between n_i and ΔU_i.

The medium's porosity at point M, $n(M)$, is defined as the limit of the ratio n_i as $\Delta U_i \to \Delta U_{0n}$:

$$n(M) = \frac{\lim}{\Delta U_i \to \Delta U_{0n}} n_i[\Delta U_i(M)] = \frac{\lim}{\Delta U_i \to \Delta U_{0n}} \frac{(\Delta U_v)_i(M)}{\Delta U_i}$$

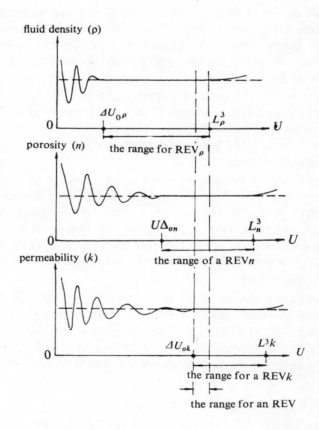

Fig. 2.   Fluctuations of same properties of the fluid and porous medium involved in the equation of motion and the REV.

For values of $\Delta U < \Delta U_{0n}$, we have to consider the actual presence of pores and solid particles; in this range there is no single value which can represent the average porosity n_i around M. The volume ΔU_{0n} is therefore the *represen-tative elementary volume of the porous medium porosity* (abbreviated as REVn) at the mathematical point M. The same procedure can be applied, theoretically, to other properties and parameters characterizing the porous medium (e.g. permeability) the fluid (e.g. density) or the flow (e.g. pressure). The magnitude of the *representative elementary volume characterizing a certain property* P (abbreviated REVP) may differ for different properties. So, for homogeneous liquids the representative elementary volume of the density (REVρ) is of an order of magnitude much smaller than a single pore, whereas the representative elementary volume of the porosity (REVn) contains a sufficient number of pores such that fluctuations in the ratio $(\Delta U_v)_i/\Delta U_i$ disappear. Another example of interest is that of a fissured porous medium. In a fissur-ed porous medium in which fissures occupy a volume much smaller than the pore space, the size of the REVn may be of an order of magnitude of the REVn of the block itself; however, generally the permeability of the fissure system considerably exceeds that of the system of pores in the individual blocks so that the size of the representative elementary volume characterizing the permeability (REVk) is given by a volume containing a number of sufficient fissures surrounding porous blocks for which fluctuations in the relationship between the permeability k_i and the volume ΔU_i disappear.

Macroscopic balances or conservation considerations lead to *partial dif-ferential equations* which express what happens to each of the dependent variables. Variables and coefficients appearing in these equations have the meaning of macroscopically averagrd values at a "point" which is the physical point; they refer simultaneously to the same "point" which is the common representative elementary volume (denoted in what follows by REV).

Actually ΔU_{0p} should be considered only as the lower limit of the range of volumes from which we may choose the elementary representative volume for a certain property P. For the case of inhomogeneities in fluid or medium properties, it is possible to define a characteristic length L (and a corresponding volume L^3) or three lengths L_i (in the directions of the coordinate system X_i, with a corresponding volume $L_1 L_2 L_3$) satisfying:

$$L = P/(\partial P/\partial X_i) \text{ or } L_i = P/(\partial P/\partial X_i)$$

where P is the considered property. The volume L^3 (or $L_1 L_2 L_3$) then serves as the upper limit of the range from which we may choose REVP (Fig 2):

$$\Delta U_{0p} \leqq \text{REVP} \leqq L^3$$

When at a (mathematical) point we consider simultaneously several properties, the REV of the medium has to be chosen from the range $\Delta U_0 \div L^3$ which is common to all properties (Fig. 2)

$$\underset{(j)}{\text{Max}} (\Delta U_{0pj}) \leqq \text{REV} \leqq \underset{(i)}{\text{Min}} [(L^3)_j]$$

read; the REV should be larger than the largest ΔU_0 of the individual properties, and smaller than the smaller value of L^3 of the individual properties. If no range exists which is common to *all* properties, we may still choose the REV out of the range common to *most* properties, with the knowledge that our treatment of the problem is no more an exact one.

Whenever a flow problem (e.g. poliphase flow) is described mathematically by a set of partial differential equations it is therefore to be understood that the same REV applies to all parameters and variables appearing in the entire set of equations.

A BRIEF LITERATURE SURVEY

In Barenblatt's (1960a, 1960b) approach to flow of a homogeneous, slightly compressible liquid, the interconnected porous and fractured medium is separated into two overlapping continuums, each filling the entire space. Flow takes place throughout each continuum, with exchange of fluid taking place between the two continuums according to a certain rule.

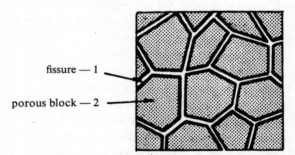

fissure — 1

porous block — 2

Fig. 3.   A fissured porous medium (after Barenblatt, 1960)

With $j = 1$ and $j = 2$ denoting the domains of fissures and porous rock, respectively (Fig. 3), the motion and mass conservation equations describing the horizontal flow in a consolidating reservoir of constant thickness are:

$$(\boldsymbol{q})_j = - \frac{k_j}{\mu} \nabla p_j; \quad j = 1, 2 \tag{3.1}$$

$$\frac{\partial (n_j \rho)}{\partial t} + \nabla [\rho(\boldsymbol{q})_j] + (-1)^j m^* = 0; \quad j = 1, 2 \tag{3.2}$$

$$n_i = (U_v)_i / U_b$$

where $\rho = \rho(p)$, and the source function $m*$ gives the transfer of fluid mass from porous blocks to fissures per unit bulk volume of the flow domain and per unit time; U_b is the bulk volume of a fractured rock sample and $(U_v)_j$ is the volume of voids of the type j contained in U_b. Barenblatt assumes that the $m* = m*(x, t)$ at a (mathematical) point x depends on the difference in pressure between the two continuums at that point;

$$m* = c\frac{k_2 A^2}{\mu}\rho_0(p_2 - p_1)$$

where c is a constant, k_2 is the permeability of the porous block, A is the area of fissures per unit volume of rock (i.e. $A = n_2/\bar{d}$, where $\bar{d}$ is the average width of the fissures), and p_1 and p_2 have the usual meaning of average values taken over REV's centered at the considered point in the two continuums.

For the slightly compressible liquid Barenblatt assumes the relationship:

$$\rho = \rho_0[1 + \beta(p - p_0)] \tag{3.4}$$

For the effect of pressure changes on the two rock porosities, Barenblatt assumes:

$$dn_1 = \alpha_1 dp_1 - \alpha_1' dp_2$$
$$dn_2 = \alpha_2 dp_2 - \alpha_2' dp_1 \tag{3.5}$$

where $\alpha_1, \alpha_2, \alpha_1'$ and α_2 are the coefficients of rock compressibility. By substituting (3.1), (3.3), (3.4) and (3.5) into (3.2), we obtain:

$$\frac{k_1}{\mu}\nabla^2 p_1 = (\alpha_1 + n_1\beta)\frac{\partial p_1}{\partial t} - \alpha_1'\frac{\partial p_2}{\partial t} - \frac{k_1\beta}{\mu}(\nabla p_1)^2 - c\frac{k_2 A^2}{\mu}(p_2 - p_1)$$

$$\tag{3.6}$$

$$\frac{k_2}{\mu}\nabla^2 p_2 = (\alpha_2 - n_2\beta)\frac{\partial p_2}{\partial t} - \alpha_2'\frac{\partial p_1}{\partial t} - \frac{k_2\beta}{\mu}(\nabla p_2)^2 + c\frac{k_2 A^2}{\mu}(p_2 - p_1)$$

as the partial differential continuity equations describing the flow in a fractured rock.

In order to enable an analytical solution of these equations for a fissured rock, Barenblatt makes the following simplifying approximations:

(i) $\alpha_1' dp_2 \ll \alpha_1 dp_1$; $\alpha_2' dp_2 \ll \alpha_2 dp_1$

(ii) $\nabla(\rho \mathbf{q}_2) \ll \partial(n_2\rho)/\partial t$ (because of the relatively low value of n_1)

(iii) $\nabla(\rho \mathbf{q}_2) \gg \partial(n_1\rho)/\partial t$ (because of the relatively low value of n_1)

He obtains:

$$k_1 \nabla^2 p_1 + ck_2 A^2 (p_2 - p_1) = 0$$

$$(\alpha_2 + n_2 \beta) \frac{\partial p_2}{\partial t} + \frac{ck_2 A^2}{\mu}(p_2 - p_1) = 0 \qquad (3.7)$$

For a *double-porosity medium* (where n_1 and n_2 are of the same order of magnitude, approximation (ii) is not valid.

By combining the two equations of (3.7), Barenblatt obtains:

$$(3.8) \qquad \frac{\partial p_1}{\partial t} - \frac{k_1}{k_2} \frac{1}{cA^2} \frac{\partial}{\partial t}(\nabla^2 p_1) = \frac{k_1}{\mu(\alpha_2 + n_2 \beta)} \nabla^2 p_1$$

where $k_1/[\mu(\alpha_2 + n_2\beta)] = x$ is called the *piezo-conductivity of the fissured rock*. If we keep reducing the dimensions of the blocks, $A \to \infty$ and the second term on the left hand side of (3.8) vanishes. As also $\alpha_2 \to 0$ and $n_2 \to n$ ($=$ overall porosity, as $n_1 + n_2 = n$ and $n_1 \ll n_2$) we obtain:

$$(3.9) \qquad \frac{\mu n \beta}{k} \frac{\partial p_1}{\partial t} = \nabla^2 p_1$$

which is the continuity equation of flow of a slightly compressible fluid in a rigid homogeneous porous medium.

In a later work (1964) Barenblatt considers the flow of two immiscible fluids (a liquid and a gas). With superscripts $(i) = (1)$ and $(i) = 2$ denoting the liquid and the gas phases, respectively, the continuity equations are:
Fluid 1 ($=$ liquid) in blocks ($=$ medium 2):

$$\frac{\partial}{\partial t}[n_2 \rho^{(1)}(p_2)s_2^{(1)}] + \nabla[\rho^{(1)}(p_2)q_2^{(1)}] = -m^{*(1)}$$

Fluid 1 in fissures:

$$\frac{\partial}{\partial t}[n_1 \rho^{(1)}(p_1)S_1^{(1)}] + \nabla[\rho^{(1)}(p_1)q_1^{(1)}] = m^{*(1)}$$

Fluid 2 in blocks:

$$\frac{\partial}{\partial t}[n_2 \rho^{(2)}(p_2)S_2^{(2)} + n_2 \rho^{(1)}(p_2)s(p_2)S_2^{(1)}] + \nabla[\rho^{2)}(p_2)q_2^{(2)}] = -m^{*(2)}$$

Fluid 2 in fissures:

$$(3.10) \qquad \frac{\partial}{\partial t}[n_1 \rho^{(2)}(p_1)S_1^{(2)} + n_1 \rho^{(1)}(p_1)s(p_1)S_1^{(1)}] + \nabla[\rho^{(1)}(p_1)q_1^{(2)}] = m^{*(2)}$$

where $m^{*(i)}$ = the transfer of mass of fluid (i) from medium 1 to medium 2, per unit bulk volume of porous medium and per unit time,

$S_j^{(i)}$ = the saturation of fluid (i) in medium $j = (U_v)^{(i)}/(U_v)_j$,

$$S_1^{(1)} + S_1^{(2)} = 1; \quad S_2^{(1)} + S_2^{(2)} = 1$$

s = the solubility of fluid 2 (gas) in fluid 1.

$q_1^{(1)}$, $q_1^{(2)}$, $q_2^{(1)}$ and $q_2^{(2)}$ are the specific discharge vectors given by Darcy's law.

Barenblatt assumes that the source functions $m^{*(i)}$ giving the transfer of the ith fluid from fissures to the block are of the form:

$$m^{*(1)} = \frac{k_2 A^2}{\mu(1)} k_{r2}(S^{(1)}) \cdot \left[\int_0^{p_2} \rho^{(1)}(p_2')dp_2' - \int_0^{p_1} \rho^{(1)}(p_1')dp_1' \right]$$

and

$$(3.11) \quad m^{*(2)} = \frac{k_2 A^2}{\mu(1)} k_{r2}(S^{(1)}) \left[\int_0^{p_2} \rho^{(1)}(p_2')s(p_2')dp_2' - \int_0^{p_1} \rho^{(1)}(p_1')dp_1' \right]$$

$$- \frac{k_2 A^2}{\mu(2)} k_{r2}(S^{(2)}) \left[\int_0^{p_2} \rho^{(2)}(p_2')dp_2' - \int_0^{p_1} \rho^{(2)}(p_1')dp_1' \right]$$

In general we would like to define two capillary pressures:

$$p_2^{(c)} = p_2^{(1)} - p_2^{(2)}$$

However, Barenblatt neglects the presence of capillary pressure which means that he assumes:

$$p_2^{(1)} = p_2^{(2)} = p_2; \quad p_1^{(1)} = p_1^{(2)} = p_1$$

The process of diplacement by pure imbibition in a horizontal water-wet reservoir is treated by Bokserman, Zheltov and Kochenshkov (1964), assuming a particular source function m^*. They assumed that at $t = 0$ the fissured porous medium is saturated with a nonwetting liquid. Starting at $t = 0$ a wetting liquid is injected at a rate $q(t)$ into the reservoir. The injected liquid advances faster along the fissures, leaving behind porous blocks saturated by nonwetting liquid and surrounded by the wetting liquid. This initiates a process of imbibibtion which displaces the nonwetting fluid from the (porous) blocks. They considered only homogeneous immiscible liquids and a nonconsolidating reservoir. As the front of the wetting liquid occupying fissures advances. the imbibition process at a given point (x_1, x_2) starts at time $t = t^*(x_1, x_2)$, instead of at $t = 0$. The source function at a point (x_1, x_2) and time t is therefore q^*, i.e. a function of the time interval $(t - t^*)$.

With superscripts $(i) = (1)$ and $(i) = (2)$ denoting the nonwetting and the wetting liquids, respectively, the set of equations describing a horizontal motion *in the fissures* (subscript $j = 1$) is:

(3.12)

$$n_1 \frac{\partial s_1^{(i)}}{\partial t} + \nabla q_1^{(i)} = (-1)^{(i)} q^* [t - t^*(x_1, x_2)] \qquad i = 1, 2$$

$$q_1 = - \frac{k_1 k_{r1}(S_i)}{\mu(i)} \nabla p_1 \qquad i = 1, 2$$

where $q^* = m^*/\rho$ is thr source function. From the results of imbibition laboratory experiments performed by Mattax and Kyte (1962), Bokserman et al, (1964), concluded that the source function q^* can be expressed in the approximate form, which is sufficiently accurate for all practical purposes:

(3.13)

$$q^* = C \cdot n_2 F(S_2^{(1)}) \, \bar{A} \left[\frac{\cos \theta (k_2/n_2)^{\frac{1}{2}}}{\mu(1)} \right] \cdot t^{-\frac{1}{2}}$$

Where C is a dimensionless constant, $F(S_2)$ a function of the saturation in the porous blocks, θ is the contact angle, and $\bar{A}$ the specific surface area of the block.

Bokserman et. al (1964) discussed one-dimensional flow in a horizontal reservoir (Fig. 4) with the boundary and initial conditions:

$$x = 0; \ t = 0: \ q_1^{(2)} = q_1^{(t)}, \ q_1 = q_1^{(1)} + q_1^{(2)};$$

$$q_1^{(1)} = 0$$

$$x \geqq 0; \ t = 0; \ S^{(1)} = S_1^{(1)} + S_2^{(1)} = 1 ; S^{(2)} = S_1^{(2)} + S_2^{(2)} = 0$$

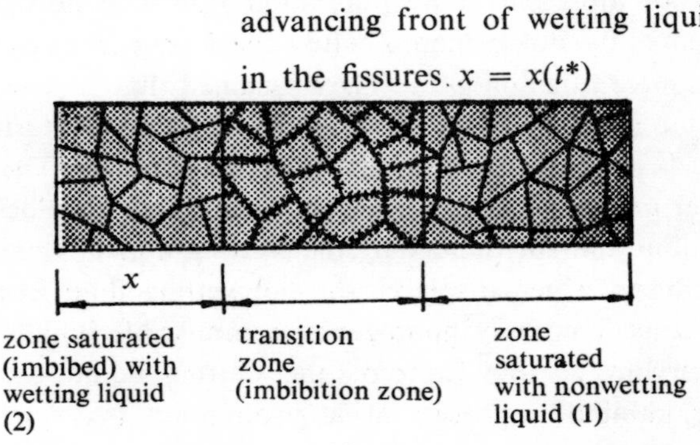

advancing front of wetting liquid

in the fissures. $x = x(t^*)$

| zone saturated (imbibed) with wetting liquid (2) | transition zone (imbibition zone) | zone saturated with nonwetting liquid (1) |

Fig. 4. The scheme of imbibition process in a fissured reservoir (after Bokserman et al. 1964).

At time t, the rate of nonwetting liquid displaced from the blocks along the reservoir, behind the advancing front, equals the rate of injection of the wetting liquid at $x = 0$:

(3.15)
$$q(t) = \int_{x=0}^{x(t=t^*)} q^*[t - t^*(x')]dx'$$

From (3.12) and (3.15) we have:

(3.16)
$$q_1^{(t)} f'(S_1)\frac{\partial S_1^{(2)}}{\partial x} + n_1 \frac{\partial S_1^{(2)}}{\partial t} + q^*[t - t^*(x)] = 0$$

where

$$f(S_1) = \frac{k_{r1}}{\mu(1)} \bigg/ \frac{k_{r1}(1)}{\mu(1)} + \frac{k_{r1}(2)}{\mu(2)}$$

If $q = $ const., the solution of the eqn. (3.16) is

$$t^* = cx^2$$

where c is a constant.

Verma (1968) solved the same problem as Bokserman et al., by using the same source function of imbibition with capillary pressure in the fissures $p_{c1} = p_{c1}(S_1^{(2)})$. For particular functions p_{c1}, and $k_{r1}{}^{(i)}$, he obtained a closed solution of the continuity equations by the method of perturbations.

A FLOW MODEL LEADING TO THE SOURCE FUNCTION q^*

The solution of a problem of flow through a fissured porous medium by Barenblatt's approach requires the knowledge of the source function q^*. As proposed by Barenblatt and derived by dimensional analysis, this function neglects the capillary pressure. Bokserman et al. (1964) consider a source function determined experimentally, for a displacement by imbibition only. In what follows, an attempt will be made to derive a source function by analyzing a conceptual model of flow through a fissured rock.

In displacement by imbibition by maintaining the water level in the fissures lower than the water level in the adjacent blocks (Fig. 5a) we obtain an ultimate maximal total recovery of oil from the reservoir (Fig. 6). However, an exploitation based on imbibition alone implies low oil ratios during a long period of time, although ultimately a higher total recovery is achieved.

It is of interest therefore to investigate what happens in a fissured reservoir during flow with gradients which are not negligible in comparison to those occurring during imbibition.

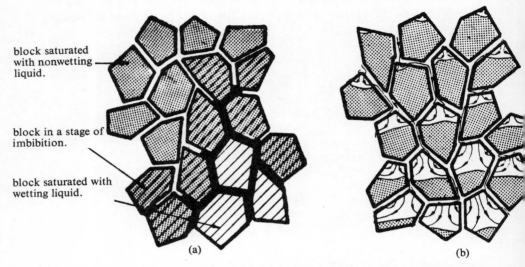

block saturated
with nonwetting
liquid.

block in a stage of
imbibition.

block saturated with
wetting liquid.

(a)   (b)

Fig. 5.   Two kinds of displacement in fissured porous rocks (a) imbibition with negligible
           gradient in the fissures (b) at low gradients.

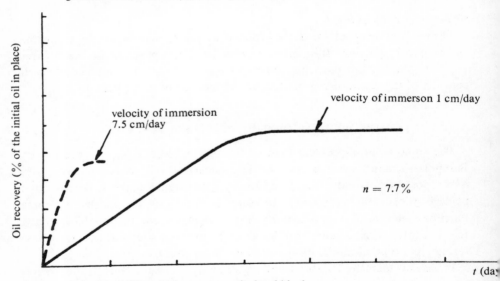

velocity of immersion
7.5 cm/day

velocity of immerson 1 cm/day

$n = 7.7\%$

Oil recovery (% of the initial oil in place)

t (day

Fig. 6.   Cheuca's (1959) experiments on an isolated block.

If the wetting liquid in the fissures by-passes the nonwetting liquid in the
adjacent blocks, a flow pattern as shown in Fig. 5b is obtained. The wetting
liquid enters the blocks and flows in the regions saturated with this liquid.

In a similar manner, the nonwetting liquid removed from the lower blocks
and flowing in the fissures may enter blocks and flow in their portions saturated

with the nonwetting liquid. At low pressure difference (between the oil in the fissure and the oil in the block) no nonwetting liquid will enter the block portions saturated with the wetting liquid at the residual saturation 'of the nonwetting liquids. Consider a model of a block made of randomly oriented capillary tubes (Fig. 7). In each of the capillary tubes, in which the nonwetting

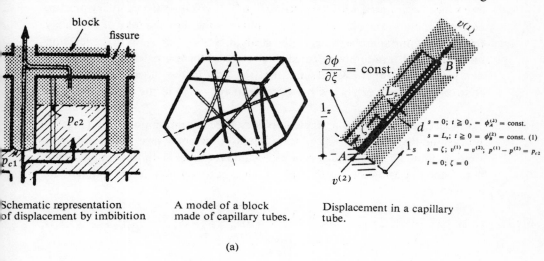

Schematic representation of displacement by imbibition

A model of a block made of capillary tubes.

Displacement in a capillary tube.

(a)

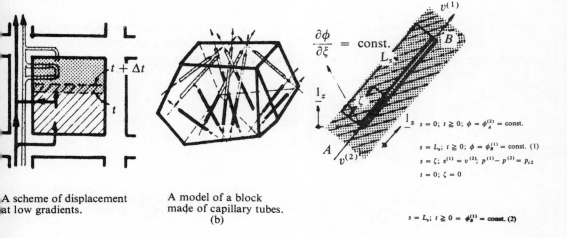

A scheme of displacement at low gradients.

A model of a block made of capillary tubes.

(b)

Displacement in a capillary tube.

$$s = L_s;\ t \geq 0 = \ \phi_B^{(1)} = \text{const.} \ (2)$$

wetting liquid
nonwetting liquid

Fig. 7. A model of (a) displacement by imbibition and (b) flow at low gradients.

liquid is displaced by the wetting one, a sharp interface exists perpendicular to the axis of the capillary tube. End effects are well represented by this model: the wetting liquid leaves the exit face of each capillary tube only at $S^{(2)} = 1$ (instead of the residual saturation of the nonwetting liquid in a porous medium) while the entry value for the nonwetting liquid when the entry face is saturated with the wetting liquid, is given by the capillary pressure corresponding to the diameter of the capillary tube. For the sake of simplicity, the capillary tubes are assumed to be straight-tubes of constant diameter, non-interconnected, randomly oriented, and each of them crosses the block from one side to the other. We assume that the flow in the blocks (medium 2) is controlled by the piezometric head gradients ($\phi_1 = \phi$) in the fissures (medium 1).

For a displacement in a single capillary tube by imbibition only (Fig. 7a) we assume that we have only wetting liquid at the entry face A and only non-wetting liquid at the exit face B of each capillary tube. With the boundary and initial conditions as shown in Fig. 7a, the mean velocity in a capillary tube is given by:

$$(4.1) \quad \langle v \rangle = \langle v_s \rangle \, \mathbf{1}_s = \frac{d^2/32}{\mu^{(1)}(L_s - \zeta) + \mu^{(2)}\zeta} (\phi_A^{(2)}\gamma^{(2)} - \phi_B^{(1)}\gamma^{(1)} + P_{c2} - \zeta\Delta\gamma\mathbf{1}_s\mathbf{1}_z)\mathbf{1}_s$$

where

$$\phi = \frac{p}{\gamma} + z \quad \text{and} \quad \Delta\gamma = \gamma^{(2)} - \gamma^{(1)}$$

$\phi_A^{(1)}$ and $\phi_B^{(2)}$ are related by the capillary pressure in the fissure. Assuming that the fissure has the form of a narrow space between two parallel walls, we obtain for a water front in the fissure at A ($z = 0$):

$$(4.2) \qquad\qquad \phi_A^{(2)}\gamma^{(2)} = \phi_A^{(1)}\gamma^{(1)} - P_{c1}$$

By substituting (4.2) in (4.1) we obtain:

(4.3)

$$\langle v_s \rangle = \langle v_s \rangle \mathbf{1}_s = \frac{d^2/32^{(1)}}{\mu^{(1)}\dfrac{(L_s - \xi)}{L_s} + \dfrac{\mu^{(2)}}{\mu^{(1)}}\dfrac{\xi}{L_s}} \left(\frac{\phi_A^{(1)} - \phi_B^{(1)}}{L_s} + \frac{P_{c2} - P_{c1}}{\gamma^{(1)}L_s} - \frac{\xi}{L_s}\frac{\Delta\gamma}{\gamma^{(1)}}\ \mathbf{1}_s\mathbf{1}_z \right)\mathbf{1}_s$$

Replacing $\phi_A^{(1)} - \phi_B^{(1)} = \Delta\phi^{(1)}$ by $\Delta\phi^{(1)} \cong \dfrac{\partial\phi^{(1)}}{\partial\xi}L_s\mathbf{1}_\xi\mathbf{1}_s$ and denoting:

$$\frac{\zeta}{L_s} = S^{(2)} = \text{wetting liquid saturation}$$

$$\frac{L_s - \zeta}{L_s} = S^{(1)} = 1 - S^{(2)} = \text{nonwetting liquid saturation}$$

$$d^2/32 = k = \text{intrinsic permeability}$$

$$k\gamma^{(i)}/\mu^{(i)} = K^{(i)}; \quad (i = 1, 2,) = \text{hydraulic conductivity}$$

$$f(S^{(2)}) = \frac{1}{(L_s - \zeta)/L_s + (\mu^{(2)}/\mu^{(1)})\zeta/L_s} = \frac{1}{s^{(1)} + (\mu^{(2)}/\mu^{(1)})s^{(2)}}$$

$$= \frac{1}{1 - S^{(2)}\Delta\mu/\mu^{(1)}} = \text{a function of saturation,}$$

one may write (4.3) as:

$$(4.4) \quad \langle \boldsymbol{v}_s \rangle = K^{(1)}f(S^{(2)}) \left[-(\nabla\phi \cdot \mathbf{1}_s + \frac{p_{c2} - p_{c1}}{\gamma^{(1)}L_s}\mathbf{1}_s - s^{(2)}\frac{\Delta\gamma}{\gamma^{(1)}}(\mathbf{1}_z \cdot \mathbf{1}_s)\mathbf{1}_s \right]$$

The liquid discharge is:

$$(4.5) \quad Q^{(i)} = \int_{A_0(i)} \langle \boldsymbol{v}_s \rangle dA; \quad (i = 1, 2)$$

where $A_0^{(i)}$ is the total area of the capillary tubes through which liquid leaves $(i = 1)$ or enters $(i = 2)$ the blocks contained in the volume U of an REV. A_0 may be related to the total surface area of the blocks A by

$$A_0^{(i)} = \alpha^{(i)}A.$$

The source function, appearing in Barenblatt's approach is given by;

$$(4.6) \quad q^{*(i)} = \frac{Q^{(i)}}{U} = \frac{\int \alpha^{(i)}A\langle \boldsymbol{v}_s \rangle dA}{U}$$

By integrating (4.6) for the nonwetting liquid (1) and expressing $\langle \boldsymbol{v}_s \rangle$ by (4.4) we obtain:

$$q^{*(1)} = -\frac{K^{(1)}}{V}\frac{\partial\phi^{(1)}}{\partial\xi}\int_{\alpha^{(1)}A} f(S^{(2)}, s)(\mathbf{1}_\xi\mathbf{1}_s)(\mathbf{1}_s\mathbf{1}_A)dA$$

$$(4.7) \quad + \frac{K^{(1)}}{v\gamma^{(1)}}\int_{(1)A} f(S^{(2)}, s) \cdot \frac{p_{c2} - p_{c1}}{L_s}(\mathbf{1}_s\mathbf{1}_A)dA$$

$$- \frac{K^{(1)}}{V}\frac{\Delta\gamma}{\gamma^{(1)}}\int_{\alpha^{(1)}A} f(S^{(2)})S^{(2)}(\mathbf{1}_z\mathbf{1}_s)(\mathbf{1}_z\mathbf{1}_A)dA$$

In an actual porous medium the capillary pressures p_{c1} and p_{c2} will be functions of the saturations in the fissure and block, respectively. We may think of such result as obtained from a porous medium model made of capillary tubes which are conic in shape instead of cylindrical. Also A_0 or α will be a function of the block's saturation (Fig. 7a), i.e. $A_0 = \alpha(S^{(2)})A$.

Consequently, the source function has the general form:

$$q^{*(1)} = -\bar{A}K_2^{(1)}F_a(S_2^{(2)})\frac{\partial\phi^{(1)}}{\partial\xi^{(1)}} + \frac{\bar{A}K_2^{(1)}}{\gamma^{(1)}}[F_b(S_2^{(2)}) - F_b'(S_1^{(2)}, S_2^{(2)})]$$

$$(4.8) \qquad\qquad - \bar{A}K_2^{(1)}\frac{\Delta\gamma}{\gamma^{(1)}}F_c(S_2^{(2)})$$

and

$$q^{*(2)} = -q^{*(1)}$$

where $\bar{A}$ is the specific surface of the blocks ($\bar{A} = A/V$) and:

$$F_a(S_2^{(2)}) = \alpha^{(1)}(S_2^{(2)})\,\overline{[f(S^{(2)}, s)(1_z \cdot 1_s)(1_s \cdot 1_A)]}$$

$$F_b(S_2^{(2)}) = \alpha^{(1)}(S_2^{(2)})\,\overline{\left[f(S_2^{(2)}, s)\frac{p_{c2}}{L_s}(1_s \cdot 1_A)\right]}$$

$$(4.9)$$

$$F_b'(S_1^{(2)}) = \alpha^{(1)}(S_2^{(2)})\,\overline{\left[f(S^{(2)}, s)\frac{p_{c1}}{L_s}(1_s \cdot 1_A)\right]}$$

$$F_c(S_2^{(2)}) = \alpha^{(1)}(S_2^{(2)})\,\overline{[f(S^{(2)}, s)(1_z \cdot 1_s)(1_s \cdot 1_s)]}$$

$F_a(S_2^{(2)}), F_b(S_2^{(2)})$ $F_b'(S_1^{(2)}S_2^{(2)})$ and $F_c(S_2^{(2)})$ are functions of saturation and must be determined experimentally. The overscore indicates average value.

For a displacement by imbibition only, the liquid in the fissures is quasi-stagnant and $\partial\phi^{(1)}/\partial\xi = 0$. Neglecting the capillary pressure in the fissures (in comparison to the capillary pressure in the blocks), and gravity, the source function may be approximated by:

$$(4.10) \qquad\qquad q^{*(1)} = \frac{\bar{A}K_2^{(1)}}{\gamma^{(1)}}F_b(S_2^{(2)})$$

If $q^* = q^*(S_2^{(2)})$ is a known function, $q^* = q^*(t)$ may be obtained by integrating the mass conservation equation in the block:

$$\frac{\partial S_2^{(2)}}{\partial t} = q^*(S_2^{(2)})$$

This is a function similar to that used by Bokserman et al. (1964) and based on the experiments of Mattax and Kyte (1962).

If the piezometric gradients are not negligible and water flows in the fissures, leaving behind blocks still containing nonwetting liquid, we consider the flow model shown in Fig. 5b and 7b. We assume that the two liquids flow in the fissures but no nonwetting liquid enters the capillary tubes saturated with the liquid at the entry face (low pressure gradients). There are capillary tubes through which only wetting or nonwetting liquid flows, and capillary tubes in which wetting liquid displaces the nonwetting liquid. At the exit face we may have the wetting or the nonwetting liquid.

The corresponding mean velocities are:

In a capillary tube in which only nonwetting liquid flows:

$$(4.11) \qquad v'^{(1)} = -K_2^{(1)} \nabla \phi^{(1)}$$

In a capillary tube in which only wetting liquid flows:

$$(4.12) \qquad v'^{(2)} = -K_2^{(2)} \nabla \phi^{(2)}$$

For a capillary tube in which the nonwetting liquid is displaced in a wetting liquid environment, with the boundary and initial conditions (1) of Fig. 7b, we have:

$$(4.13) \qquad \langle v_s \rangle = K^{(1)} f(S^{(2)}) \frac{\gamma^{(2)}}{\gamma^{(1)}} \left(-\nabla \phi^{(2)} + \frac{p_{c2} - p_{c1}}{L_s \gamma^{(2)}} \mathbf{1}_s + \frac{\Delta \gamma}{\gamma^{(2)}} S^{(1)} \mathbf{1}_z \right)$$

where p_{c1} is the capillary pressure in the fissure. It equals the capillary pressure at the exit phase (B) of the nonwetting liquid in the capillary tube and the wetting liquid in the fissure.

For a capillary tube in which the nonwetting liquid B is displaced in a nonwetting liquid environment, a similar relationship is obtained.

The source function is given by:

$$(4.14) \qquad q^{*(i)} = \frac{\int_A \langle v^{(i)} \rangle dA}{U}$$

The capillary tubes for which the saturation $S^{(i)} = 1$ (i.e. containing a single liquid only) give a zero contribution to the source function. Then (4.13) alone determines the value of the source function in (4.14).

By integrating (4.14) for $\langle vs^{(i)} \rangle$ given by (4.13), we obtain a source function similar to that given by (4.8):

$$q^{*(1)} = -\bar{A}K_2^{(1)}F_a(S_2^{(2)}, \mathbf{1}_s)\frac{\gamma^{(2)}}{\gamma^{(1)}}\frac{\partial\phi^{(2)}}{\partial\xi} + \frac{\bar{A}K_2^{(1)}}{\gamma^{(1)}}[F_b(S_2^{(2)}) - F_b{}'(S_1^{(2)}, S_2^{(2)})] -$$

(4.15)
$$-\frac{\bar{A}K_2^{(1)}}{\gamma^{(1)}}F_c(S_2^{(2)})$$

and $q^{*(2)} = -q^{*(1)}$

where $F_a(S_2^{(2)})$, $F_b(S_2^{(2)})$ and $F_c(S_2^{(2)})$ have the forms similar to those given by (4.9) with $\alpha(S_2^{(2)})$ corresponding to one of the capillary tubes of our porous medium model, in which water displaces oil.

Altogether we have four different functions of the various saturations, contributing to the source function q^*:
$F_a(S_2^{(2)}, \mathbf{1}_\xi)$ which expresses the effect of the average gradient of the wetting phase (2), $F_b(S_2^{(2)})$ expresses the effect of capillary pressure in the blocks $F_b'(S_1^{(2)}, S_2^{(2)})$ expresses the effect of capillary pressure in the fissutes, and $F_c(S_2^{(2)})$ expresses the effect of gravity. It is of interest to note that F_a depends on the direction $\mathbf{1}_\xi$ of the gradient. The influence of the pressure gradients on the displacement process can be observed in the Graham and Richardson (1959) experiments (Fig. 8).

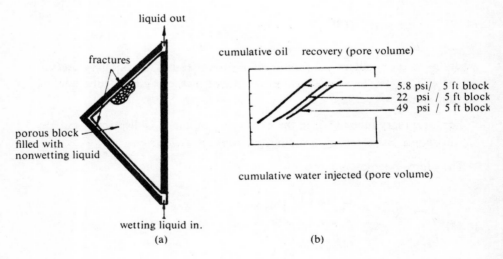

Fig. 8.   Graham and Richardson's (1959) experiments: (a) the model (b) the result of the experiments.

As follows from (4.15) which was obtained from a conceptional model, and also from Mattax and Kyte's (1962) experiments, the contribution to the source function $q*$ due to imbibition is inversely proportional to the square characteristic length of the block (L^2). Thus the influence of this term diminishes as L increases and for a certain characteristic length of the blocks the contribution to $q*$ due to piezometric gradients of the wetting liquid may be of the same order of magnitude as the imbibition term even at low gradients. Therefore the consideration of a source function $q*$ with a term due to piezometric gradients is of practical interest.

THE FORMULATION OF THE FLOW IN FISSURED POROUS MEDIA

In the flow model to be considered below, only the fissures are regarded as a continuous space; the flow in the blocks is controlled by the gradients in the fissures. These assumptions are different from those assumed by Barenblatt who regarded both the blocks and the fissures as two overlapping continuum with a different piezometric head and head gradient in each of them. Each liquid may pass from blocks to fissures and then return to blocks downstream. Consequently each liquid flows in a parallel-series system of the two media, fissures and blocks; the connectiontween the two media in a randomal arrangement of blocks is assumed a function of saturation, independent of the direction of flow. A source-sink function given by (4.15) is introduced.

Neglecting the capillary pressure in the fissures, $F_{b'}$ in (4.15) vanishes and following the schematic representation of the flow as shown in Fig. 7b, the equations characterizing the flow are:

The equations of motion with an effective hydraulic conductivity K_e or a relative permeability $k_{r'}$ ($K_e = Kk_r$) as a function of both saturations, in the fissures and blocks (Fig. 9):

$$(5.1) \qquad q^{(i)} = - K_e^{(i)}(S_1^{(2)}, S_2^{(2)}) \nabla \phi^{(i)}; \quad \phi^{(i)} = z + \frac{p}{\gamma^{(1)}}; \ (i = 1, 2);$$

continuity equation in the fissures:

$$(5.2) \qquad n_1 \frac{\partial S_1^{(i)}}{\partial t} + V q^{(i)} = -(-1)^{i+1} q*$$

$$(5.3) \qquad S_1^{(1)} + S_1^{(2)} = 1$$

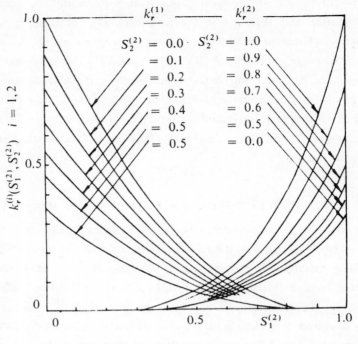

Fig. 9. $k_r^{(i)} = k_r^{(i)}(S_1^{(2)}, S_2^{(2)})$

The continuity equations in the blocks:

(5.4) $n_2 \dfrac{\partial S_2^{(i)}}{\partial t} = -(-1)^i q^*$

(5.5) $S_2^{(1)} + S_2^{(2)} = 1$

The source function given by (4.15) is:

(5.6) $q^* = -AF_a(S_2^{(2)}) \dfrac{\partial \phi^{(1)}}{\partial \xi} + AF_b(S_2^{(2)}) - AF_c(S_2^{(2)})$

where $A = \bar{A} K_2^{(1)}$ is a constant.

If $K^{(i)}(S_1^{(2)}, S_2^{(2)})$, $F_a(S_2^{(2)})$, $F_b(S_2^{(2)})$ and $F_c(S_2^{(2)})$ are known functions of the block saturation, (5.1) to (5.6) constitute a system of 14 equations with 14 unknown parameters $q^{(i)}$, $\phi^{(i)}$, p, $S_1^{(i)}$, $S_2^{(i)}$ and q^*.

ONE DIMENSIONAL HORIZONTAL FLOW

For a one-dimensional horizontal flow in the x-direction, the system (5.1) through (5.6) reduces to:

$$(5.7) \qquad q^{(i)} = - \frac{K^{(i)}(S_1^{(2)}, S_2^{(2)})}{\gamma^{(i)}} \frac{\partial p}{\partial x} \quad (i = 1, 2)$$

$$(5.8) \qquad n_1 \frac{\partial S_1^{(i)}}{\partial t} + \frac{\partial q^{(i)}}{\partial x} = (-1)^{i+1} q^*$$

$$(5.9) \qquad S_1^{(1)} + S_1^{(2)} = 1$$

$$(5.10) \qquad n_2 \frac{\partial S_2^{(2)}}{\partial t} = + q^*$$

$$(5.11) \qquad q^* = - AF_a(S_2^{(2)}) \frac{\partial p}{\partial x} + AF_b(S_2^{(2)})$$

If $K^{(i)}$, F_a and F_b are known functions of saturation (5.7) through (5.11) is a system of 7 equations with 7 unknown parameters $q^{(i)}$, $S_1^{(1)}$, $S_2^{(2)}$, p and q^*. By adding equations (5.8) for $i = 1$ and $i = 2$, we obtain:

$$(5.12) \qquad \frac{\partial}{\partial x}(q^{(1)} + q^{(2)}) = 0$$

$$(5.13) \qquad \frac{\partial}{\partial x} q = 0; \quad q = q^{(1)} + q^{(2)}$$

where q is the total specific discharge through the fractures rock system. From (5.13) it follows that:

$$(5.14) \qquad q = q(t)$$

If the displacement of the oil by the water is taking place at a constant $q \equiv q^{(2)}$ with the boundary and initial conditions:

$$x = 0, \quad t \geq 0 \quad q^{(2)} = \text{const.}$$

$$(5.15) \qquad q^{(1)} = 0$$

$$t \leq 0, \ 0 \leq x \leq L; \ S^{(1)} = 1$$

we obtain:

$$(5.16) \qquad q = -\left(\frac{K^{(1)}(S_1^{(2)}, S_2^{(2)})}{\gamma^{(1)}} + \frac{K^{(2)}(S_1^{(2)}, S_2^{(2)})}{\gamma^{(2)}} \right) \frac{\partial p}{\partial x}$$

or:

(5.17)
$$\frac{\partial p}{\partial x} = \frac{-q}{K^{(1)}(S^{(2)}S^{(2)})/\gamma^{(1)} + K^{(2)}(S^{(2)}S^{(2)})/\gamma^{(2)}}$$

By substituting (5.7), (5.11) and (5.17) in (5.8) for $i = 2$, we obtain:

$$n_1 \frac{\partial S_1^{(2)}}{\partial t} + q \frac{\partial}{\partial x} \left[\frac{K^{(2)}(S_1^{(2)}, S_2^{(2)})/\gamma^{(2)}}{K^{(1)}(S_1^{(2)}, S_2^{(2)})/\gamma^{(1)} + K^{(2)}(S_1^{(2)}, S_2^{(2)})/\gamma^{(2)}} \right]$$

(5.18)
$$= - AF(S_2^{(2)}) \frac{q}{K^{(1)}(S_1^{(2)}, S_2^{(2)})/\gamma^{(1)} + K^{(2)}(S_1^{(2)}, S_2^{(2)})/\gamma^{(2)}} - AF_b(S_2^{(2)})$$

From (5.10) we obtain:

$$n_2 \frac{\partial S_2^{(2)}}{\partial t} = + AF_a(S_2^{(2)}) \frac{q}{K^{(1)}(S_1^{(2)}, S_2^{(2)})/\gamma^{(1)} + K^{(2)}(S_1^{(2)}, S_2^{(2)})/\gamma^{(2)}}$$

(5.19)
$$+ AF_b(S_2^{(2)})$$

(5.18) and (5.19) are two equations with two unknown functions $S_1^{(2)}$ and $S_2^{(2)}$ of x and t. Or:

$$n_1 \frac{\partial S_1^{(2)}}{\partial t} + q \frac{\partial}{\partial x} F^{(2)}(S_1^{(2)}, S_2^{(2)}) = - Af^{(2)}(S_1^{(2)}, S_2^{(2)})$$

(5.20)

and:
$$n_2 \frac{\partial S_2^{(2)}}{\partial t} = Af^{(2)}(S_1^{(2)}, S_2^{(2)})$$

where:

(5.21)
$$F^{(2)}(S_1^{(2)}, S_2^{(2)}) = \frac{1}{1 + \dfrac{k_r^{(1)}(S_1^{(2)}, S_2^{(2)})\mu^{(2)}}{k_r^{(2)}(S_1^{(2)}, S_2^{(2)})\mu^{(1)}}}$$

has the general form shown in Fig. 10.

and:

(5.22)
$$f(S_1^{(2)}, S_2^{(2)}) = F_a(S_2^{(2)}) \frac{q}{\dfrac{K^{(1)}(S_1^{(2)}, S_2^{(2)})}{\gamma^{(1)}} + \dfrac{K^{(2)}(S_1^{(2)}, S_2^{(2)})}{\gamma^{(2)}}} + F_b(S_2^{(2)})$$

If $F(S_1^{(2)}, S_2^{(2)})$ and $f(S_1^{(2)})$ are known functions, equation (5.20) may be solved for the unknown variables $S_1^{(2)}$ and $S_2^{(2)}$.

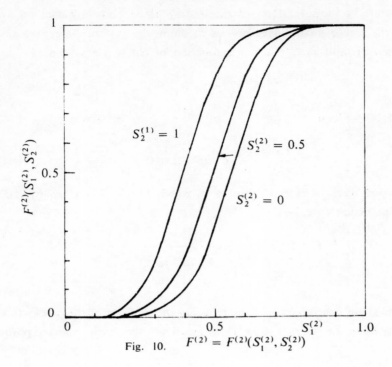

Fig. 10. $F^{(2)} = F^{(2)}(S_1^{(2)}, S_2^{(2)})$

For an ordinary porous medium, (5.20) gives the known Buckly-Leverett equation. Indeed, by adding the two equations in (5.20) we obtain:

$$(5.23) \qquad n_1 \frac{\partial S_1^{(2)}}{\partial t} + n_2 \frac{\partial S_2^{(2)}}{\partial t} + q \frac{\partial}{\partial x} F^{(2)} S_1^{(2)}, S_2^{(2)}) = 0$$

Or:

$$(5.24) \qquad n \frac{\partial S^{(2)}}{\partial t} + q \frac{\partial}{\partial x} F^{(2)}(S_1^{(2)}, S_2^{(2)}) = 0$$

where:

$$n \frac{\partial S^{(2)}}{\partial t} = n_1 \frac{\partial S_1^{(2)}}{\partial t} + n_2 \frac{\partial S_2^{(2)}}{\partial t}$$

follows from the definitions of n_1, n_2, S_1 and S_2. Eqn. (5.24) differs from the Buckly-Leverett equation only by the fact that

$$F(S_1^{(2)}, \quad S_2^{(2)})$$

is a function of the saturations of the wetting liquid both in medium 1 and in medium 2. For an ordinary porous medium, the two media reduce to a single medium and $F(S_1^{(2)}, S_2^{(2)})$ is a function of the single saturation

$$S^{(2)} = (S_1^{(2)} + S_2^{(2)})$$

From (5.24) we obtain:

$$n \frac{\partial S^{(2)}}{\partial t} + \frac{\partial}{\partial x} F^{(2)}(S^{(2)}) = 0$$

which is the Buckly-Leverett equation in which $F^{(2)}(S^{(2)})$ is given by (5.21) for this particular case, i.e.:

$$(5.26) \qquad F^{(2)}(S^{(2)}) = \cfrac{1}{1 + \cfrac{k_r^{(1)}(S^{(2)})\mu^{(2)}}{k_r^{(2)}(S^{(2)})\mu^{(1)}}}$$

The same result can be obtained by reducing the medium of the blocks to an impervious one, i.e. $n_2 = 0$, $S_2 = 0$, as again we obtain an ordinary porous medium.

CONCLUSIONS

The system of equations and the flow parameters characterizing the flow in a fissured porous medium were established, and thus the objectives of the present paper were achieved.

Special attention was given to determine the parameters affecting the source function q^*, the piezometric head gradients, the capillary pressure in the blocks (imbibition) and fissures, and gravity.

It follows that a source function q^* with a term due to piezometric gradients of the wetting liquid is of practical interest. A large characteristic length L of the blocks may reduce the importance of the term due to imbibition in the q^* function and the term due to piezometric gradients must be considered even at low gradients.

Flow of a single phase incompressible fluid in a nonconsolidating fractured porous medium may be treated without difficulties by the usual approach considering the fractured rock as a single continuum.

In two phase flow, the relative permeability of the fissured porous medium is a function of the total saturation and the fissure – block saturation distribution, i.e.,

$$k_r = k_r(S_1^{(2)}, S_2^{(2)})$$

At this stage, although we have established the system of equations and the various parameters appearing in them, deriving solution to problems of interest by solving these equations is impractical, as there is no way to derive experimentally the various parameters. Nevertheless, the authors believe that the analysis presented here gives some insight into the flow pattern and the various forces active in two phase flow in fractured rocks. Experiments are being carried out at the present at the Hydrodynamic and Hydraulic Engineering Laboratory of the Technion to verify some of the results of the theory presented above.

REFERENCES

BARENBLATT, G. I. and ZHELTOV IU. P. (1960a), On the basic equations of seepage of homogeneous liquids in fissured rocks, *Dokl. Akad. Nauk SSSR, 132*, 3, 545–548.

BARENBLATT, G. I., ZHELTOV, IU. P. and KOCHINA, I. N. (1960b), Basic concepts in the theory of seepage of homogeneous liquids in fissured rocks, *PMM, 24*, 5, 1286–1303.

BARENBLATT, G. I. (1964), *O dvizhenii gazozhidkostn'kh smesei v treshchinovato — porist'kh porodakh*, Izvestia Akademii Nauk SSSR, Mekhanika i Mashinostroienie, No. 3, 47–50.

BOKSERMAN, A. A., ZHELTOV, IU. P. and KOCHESHKOV, A. A. (1964), Motion of immiscible liquids in a cracked porous medium, *Soviet Physics Doklady, 9*, 4, 285–287.

BEAR, J. (1969), Dynamics of fluids in porous media (under publication).

CHEUCA, A. (1959), Répartitions des charactéristiques et facteurs influensant la récupération finale dans un reservoir carbonaté hétérogène, *Revue de l'Institut Francais du Petrole, 14*, No. 11, pp. 1468–1511.

DANIEL, E. I. (1954), Fractured Reservoirs of the Middle East, *Bulletin of American Assoc. of Petroleum Geologists, 38*, 5, 778–815.

FREEMAN, H. A. and NATANSON, S. G. (1959), *Recovery problems in a fractured-pore system, Kirkuk Field*, Fifth World Petroleum Congress Sec. II, paper 24 pp. 1–19.

GIBSON, H. S. (1948), The production of oil from the fields of South Western Iran, *J. Inst. Petroleum, 34*.

GRAHAM, J. W. and RICHARDSON, J. G. (1959), *Theory and application of imbibition phenomena in recovery of oil*, Petroleum Transaction AIME, *216*, 377–381.

MATTAX, C. C. and KYTE, J. R. (1962), Imbibition oil recovery from fractured water-drive reservoir, *Society of Petroleum Engineering Journal* 177–184.

PIRSON, S. J. (1953), Performance of fractured oil reservoires, *Bulletin of American Assoc. Petrol Geologists, 37*, 2, 232–244.

PIRVU, G. (1965), *Roci fisurate colectoare de petrol*. Ed. Tehnica Bucuresti.

PRANDTL, L. and TIETJENS, O. G. (1934), *Fundamentals of hydro- and aeromechanics*. New York, McGraw-Hill, 270 p.

RATS, M. V. and CHERNYASHOV, S. N. (1965), *Statistical aspect of the problem on the permeability of the joint rocks*, Proceedings of the Dubrovnik Symposium — Hydrology of fractured rocks, vol. I, pp. 227–236.

VERMA, A. P. (1968), Motion of imbiscible liquids in a cracked heterogeneous porous medium with capillary pressure, *Rev. Roumaine Sci. Techn — Mé Appl., 13*, 2, 277–292.

WILKINSON, W. M. (1953), Fracturing in Spraberry Reservoir, West Texas, *Bulletin of the American Assoc. of Petroleum Geologists, 37*, 2, 250–265.

TECHNION—ISRAEL INSTITUTE OF TECHNOLOGY,
        HAIFA, ISRAEL

SUR LES ÉQUATIONS DE LA MAGNÉTOHYDRO-DYNAMIQUE DES MILIEUX POREUX

HORIA I. ENE

1. INTRODUCTION

Le problème du mouvement de filtration d'un fluide sous l'influence d'un champ magnétique n'a pas encore été traité. On a étudié le problème du mouvement sous l'influence d'électro-osmose [1], [4], mais alors il n'y a pas de forces électromagnétiques.

Les équations que nous déduisons ici, contiennent l'hypothèse essentielle que la portion solide du milieu est neutre du point de vue électrique.

Le cas le plus général, quand la portion solide du milieu est un bon conducteur d'électricité, va faire l'objet d'un autre article.

2. LES ÉQUATIONS DU PROBLÈME

Les équations de la magnétohydrodynamique sont [2]:

— l'équation de conservation de la quantité de mouvement:

$$(1) \qquad \frac{\partial u}{\partial t_1} + (u \cdot \mathrm{grad})u = F + \frac{1}{\rho}\Phi - \frac{1}{\rho}\mathrm{grad}\, p$$

— l'équation de conservation de la masse:

$$(2) \qquad \frac{\partial \rho}{\partial t} + \mathrm{div}(\rho u) = 0$$

— les équations de Maxwell:

$$(3) \qquad \mathrm{rot}\, H = 4\pi J + \frac{\partial D}{\partial t}$$

$$(4) \qquad \mathrm{rot}\, E = -\frac{\partial B}{\partial t}$$

$$(5) \qquad \mathrm{div}\, D = 0$$

$$(6) \qquad \mathrm{div}\, B = 0$$

— la loi d'Ohm:

$$J = \sigma(E + u \times B)$$

où u — la vitesse du fluide, ρ — la densité, p — la pression, F — les forces massiques, H — le champ magnétique, J — la densité du courant électrique, E — le champ électrique, B — l'induction magnétique, D — l'induction électrique, σ — le coefficient de conducttivité électrique, et Φ:

$$(8) \qquad \Phi = J \times B$$

la force extérieure qu'exerce le champ électromagnétique sur le milieu.

Entre les vecteurs D, E, B et H existent les relations suivantes:

$$(9) \qquad D = \varepsilon E; \quad B = \mu H$$

où ε — le coefficient diélectrique, μ — la perméabilité magnétique.

Pour définir le mouvement dans un milieu poreux, il est nécessaire [3] d'introduire la notion de vitesse de filtration, définie par:

$$(10) \qquad v = mu$$

où m — la porosité du milieu.

En faisant l'hypothèse que la portion solide du milieu poreux est neutre du point de vue électrique, nous pouvons appliquer aux équations (1)–(7) un raisonnment analogue à celui de la théorie de filtration.

En introduisant (10) en (1) nous avons:

$$(11) \qquad \frac{1}{m}\frac{\partial v}{\partial t} + \frac{1}{m^2}(v \cdot \mathrm{grad})v = F + \frac{1}{\rho}\Phi - \frac{1}{\rho}\mathrm{grad}\, p.$$

La vitesse de filtration est petite, et le produit entre elle et ses dérivées peut être négligé:

$$(11') \qquad \frac{1}{m}\frac{\partial v}{\partial t} = F + \frac{1}{\rho}\Phi - \frac{1}{\rho}\mathrm{grad}\, p.$$

Si nous considérons le cas stationnaire l'équation (11′) devient:

$$(12) \qquad F + \frac{1}{\rho}\Phi - \frac{1}{\rho}\mathrm{grad}\, p = 0.$$

Les forces massiques F ont en général deux composantes, une composante extérieure $X^{(e)}$ et l'autre intérieure $X^{(i)}$, c'est-à-dire $F = X^{(e)} + X^{(i)}$.

En l'absence de forces extérieures et de forces électromagnétiques, on a la loi du Darcy:

$$(13) \qquad \iota = -\frac{k}{\rho g}\mathrm{grad}\, p$$

où k — coefficient de filtration. Dans le même hypothèse l'équation (12) nous donne:

$$(14) \qquad X^{(i)} = \frac{1}{\rho} \operatorname{grad} p \,.$$

De (13) et (14) il résulte:

$$(15) \qquad v = - \frac{k}{g} X^{(i)} \,.$$

En introduisant (15) dans (12), nous avons:

$$(16) \qquad v = - \frac{k}{\rho g} (\operatorname{grad} p - \rho X^{(e)} - \mathbf{\Phi})$$

ou en projection sur les axes de coordonnées:

$$(16') \qquad v_j = - \frac{k}{\rho g} \left(\frac{\partial p}{\partial x_j} - \rho X_j^{(e)} - \Phi_j \right) \qquad (j = 1, 2, 3) \,.$$

En général les forces massiques extérieures sont données par la force de gravitation, donc $X_1^{(e)} = X_2^{(e)} = 0$ et $X_3^{(e)} = -g$, et:

$$v_1 = - \frac{k}{\rho g} \left(\frac{\partial p}{\partial x_1} - \Phi_1 \right)$$

$$v_2 = - \frac{k}{\rho g} \left(\frac{\partial p}{\partial x_2} - \Phi_2 \right)$$

$$v_3 = - \frac{k}{\rho g} \left(\frac{\partial p}{\partial x_3} + \rho g - \Phi_3 \right) \,.$$

Si nous faisons abstraction des forces massiques extérieures, c'est-à-dire $X^{(e)} = 0$, il résulte que:

$$(17) \qquad v = - \frac{k}{\rho g} (\operatorname{grad} p - \mathbf{\Phi}) \,.$$

On observe que (16) ou (17) réprésentent la généralisation de la loi de Darcy pour le cas du mouvement d'un fluide électroconducteur dans un millieu poreux. On voit que si nous faisons $\mathbf{\Phi} = 0$, c'est-à-dire lorsque l'action du champ magnétique est nulle, nous retrouverons exactement la loi classique de Darcy.

Avec (10), l'équation de continuité (2) devient:

$$\frac{\partial(m\rho)}{\partial t} + \operatorname{div}(\rho v) = 0 \,,$$

ou dans le cas du mouvement stationnaire:

$$(18) \qquad \operatorname{div}(\rho v) = 0 \,.$$

Enfin, la loi d'Ohm (7) devient:

$$(19) \qquad J = \sigma\left(E + \frac{1}{m}v \times B\right).$$

Les équations de la magnétohydrodynamique du mouvement stationnaire en milieux poreux sont (17), (18) et (19), auxquelles se rattachent les équations de Maxwell.

Dans le cas d'un fluide parfait incompressible, ces équations deviennent:

$$v = -\frac{k}{\rho g}(\operatorname{grad} p - J \times \mu H)$$

$$J = \sigma\left(E + \frac{1}{m}v \times \mu H\right)$$

$$(20) \qquad \operatorname{rot}\ H = 4\pi J$$

$$\operatorname{rot}\ E = 0$$

$$\operatorname{div}\ H = 0$$

$$\operatorname{div}\ v = 0$$

En éliminant J entre la deuxième et la troisième équation (20):

$$(21) \qquad \operatorname{rot} H = 4\pi\sigma\left(E + \frac{1}{m}v \times H\right).$$

3. CAS DE RÉDUCTION DES ÉQUATIONS

a) On peut éliminer J et E du système (20). La première équation (20) s'écrit:

$$(22) \qquad v = -\frac{k}{\rho g}\left(\operatorname{grad} p - \frac{\mu}{4\pi}\operatorname{rot} H \times H\right).$$

En appliquant à (21) l'opérateur rot, nous avons:

$$(23) \qquad \Delta H = -\frac{4\pi\sigma\mu}{m}\operatorname{rot}(v \times H)$$

où Δ est l'opérateur de Laplace.

Nous avons aussi:

$$(24) \qquad \operatorname{div} v = \operatorname{div} H = 0$$

b) Écoulements à champ magnétique aligné.

Dans ce cas $H = \lambda v$. La deuxième équation (24) nous donne:

$$\operatorname{div}(\lambda v) = \lambda \operatorname{div} v + v \cdot \operatorname{grad} \lambda = 0$$

ou
$$\boldsymbol{v} \cdot \operatorname{grad} \lambda = 0$$

donc
$$\lambda = \text{const.}$$

Dans (23) nous avons rot $(\boldsymbol{v} \times \lambda \boldsymbol{v}) = 0$, et donc:

(25)
$$\Delta \boldsymbol{H} = 0.$$

En appliquant à (22) l'opérateur div et en tenant compte de la premiére équation (24) nous avons:

$$\Delta p + \frac{\mu}{4\pi}\left[\boldsymbol{H} \cdot \Delta \boldsymbol{H} + (\operatorname{rot} \boldsymbol{H})^2\right] = 0$$

mais $\Delta \boldsymbol{H} = 0$, et

$$\operatorname{rot} \boldsymbol{H} = \operatorname{rot}(\lambda \boldsymbol{v}) = \lambda \operatorname{rot} \boldsymbol{v}$$

donc

(26)
$$\Delta p + \frac{\mu \lambda^2}{4\pi}(\operatorname{rot} \boldsymbol{v})^2 = 0.$$

c) Écoulements à champ magnétique orthogonal.

Nous supposerons que le movement est à vitesse parallèle à un plan fix $x0y$ et que nous avons une seule composante du champ magnétique dans la direction $0z$:

$$\boldsymbol{v} = \boldsymbol{i}u(x, y) + \boldsymbol{j}v(x, y)$$

$$\boldsymbol{H} = \boldsymbol{k}H(x, y).$$

i, j, k — les vecteurs unitaires des axes.

Dans ce cas (22) devient:

(27)
$$\boldsymbol{v} = -\frac{k}{\rho g} \operatorname*{grad}_{(x,y)}\left(p + \frac{\mu H^2}{8\pi}\right)$$

et (23) s'écrit:

(28)
$$\operatorname*{\Delta}_{(x,y)} H = \frac{4\pi \sigma \mu}{m}\left[\frac{\partial(uH)}{\partial x} + \frac{\partial(vH)}{y}\right].$$

4. CAS D'UNE SPHÈRE AIMANTÉE ENTOURÉE D'UN MILIEU POREUX INFINI

Nous supposerons qu'une sphère aimantée de rayon a est située dans un milieu poreux infini homogène de coefficient de filtration $k = \text{const.}$ Un fluide parfait incompressible, de masse spécifique ρ_0, de conductivité électrique σ, est animé à l'infini d'une vitesse uniforme U. La perméabilité magnétique μ du fluide est supposée constante. La sphère est un dipôle de moment magnétique M de conductivité électrique σ' constante et de perméabilité μ' constante.

On suppose qu'un mouvement stationnaire est établi d'équations:

$$v = -\frac{k}{\rho g}\left(\operatorname{grad} p - \frac{\mu}{4\pi}\operatorname{rot} H \times H\right)$$

(29)
$$\operatorname{rot} H = 4\pi\sigma\left(E + \frac{\mu}{m}v \times H\right)$$

$$\operatorname{rot} E = 0$$

$$\operatorname{div} H = 0$$

$$\operatorname{div} v = 0\,.$$

Par suite de la symétrie des données, l'écoulement est dans le méridien de révolution. En coordonnées sphèriques: $x = r\,\sin\theta\cos\phi$, $y = r\sin\theta\sin\phi$, $z = r\cos\theta$. En designant par e_1, e_2, e_3 les vecteurs unitaires des directions (r,θ,ϕ), nous avons:

$$v = e_1 v_r(r,\theta) + e_2 v_\theta(r,\theta)$$

$$H = e_1 H_r(r,\theta) + e_2 H_\theta(r,\theta)$$

$$\operatorname{rot} H = -\frac{e_3}{r}\left\{\frac{\partial H_r}{\partial \theta} - \frac{\partial(rH_\theta)}{\partial r}\right\}\,.$$

Le produit $v \times H$ a une seule composante dans la direction e_3; la deuxième équation (29) indique que la seule composante non-nulle du champ électrique est E_ϕ. La troisième équation (29) donne:

$$\frac{\partial}{\partial\theta}(r\sin\theta\,E_\phi) = 0 \quad ; \quad \frac{\partial E_\phi}{\partial r} = 0$$

Dans le plan du mouvement $E_\phi = 0$.

En introduisant les grandeurs sans dimensions:

$$v^* = \frac{v}{U} \qquad H^* = \frac{H}{h} \text{ avec } h = \frac{M}{a^3}$$

$$p^* = \frac{p}{\rho_0 U^2} \quad r^* = \frac{r}{a}$$

les équations (29) s'écrivent:

$$v^* = K(\operatorname{grad} p^* - \beta\operatorname{rot} H^* \times H^*)$$

$$\operatorname{rot} H^* = R_m v^* \times H^*$$

(30)
$$\operatorname{div} v^* = 0$$

$$\operatorname{div} H^* = 0$$

avec

$$K = -\frac{kU}{ag}$$

$$u = \frac{\mu h^2}{4\pi\rho_0 U^2}$$

et

$$R_m = \frac{4\pi\sigma\mu Ua}{m}.$$

R_m est le nombre de Reynolds magnétique. Il faut remarquer que dans ce cas R_m contient le facteur sans dimensions $1/m$ qui caractérise le milieu poreux. Pour $m = 1$, c'est-à-dire en absence de milieu poreux, le fluide étant libre, R_m devient le nombre de Reynolds classique.

Pour la résolution du problème nous pouvons adopter le même procédé que dans le cas de la magnétohydrodynamique classique [2], c'est-à-dire nous chercherons la solution de l'équation (30) sous forme de dévelopements:

$$v^* = v_0(r,\theta) + \beta v_1(r,\theta) + \cdots$$
(31)
$$p^* = p_0(r,\theta) + \beta p_1(r,\theta) + \cdots$$
$$H^* = H_0(r,\theta) + \cdots$$

En portant (31) dans (30), on obtient:

$$\operatorname{rot} H_0 = R_m v_0 \times H_0$$
(32)
$$\operatorname{div} H_0 = 0$$

$$v_1 = K(\operatorname{grad} p_1 - \operatorname{rot} H_0 \times H_0)$$
(33)
$$\operatorname{div} v_1 = 0.$$

a) Calcul du champ magnétique.

Connaissant v_0, solution du problème du mouvement en absence de champ magnétique, il est possible de déterminer $H_0(r,\theta)$ à partir de (32). v_0 a l'expression [3]:

$$v_0 = K\left\{e_1\left(1 - \frac{1}{r^3}\right)\cos\theta - e_2\left(1 + \frac{1}{2r^3}\right)\sin\theta\right\}.$$

Nous avons donc un nouveau paramètre sans dimensions $\delta = R_m K$. A défaut d'une solution exacte, on peut chercher une solution développable en puissances croissantes de δ.

La seconde équation (32) est vérifiée par la fonction $A(r,\theta)$ définie par:

$$H_{or} = \frac{1}{r^2\sin\theta}\frac{\partial A}{\partial\theta} \qquad H_{0\theta} = -\frac{1}{r\sin\theta}\frac{\partial A}{\partial r}.$$

La première équation (32) devient donc:

$$(34)\quad \frac{\partial^2 A}{\partial r^2} + \frac{\sin\theta}{r^2}\frac{\partial}{\partial\theta}\left(\frac{1}{\sin\theta}\frac{\partial A}{\partial\theta}\right) = \delta\left\{\left(1 - \frac{1}{r^3}\right)\cos\theta\frac{\partial A}{\partial r} - \frac{1}{r}\left(1 + \frac{1}{2r^3}\right)\sin\theta\frac{\partial A}{\partial\theta}\right\}.$$

La seule différence entre (34) et l'équation du cas classique est qu'ici δ remplace R_m. A l'aide d'un procédé analogue à celui du cas de la magnéto-hydrodynamique classique et en imposant les conditions de continuité pour la composante normale de l'induction magnétique et pour la composante tangentielle du champ magnétique, nous avons la solution:

$$A_e = \exp\left\{-\delta r\frac{1-\cos\theta}{2}\right\}\left\{\frac{a_1}{2}\sin\theta + \delta\left(\frac{a_1}{2} + \frac{b_2\cos\theta}{r}\right)\sin^2\theta\right\}$$

$$A_i = \frac{\sin^2\theta}{r} + c_2 r^2\sin^2\theta + c_3 r^3\sin^2\theta\cos\theta$$

avec:

$$a_1 = \frac{3\mu^1}{2\mu + \mu^1} \qquad\qquad b_2 = -\frac{9\mu\mu^1}{2(2\mu + \mu^1)(3\mu + 2\mu^1)}$$

$$c_2 = \frac{\mu - \mu^1}{2\mu + \mu^1} \qquad\qquad c_3 = \frac{3\mu\mu^1\delta}{(2\mu + \mu^1)(3\mu + 2\mu^1)}.$$

b) Calcul de la vitesse.

La seconde équation (33) nous indique qu'il existe une fonction $\Psi(r,\theta)$, telle que:

$$(35)\qquad\qquad v_{1r} = \frac{1}{r^2\sin\theta}\frac{\partial\Psi}{\partial\theta} \qquad v_{1\theta} = -\frac{1}{r\sin\theta}\frac{\partial\Psi}{\partial r}.$$

La première équation (33) nous donne:

$$\text{rot}\,v_1 = -K\,\text{rot}(\text{rot}\,H_0 \times H_0)$$

donc nous connaissons $\omega = \text{rot}\,v$. Pour Ψ nous avons l'équation:

$$(36)\qquad\qquad \frac{\partial^2\Psi}{\partial r^2} + \frac{\sin\theta}{r^2}\frac{\partial}{\partial\theta}\left(\frac{1}{\sin\theta}\frac{\partial\Psi}{\partial\theta}\right) = r\sin\theta\omega.$$

Cette équation est formellement analogue à celle du cas de la magnétohydro-dynamique classique. Donc nous pouvons développer en série de polynômes $P_n'(\cos\theta)$:

$$\omega = \sum_{n=1}^{\infty}\omega_n(r)P_n'(\cos\theta)$$

$$\Psi = \sum_{n=1}^{\infty}\Psi_n(r)P_n'(\cos\theta).$$

Nous tirons de (36):

$$\Psi_n'' - \frac{n(n+1)}{r^2}\Psi_n = -r\omega_n$$

On peut donc calculer v_{1r} et $v_{1\theta}$ à l'aide de (35).

BIBLIOGRAPHIE

1. BURDAC, N. M. (1953), Trudi Mosc. energ. instit. nr. 14, 144 (en russe).

2. CABANNES, H. (1965), Introduction à la magnétodynamique des fluides, Cours à la Faculté des Sciences de Paris, Paris.

3. GHEORGHITA, ST. I. (1967) *Methode matematice in hidrogazodinamica subterana*, Ed. Acad. Bucarest, (en roumain).

4. NETUSIL, A. V. et K. M. POLIVANOV (1953), *Dok. Akad. Nauk*, *89*, 5, 845 (en russe).

INSTITUT DE MATHÉMATIQUES DE L'ACADÉMIE DE LA R.S. ROUMANIE,
RUE M. EMINESCU 47,
BUCAREST 3, ROUMANIA

TRANSFER PROPERTIES AND FRICTION COEFFICIENTS FOR SALT AND WATER FLOW THROUGH CLAYS

A. Banin

ABSTRACT

The friction model was applied to clay systems in which steady flow of salt and water, in response to salt concentration gradients, was taking place. The necessary equations were presented, and a detailed set of data, from flow experiments in Na-montmorillonite was used to calculate the friction coefficients of the moving species with the clay surface. The magnitude and mode of variation of the friction coefficients was correlated with the extent of electrical double layer interaction in the pores of the clay plug, and was shown to affect the over all transfer properties of the system.

INTRODUCTION

The process of salt and water flow through clay and soil plugs was extensively studied recently [1, 7, 8, 11]. It was found that this process is phenomenologically well described by the thermodynamic theory of irreversible processes in its macroscopic approach [1, 11]. According to this approach the overall flows are expressed as a function of the over-all forces operating on the system, using a set of mobility or conductence coefficients (L_{ij}'s).

However, a microcsopic approach is also possible. According to it, the local forces inside the plug are expressed as a function of local velocities and frictional coefficients [6, 10].

The microscopic approach was not applied yet to soil or clay systems mainly due to lack of the experimental information needed. In this article the results of a comprehensive experimental study of salt and water flow through clay plugs, reported in detail elsewhere [2, 8], are used to calculate the friction coefficients of the moving species and the physical meaning of these coefficients is related to the plug properties.

THE FRICTION MODEL

Consider a clay plug separating two aqueous solutions having different concentrations of a mono-monovalent salt. Water and ions flow through the clay, driven by the free energy gradients. At the steady state the thermodynamic force, acting on each species, is counterbalanced by several frictional hydrodynamic forces arising from the interaction of the moving particle with its surroundings [3, 6, 10]. This can be written as follows:

(1)
$$-\frac{dF_i}{dx} = \sum_{k=1}^{j} - G_{ik}$$

where dF_i/dx is the gradient of the partial moalr free energy in the direction of flow and G is a counteracting frictional force.

Every frictional force is given by

(2)
$$G_{ik} = -f_{ik}(V_i - V_k)$$

where f is the friction coefficient and V the velocity of the species. On the other hand

(3)
$$J_i = c_i V_i$$

where J is the flow of the species and c its local concentration. Introducing from (3) and (2) into (1) we get

(4)
$$-\frac{dF_i}{dx} = \sum_{k=1}^{j} f_{ik} \left(\frac{J_i}{c_i} - \frac{J_k}{c_k} \right)$$

Principally, Eq. (4) can be used to calculate the friction coefficients if dF_i/dx, c and J are measured experimentally. This usually needs, for a system like the one discussed here, at least six independent measurements, some of them rather complicated technically. However, in the simplest case, when the electrical current is zero, and therefore

(5)
$$J_1 = J_2 = J_s$$

a smaller and less involved set of measurements is sufficient [6, 10]. (Subscripts 1, 2 and s denote the cation, anion and salt, respectively). This case is to be discussed here.

By writing a set of equations similar to (4) for this simple case, and by using Einstein's equation for the relationship between the friction coefficient and the diffusion coefficient $\left(f_{io} = \dfrac{RT}{D_i} \right)$, the following set of equations can be developed for the calculation of the friction coefficients:

(6)
$$f_{12} = \frac{RT}{2} \left[\frac{1}{c_1 D_1} + \frac{1}{c_2 D_2} - \frac{1}{c_s D_s} \right] c_2$$

(7)
$$f_{21} = \frac{c_1}{c_2} f_{12}$$

$$(8) \qquad f_{1o} = \frac{c_0}{J_o}\left[J_s\left(\frac{RT}{D_1} - \frac{f_{12}}{c_2}\right) + \frac{dF_1}{dx}\right]$$

$$(9) \qquad f_{2o} = \frac{c_0}{J_o}\left[J_s\left(\frac{RT}{D_2} - \frac{f_{21}}{c_1}\right) + \frac{dF_2}{dx}\right]$$

$$(10) \qquad f_{1m} = \frac{RT}{D_1} - f_{12} - f_{1o}$$

$$(11) \qquad f_{2m} = \frac{RT}{D_2} - f_{21} - f_{2o}$$

$$(12) \qquad f_{sm} = f_{1m} + f_{2m}$$

where R denotes the gas constant, T the absolute temperature, D the diffusion coefficient and subscripts o and m, the water and clay surface, respectively.

EXPERIMENTAL

The experimental set-up consisted of a 14 cm long clay-water-salt plug confined in a Teflon tube by two porous end plates. Salt concentration gradient was maintained across the plug. The clay was montmorillonite (Wyoming Bentonite) in the Na state and the salt was NaCl. In five equally spaced locations along the plug various transducers were inserted. When the system reached the steady state the following measurements were taken: Across the plug, the water flow, J_o, and salt flow, J_s. In each location: the sodium and chloride ion "activities" (a_1 and a_2, respectively) and the salt activity (a_s); the swelling pressure (P_w) and the hydrostatic pressure in the equilibrium solution (P_s); the salt concentration inside the clay paste (c'_s) and in the equilibrium solution (c''_s), and the clay content. Clay samples were then taken from each location and self-diffusion coefficients of the sodium ion (D_1^*) and the chloride ion (D_2^*) were measured. Finally, the hydraulic conductivity of the clay was measured, in separate experiments, on plugs having a similar range of clay and salt solution concentrations. A detailed description of the methods used is given elsewhere [2, 8].

From the measured activities and pressures, the relative partial molar free energies (F_i) of the salt and water in the various locations were calculated. The thermodinamic driving forces $-dF_i/dx$, in each location were found from graphs of the relative partial molar free energy versus the distance along the plug, x being considered positive in the direction from the concentrated to the dilute end solution. The salt self-diffusion coefficient, D_s^*, was calculated from the ionic self-diffusion coefficients as described previously [2, 8].

RESULTS AND DISCUSSION

The experiment was repeated using twice two clay plugs. A collection of the data from one experiment (II) is presented in Table 1. In general, they replicated the results of the other plug (I). One noticeable exception was found, however, in the salt flow, which was about four times larger in plug II than in plug I. $[J_s(I) = 0.342 \times 10^{-11}, J_s(II) = 1.278 \times 10^{-}e^{11}$ mole.cm^{-2}. sec$^{-1}]$.

Using the frictional model equations (6–12) and the experimental data, the various friction coefficients were calculated. Out of the large set of friction coefficients we presnt in Fig. 1 the f_{im}'s, i.e., the frricon oefficients of the ions andthe salt with the clay surface ("membrane matix"), as they vary along the plug.

The coefficients are of the same order of magnitude, or one order of magnitude different from those found for the over-all friction coefficients of non-electrolyte solutes flowing thrhough dialysis tubing and wet-gel membranes [5].

Considering the variation along the plugs we find that in plug I the friction coefficient of the sodium ion attains larger positive values and that of the chloride ion attains larger negative values in the direction from the concentrated salt solution to the dilute salt solution. The salt friction coefficient fol ows mainly the chloride friction coefficient. In plug II, much smaller variation in the friction coefficients was observed.

The two plugs were intended to be replicates having similar properties. The clay type, the salt concentration-gradient and the general procedures were the same for both. However, we noted already that their macroscopic salt flow differed and we note now that microscopically, the friction coefficients differ too, showing more interaction between the salt and the surface for the plug with the smaller salt flow (I). These two observations are in agreement and may result from the same basic difference between the two plugs. In the following we will attempt to analyse the nature of this difference.

The clay surface area is negatively charged, and a diffuse layer (e.d.l.) of cations is developed adjacent to it. In a pore, limited by two parallel clay plates, there is overlapping of the opposing e.d.l. which form a space in which higher elecurostatic interaction of moving ions will occur. The average amount of e.d.l. overlaping in a given clay-water paste depends on the double layer thickness and on the plate separation.

The e.d.l. thickness, D, in Å, can be estimated using Schofield's equation [9] for the depth of anion exclusion near charged surfaces

$$(13) \qquad D = \frac{q \cdot 10^8}{\sqrt{\beta z n}} - \frac{4 \cdot 10^8}{\beta z \sigma}$$

where q is a factor depending on the cation : anion valency ratio (2 for NaCl); $\beta = 8\pi F^2/\varepsilon RT = 1.06 \times 10^{-15}$ cm/meq for water at 25°C, z is the cationic

TABLE I

Properties of clay plug (II) in the steady state of a salt and water flow process.

Flows (mole . cm^{-2} . sec^{-1}): $J_o = -6.17 \times 10^{-9}$

$J_s = 1.278 \times 10^{-11}$

Location No.	Forces		Clay content	Local salt concentration c'_s	Na ion concentration c_1	Self Diffusion Coefficients			Hydraulic conductivity k
	$-d\bar{F}_o/dx$	$d\bar{F}_s/dx$				D_1^*	D_2^*	D_s^*	
	dyne . mole^{-1}		g per. 100 g paste	mole . cm^{-3}	mole . cm^{-3}	cm^2 . sec^{-1}			cm^2 . sec^{-1} milibar^{-1}
Conc.				0.0500×10^{-3}					
1	-3.16×10^6	4.16×10^9	25.82	0.0234	0.381×10^{-3}	1.65×10^{-6}	4.53×10^{-6}	4.16×10^{-6}	1.59×10^{-9}
2	-2.71	4.00	26.97	0.0195	0.399	1.57	4.33	3.99	1.18
3	-2.72	6.27	28.27	0.0140	0.419	1.48	4.06	3.81	0.95
4	-2.93	9.36	27.15	0.0095	0.392	1.56	4.29	3.96	1.05
5	-2.95	18.04	23.88	0.0052	0.330	1.80	5.00	4.50	1.91
Dil.				0.0001					

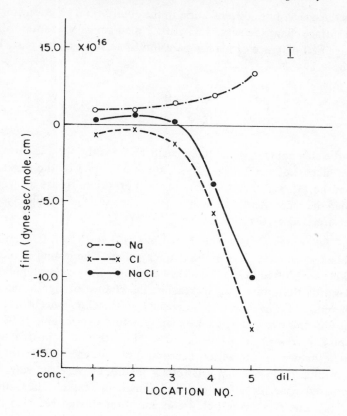

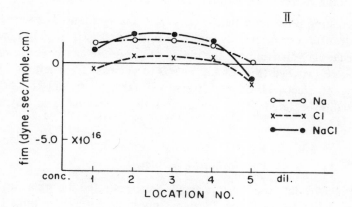

Fig. 1. Variation with location along the plug, of the friction coefficients of the ions and the salt with the surface, during the steady-state of a salt and water flow process through clay plugs. [I and II denote two different plugs].

valency, n is the electrolyte concentration in the equilibrium solution, meq/cm^3 and σ is the surface charge density, 1.5×10^{-7} meq/cm^2 for montmorillonite.

The average half distance of plate separation, d in Å, in a clay paste can be calculated by

$$(14) \qquad\qquad d = \frac{100 - W}{W \cdot S \cdot 10^{-4}}$$

where W is the clay content, g per 100 g paste, and S is the specific surface area of the clay, taken to be 560 m^2/g. This value was found by the negative adsorption method [4], and therefore it is more appropriate to use it in our calculations than the theoretical area of 800 m^2/g. The extent of double layer overlap is then estimated by the difference $d - D$.

Using the experimental data and Equations (13) and (14), D, d and $d - D$ were calculated at each location in the two clay plugs. The results of this series of calculations are plotted in Fig. 2. In the two plugs, D increases regularly and similarly towards the dilute end, increasing the chance of significant double layer overlapping. However, d, which depends on the clay concentration, varies differently in the two plugs. Consequently actual overlapping $[(d - D) < 0]$ occurs in plug I between locations 3 and 4. Note that it is at this location that the great change in interaction between ions and the surface was found (Fig. 1, I). In plug II no overlap was found, mainly because the clay concentration decreased near the dilute end and thus d increased. Thus, although only slight differences in overall clay concentration existed between the two plugs, their salt transfer properties differed considerably, because of a critical difference in the charge distribution in the pores of the clay plug.

One interesting possibility that results from the mechanism described here relates to transfer coefficients asymmetry. For example, if in plug I, with its given internal clay distribution, the places of the dilute and concentrated salt solutions were switched (Fig. 2), causing D to decrease from left to right, no overlapping of electrical double layers will take place and the salt flow would possibly be larger. The clay plug is thus behaving as an asymmetric membrane.

ACKNOWLEDGMENT

The experimental data used in this article were collected during a stay in the laboratory of Prof. P. F. Low, Purdue University, Lafayette, Ind., U.S.A.; the author gratefully acknowledges the valuable discussions and guidance offered by Prof. Low. Thanks are due also to Prof. O. Kedem for helpful discussions.

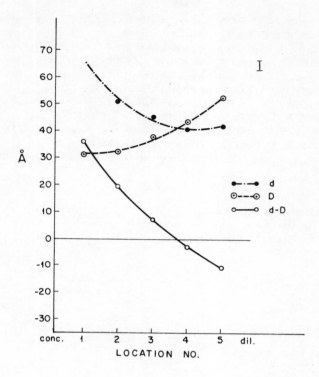

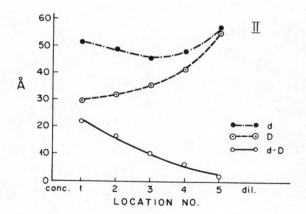

Fig. 2. Electrical double layer thickness ("Depth of anion exclusion", D), average half distance of plate separation (d) and estimated extent of double layer overlap ($d-D$) along clay plugs, during the steady state of a salt and water flow process.

REFERENCES

1. ABD-el-AZIZ, M. H. AND S. A. TAYLOR. (1965). *Soil. Sci. Soc. Amer. Proc.*, *29*, 141.
2. BANIN, A. AND P. F. LOW. *Soil. Sci.*, in press.
3. DORST, W., A. J. STAVERMAN AND R. CARAMAZZA. (1964) *Rec. Trav. Chim. 83*, 1329.
4. EDWARDS, D. G. AND J. P. QUIRK. (1962) *J. Colloid Sci. 17*, 872.
5. GINZBURG, B. Z. AND A. KATCHALSKY. (1963) *J. Gen. Physiol.*, *47*, 403.
6. KEDEM, O. AND A. KATCHALSKY. (1961) *J. Gen. Physiol. 45*, 143.
7. KEMPER, W. D. AND J. B. ROLLINS, (1966) *Soil Sci. Soc. Amer. Proc. 30*, 529.
8. MOKADY, R. S. AND P. F. LOW. (1968) *Soil Sci.*, *105*, 112.
9. SCHOFIELD, R. K. AND O. TALIBUDDIN. (1948) *Disc. Farady Soc.*, *3*, 51.
10. SPIEGLER, K. S. (1958) *Trans. Farady Soc.*, *54*, 1408.
11. TAYLOR, S. A. AND J. W. CARY. (1960) *Int. Congr. Soil Scï., Trans.*, 7th (Madison) *1*, 80.

DEPARTMENT OF SOIL SCIENCE,
   THE HEBREW UNIVERSITY OF JERUSALEM,
      P.O.B. 12, REHOVOT, ISRAEL

ON STABILIZATION OF FINGERS
IN A SLIGHTLY CRACKED
HETEROGENEOUS POROUS MEDIUM

A. P. VERMA

ABSTRACT

The stabilization problem of fingers in a specific oil-water displacement process has been examined from a statistical viewpoint for a slightly cracked heterogeneous porous medium. A perturbation procedure has been employed for the analytical solution of the equation of motion, and experimental results of Evgenev, Mattax and Kyte, Bokserman et al. are taken into consideration. It is shown that the perturbation solution does produce "stable fingers" in at least one special case corresponding to the investigated problem.

1. INTRODUCTION

The growth of fingers in displacement processes for homogeneous porous media was statistically examined by Scheidegger and Johnson (S and J) [1], and they found that no stabilization of fingers is possible in the statistical theory. Bokserman, Zheltov and Kocheshkov (BZK) [2] have recently discussed the physics of oil–water motion in a cracked porous medium. Some special problems with additional physical phenomena have been investigated by the author [3, 4, 5]. The present paper discusses the stabilization problem for a well developed finger flow in a slightly cracked heterogeneous porous medium from the S and J viewpoint, and using BZK description of a cracked medium. The finger flow is furnished by water displacing oil from an underground formation. It is shown, by using a perturbation technique, that the stabilization of fingers may occur in this particular case.

Thus we have shown that, by changing the basic assumptions of S and J theory, it is possible to find one case in which "stable fingers" may be produced.

2. STATEMENT OF PROBLEM

Water is injected with constant velocity V into an underground oil saturated seam which consists of a slightly cracked heterogeneous porous medium. The displacement of oil by water gives rise to a well developed finger flow. It is assumed that the entire oil on the initial boundary, $x = 0$ (x is measured in the direction of displacement) is displaced through a small distance due to the impact of the injecting water. For definiteness, the laws of variation in the characteristics of the medium are taken in standard forms (Section 4) (as in [3, 5]).

The particular interest of the present investigation is to determine the stabilization of fingers under the special conditions of this problem

3. STATISTICS OF FINGERS, AND CRACKED MEDIUM

In the statistical treatment of fingers [1], only the average cross-sectional area occupied by the fingers is taken into consideration, their individual size and shape are disregarded (see Fig. 1). With the introduction of the notion of fictitious relative permeability [1], this treatment of fingers becomes formally identical to the Buckley-Leverett description of immiscible fluids flow in porous media. The saturation of ith fluid (S_1) is then defined as the average cross sectional area occupied by the ith fluid at the level x i.e., $S_1 = S_1(x, t)$. Thus the saturation of the displacing fluid in the porous medium represents the average cross sectional area occupied by fingers.

The following relationship suggested by S and J has been considered.

$$(3.1) \qquad \tilde{K}_\omega = S_\omega, \quad \tilde{K}_0 = S_0 = 1 - S_\omega,$$

where $\tilde{K}_\omega$ and $\tilde{K}_0$ denote the fictitious relative permeabilities of water and oil, while S_ω, S_0 are their saturations.

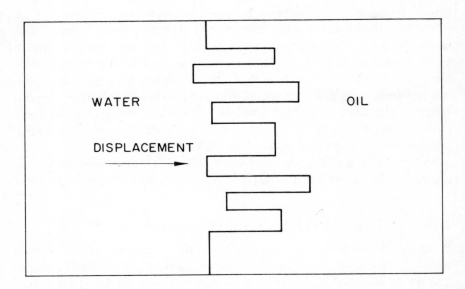

Fig. 1. Schematic representation of fingers at level "x"

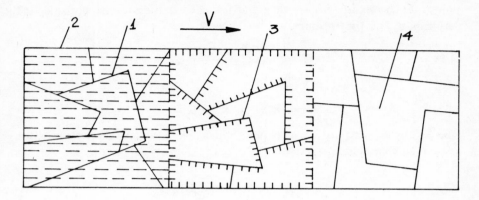

Fig. 2. Diagram showing the impregnation of a cracked one dimensional porous seam with water. The numbers denote:
1. Cracks   2. Completely impregnated blocks   3. Blocks being impregnated
4. Non-impregnated blocks

For flow in a cracked porous medium [3] (see Fig. 2), the volume of water entering the blocks in an elementary volume of seam (under the capillary action) is called the Impregnation function $\Phi(t)$, which is defined by Mattax and Kyte [6] (also, by Vezirov and Kocheshkov [7]) as below

$$(3.2) \qquad \Phi(t) = D(\varepsilon t)^{-\frac{1}{2}}$$

where D and ε, depending on the nature of cracked medium, are constants (values are given in [3]); while t denotes the time. Ryzhik (as in [2]) has pointed out that the equations of continuity for flowing phases in cracked porous medium include an additional term, the capillary suction function $\Phi\{T - \tau(\xi)\}$ which is defined as

$$\Phi\{T - \tau(\xi)\} = D[T - \tau(\xi)]^{-\frac{1}{2}}$$

(3.3)

$$T = \varepsilon t$$

$$\tau(\xi) = \bar{a}\xi^2 \quad (\bar{a} \text{ is constant})$$

(3.3)

$$\xi = \frac{x}{l} \quad (l \text{ is the mean blocksize})$$

Since we are considering a slightly cracked porous medium therefore, the capillary suction function is assumed small in the discussion.

4. LAWS OF VARIATION IN THE CHARACTERISTICS OF POROUS MEDIUM

Folowing Oroveanu [8], we take the laws of variation in the porosity and permeability of the medium as

$$(4.1) \qquad m = m(x) = \frac{1}{a - bx},$$

and,

$$(4.2) \qquad K = K(x) = K_0(1 + a_1 x),$$

where a, b, k_0 and a_1 are constants. Since $m(x)$ cannot exceed unity, we assume further that $a - bx \geqq 1$ or, $x \leqq (a-1)/b$.

5. FUNDAMENTAL EQUATIONS

The seepage velocity of water (v_w) and oil (v_0) may be written, from Darcy's law, as

$$(5.1) \qquad v_w = - \frac{\tilde{K}_w}{\mu_w} K \frac{\partial p}{\partial x}$$

$$(5.2) \qquad v_0 = - \frac{\tilde{K}_0}{\mu_0} K \frac{\partial p}{\partial x}$$

where $K = K(x)$ is the variable permeability of the cracked medium, p is the common pressure of the phases (capillary pressure is neglected), and μ_w, μ_0 are the constant viscosities of water and oil respectively.

Following Ryzhik (as in [2]), the equations of continuity for flowing phases (phase densities regarded constant) are written as

$$(5.3) \qquad m \frac{\partial S_\omega}{\partial t} + \frac{\partial v_\omega}{\partial x} + \Phi[T - \tau(\xi)] = 0$$

$$(5.4) \qquad m \frac{\partial S_0}{\partial t} + \frac{\partial v_0}{\partial x} - \Phi[T - \tau(\xi)] = 0$$

where $m = m(x)$ is the variable porosity of the medium and remaining symbols are already defined in Section 3.

Lastly, the definition of phase saturation [9] gives

$$(5.5) \qquad S_\omega + S_0 = 1$$

6. EQUATION OF MOTION FOR SATURATION

Putting the value of v_ω and v_0 from equations (5.1) and (5.2) in (5.3) and (5.4) respectively, we get

$$(6.1) \qquad \frac{\partial}{\partial x}\left(\frac{\tilde{K}_\omega}{\mu_\omega} K \frac{\partial P}{\partial x}\right) = m \frac{\partial S_\omega}{\partial t} + \Phi\{T - \tau(\xi)\}$$

(6.2)
$$\frac{\partial}{\partial x}\left(\frac{\tilde{K}_0}{\mu_0}K\frac{\partial P}{\partial x}\right) = m\frac{\partial S_0}{\partial t} - \Phi\{T - \tau(\xi)\} .$$

Combining equations (6.1) and (6.2), we get

(6.3)
$$\frac{\partial}{\partial x}\left\{\left(\frac{\tilde{K}_\omega}{\mu_\omega} + \frac{\tilde{K}_0}{\mu_0}\right)K\frac{\partial p}{\partial x}\right\} = 0 .$$

Integrating equation (6.3) with respect to x, and evaluating the constant of integration by using the condition

(6.4)
$$v_\omega(0,t) = V, \quad \text{and} \quad \tilde{K}_0(0,t) = 0$$

(the later condition becomes obvious when we note that the entire oil from the initial boundary of the seam is displaced for all time by the injected water) we get

(6.5)
$$\frac{\partial P}{\partial x} = -\frac{V}{\dfrac{\tilde{K}_\omega}{\mu_0}K + \dfrac{\tilde{K}_0}{\mu_0}K} .$$

From equations (6.5) and (6.1), we get

(6.6)
$$V\frac{\partial}{\partial x}\left[\frac{\dfrac{\tilde{K}_\omega}{\mu_\omega}K}{\dfrac{\tilde{K}_\omega}{\mu_\omega}K + \dfrac{\tilde{K}_0}{\mu_0}K}\right] = -m(x)\frac{\partial S_\omega}{\partial t} - \Phi\{T - \tau(\xi)\} .$$

Using equation (3.1), we may write

(6.7)
$$\gamma(S_w) = \left[\frac{\dfrac{\tilde{K}_w}{\mu_w}}{\dfrac{\tilde{K}_w}{\mu_w} + \dfrac{\tilde{K}_0}{\mu_0}}\right] = \frac{P}{P - 1 + \dfrac{1}{S_w}}, \quad P = \frac{\mu_0}{\mu_w} .$$

Differentiation gives

(6.8)
$$\gamma'(S_\omega) = \frac{P}{[S_\omega(P-1) + 1]^2} .$$

Rewriting equation (6.6) with the help of equation (6.7)–(6.8), we have

(6.9)
$$m\frac{\partial S_\omega}{\partial t} + V\left\{\frac{P}{[S_\omega(P-1) + 1]^2}\right\}\frac{\partial S_\omega}{\partial x} + \Phi\{T - \tau(\xi)\} = 0.$$

This is the equation of motion for the saturation S_w.

7. SOLUTION BY THE PERTURBATION METHOD

Since the capillary suction function $\Phi\{T - \tau(\xi)\}$ is small in our problem (see remarks at the end of Section 3) therefore we use a perturbation procedure

[10] to solve equation (6.9). Thus neglecting $\Phi\{T - \tau(\xi)\}$, in equation (6.9), we have

$$(7.1) \qquad m\frac{\partial S_\omega}{\partial t} + V\left\{\frac{P}{[S_\omega(P-1)+1)^2}\right\}\frac{\partial S_\omega}{\partial x} = 0 .$$

The characteristic equations of (7.1) are

$$(7.2) \qquad \frac{dx}{dt} = \frac{P}{[S_\omega(P-1)+1]^2}\frac{V}{m(x)}$$

$$(7.3) \qquad \frac{dS_\omega}{dt} = 0 .$$

Substituting the value of $m(x)$ from equation (4.1) in equation (7.2), integrating and applying the condition

$$(7.4) \qquad x = 0, \; t = 0$$

for all saturations (since all saturations start intially from the initial boundary), we have

$$(7.5) \qquad x = \frac{a}{b}\left[1 - \exp\left\{-\frac{bPVt}{[S_\omega(P-1)+1]^2}\right\}\right] .$$

On transforming t into T, and substituting the value of $\Phi\{T - \tau(\xi)\}$ from equation (3.3), we may write equation (6.9) as

$$(7.6) \qquad \varepsilon m\frac{\partial S_\omega}{\partial T} + V\left\{\frac{P}{[S_\omega(P-1)+1]^2}\right\}\frac{\partial S_\omega}{\partial x} + \frac{D}{\sqrt{T - Rx^2}} = 0 .$$

Recently, Evgenev [11] has pointed out that the value of P, in most cases of practical interest, is large, and therefore we may consider $1/P$ to be a small quantity. Keeping these remarks in the mind, and substituting the value of $T(=\varepsilon t)$ from equation (7.5) in the last term of equation (7.6), we may write equation (7.6), after simplification, as

$$(7.7) \qquad \varepsilon m\frac{\partial S_\omega}{\partial T} + V\left\{\frac{P}{[S_\omega(P-1)+1]^2}\right\}\frac{\partial S_\omega}{\partial x} + D\left(\frac{PVa}{\varepsilon}\right)^{\frac{1}{2}}\frac{x^{-\frac{1}{2}}}{S_\omega(P-1)+1} = 0.$$

The characteristic equations of this quasilinear equation are

$$(7.8) \qquad \frac{dt}{m(x)} = \frac{dx}{V\left\{\dfrac{P}{[S_\omega(P-1)+1]^2}\right\}} = \frac{dS_\omega}{D\left(\dfrac{PVa}{\varepsilon}\right)^{\frac{1}{2}}\dfrac{x^{-\frac{1}{2}}}{[S_\omega(P-1)+1]}} .$$

This may be written in the equivalent form as

$$(7.9) \qquad D\left(\frac{a}{PV\varepsilon}\right)^{\frac{1}{2}}x^{-\frac{1}{2}}dx = \frac{1}{[S_\omega(P-1)+1]}dS_\omega$$

(7.10) $$x^{-\frac{1}{2}}(a-bx)dt = \cfrac{1}{D\left(\cfrac{PVa}{\varepsilon}\right)^{\frac{1}{2}}\cfrac{1}{[S_\omega(P-1)+1]}}dS_\omega .$$

Integration of equation (7.9) gives

(7.11) $$2D\left(\frac{a}{PV\varepsilon}\right)^{\frac{1}{2}}x^{\frac{1}{2}} = \frac{1}{(P-1)}\log[S_\omega(P-1)+1] + E,$$

where E is the constant of integration.

Integrating equation (7.10) with the help of equation (7.11), we get

$$Da^{\frac{1}{2}}\left(\frac{PV}{\varepsilon}\right)^{\frac{1}{2}}t = \frac{\sqrt{PV\varepsilon}}{2D\sqrt{a(P-1)^2}}\left[1 + \frac{b}{a}\left(\frac{\sqrt{PV\varepsilon}}{2D\sqrt{a(P-1)}}\right)^2\right.$$

$$\left(3 + \frac{6E}{(P-1)^2} + \frac{3E^2}{(P-1)}\right)\right]$$

(7.12) $$\left[\frac{\{S_\omega(P-1)+1\}^2}{2}\log\{S_\omega(P-1)+1\} - \frac{1}{4}\{S_\omega(P-1)+1\}^2\right]$$

$$+ \frac{\sqrt{PV\varepsilon}}{2D\sqrt{a(P-1)}}E\left[1 + \frac{b}{a}\left(\frac{\sqrt{PV\varepsilon}}{2D\sqrt{a}}\right)^2 + E^2\right][S_\omega(P-1)+1]^2 + F,$$

where F is the constant of integration. An arbitrary functional relation between the two integrals given by equations (7.11) and (7.12) gives a solution of equation (7.7).

Since the saturation S_w is defined as the average cross sectional area occupied by the fingers (Section 3), therefore $S_w = 0$ may be regarded as a criteria for investigating their stabilization. It is observed from equations (7.11) and (7.12) that definite value of x and t correspond to the zero value of the saturation S_w, and this, in turn, implies that the stabilization of fingers is possible in the specific problem investigated.

It may be mentioned here that the conclusion depends on a perturbation procedure. Notwithstanding the difficulty in using such a procedure for studying the long term behaviour of the solution (cf. [12]), we have employed it due to the special nature of the medium, and the particular interest of the present investigation viz., showing the occurrence of the "stable" fingers in at least one case of queer permeability-heterogeneoity and capillary suction term.

8. PARTICULAR CASES

The following particular cases are discussed.

(i) *Heterogeneous medium without cracks*

For discussing this case, we put $\Phi = 0$ in equation (6.9) so that the equation of motion reduces to equation (7.1), and hence the solution may be written as

$$(8.1) \qquad x = \frac{a}{b} \left[1 - \exp\left\{ -\frac{bPVt}{[S_\omega(P-1)+1]^2} \right\} \right].$$

It follows from this equation that x tends to a definite limit a/b as time increases indefinitely. However, we cannot immediately infer the stabilization of fingers from the definiteness of a/b, because our discussion is restricted to $x \leqq (a-1)/b$ (Section 4).

(ii) *Homogeneous medium without cracks*

For discussing this case, we put $b = 0$ and $\Phi = 0$ in equation (6.9) so that the equation of motion becomes the same as that of S and J [1], and therefore no stabilization of fingers is possible.

(iii) *Slightly cracked homogeneous medium*

For discussing this case, we put $b = 0$ and $\Phi \neq 0$ in equation (7.7) so that the solution is given by equations (7.11) and (7.12) with $b = 0$, and therefore the fingers may stabilize, in this case.

9. CONCLUSION

We have shown that a perturbation solution does produce "stable" fingers in one special case corresponding to the conditions of the investigated problem. Three particular cases viz. heterogebeous medium without cracks, and homogoneous medium with or without cracks, are also discussed.

REFERENCES

1. SCHEIDEGGER, A. E. AND E. F. JOHNSON (1961), The statistical behaviour of instabilities in displacement processes in porous media. *Canadian J. Physics, 39*, 326.
2. BOKSERMAN, A. A., P Yu. ZHELTOV AND A. A. KOCHESHKOV, (1964), Motion of immiscible liquids in a cracked porous medium, *Soviet Physics Doklady, 9*, 4, 285.
3. VERMA, A. P. (1968), Motion of immiscible liquids in a cracked heterogeneous porous medium with capillary pressure, *Rev. Roum. des Sci. techn. Mec. Appli., Rumania, 13*, 2, 277.
4. VERMA, A. P. Motion of immiscible liquids in a cracked hetrogeneous porous medium with pressure dependcnt densities, *Proc. National Institute of Sciences, India, Part A 35*, 8, 458 (1969).
5. VERMA, A. P., Statistical behaviour of fingering in a displacement process in heterogeneous porous medium with capillary pressure, *Canadian J. Phys., 47*, 3, 319 (1969).
6. MATTAX, C. C. AND I. R. KYTE (1962), *Petroleum Engg. J.*, 2.
7. VEZIROV, D. SH. AND A. A. KOCHESHKOV, *Izvestiya An SSR, Mekh. i mashinost*, 6.
8. OROVEANU, T (1963), *Scurgerea Fluidelor prin medii poroase neomogene*, Editura Academiei R.P.R., Rumania, pp. 92 and 328.
9. SCHEIDEGGER, A. E. (1960), *The Physics of Flow Through Porous Media*, University of Toronto Press, p. 216.
10. MORSE, P. M. AND FESHBACK, H. (1953), *Methods of Theoretical Physics*, McGraw Hill, p. 1001.
11. EVGENEV, A. E. (1965), Phase permeabilities in flltration of a two phase system through a porous medium, *Soviet Physics Doklady, 10*, 5, 111.
12. SCHEIDEGGER, A. E. (1960, Growth of instabilities on displacement fronts in porous media, *The Physics of Fluids, 3*, 1.

DEPARTMENT OF MATHEMATICS,
FACULTY OF TECHNOLOGY AND ENGINEERING,
M.S. UNIVERSITY OF BARODA,
BARODA, INDIA

SOME PROBLEMS CONNECTED WITH THE USE OF CLASSICAL DESCRIPTION OF FLUID/FLUID DISPLACEMENT PROCESSES

WALTER D. ROSE

NOTATIONS

A	specific surface area (L^{-1})
D	transport coefficient (T)
E	energy per unit volume (M/LT^2)
f	fractional porosity
F, F_1	arbitrary saturations function
g	acceleration constant (L/T^2)
h	height of capillary rise (L)
i	subscript denoting fluid phase as N, W
k, k_i	specific, effective permeability (L^2)
M	mobility ratio
N, W	subscripts denoting nonwetting wetting, fluids
p_i	pressure in i-th fluid (M/LT^2)
Pc, Pc^*	dynamic, static capillary pressure, (M/LT^2)
q_i	flux of i-th fluid (L/T)
q_T	summed flux (cf. Eq. 2.6) (L/T)
R, R_1, R_2	interfacial radii of curvature (L)
S	subscript denoting solid phase
S_i	fractional saturation of i-th fluid
v_i	pore fluid velocity (L/T)
$x, y, z, t,$	independent spatial and time coordinates (L, T)
γ	interfacial tension (M/T^2)
μ	fluid viscosity (M/LT)
θ	contact angle
Φ	fluid potential (L^2/T^2)
$\rho, \Delta\rho$	density and density difference (M/L^3)
$\dot{\sigma}$	entropy production (L^2/T^3)

1. INTRODUCTION

A porous medium can be thought of as the composite of a two-region domain, comprised of a pore space of complicated shape and structure that both surrounds and is surrounded by the supporting solid phase skeleton. For the purpose of the analysis here, it matters little whether or not contiguous elements of the solid skeleton are linked rather than just touching, undeformable, and chemically homogeneous and unreactive; therefore, in what follows these constraints are applied for simplicity.

We start by imagining such a porous medium system to have the pore space region itself initially partitioned into more-or-less continuous sub-space domains, each occupied by one of the several immiscible fluid phases present that together completely fill the interstices. Without loss in generality, we shall limit our attention in what follows to two-phase saturated systems, where a wetting and a nonwetting fluid fill the pore space, each of which in part microscopically is bounded by the common fluid/fluid interface, and elsewhere bounded by parts of the interstitial solid surfaces that define the geometry of the pore space. And, as a further simplification, we choose to regard these immiscible fluids as imcompressible, chemically homogeneous and unreactive, and of a composition that stays fixed during the course of the ensuing displacement process.

Of course, we must provide source and sink (macroscopic) boundaries for the system space, across which displacing and displaced fluids can flow, and we shall take note of the idea that the other impermeable (macrocsopic) boundaries of the open system locate the limiting streamlines of the consequent macroscopic fields of flow.

In the usual case, that is except when very special initial and boundary conditions hold, the consequence of letting fluids flow across the (source and sink) boundaries will be to induce changes in time and space of the location and microscopic configuration of the fluid/fluid interfaces. This aspect, in fact constitutes the basic feature of a displacement process.

Thus, as the partitioning of the pore space between the immiscible wetting and nonwetting fluids changes in time and space during the displacement process, somewhere at some time (and perhaps even everywhere and continuously) the saturation of the entering fluid will increase, causing a corresponding decrease in the saturation of the displaced immiscible fluid(s).

Obviously

$$(1.1) \qquad\qquad \Sigma\, S_i = 1 \qquad i = W, N, \cdots$$

where the saturation of the i-th phase is simply the extrapolated limit (in the sense used by Hubbert, 1956) at the system time and point coordinate in question, of the ratio of the local volume of the partitioned pore space occupied by the i-th phase to the total volume of the local pore space.

The classical method of describing such displacement processes can be traced back as far as the pioneereing works of Richards (1931) and Buckley and Leverett (1942). The modern points of view, contained in what now has grown to be a voluminous literature, are adequately suggested in the monograph of Collins (1961). It is the aim in this paper to show that the theory as originally conceived, and even as applied by the later workers, in several essential respects is an unsatisfactory one resulting in indetermination.

In the next section we start by giving a more general derivation of the classical displacement equations than has previously appeared. In this way we shall expose just where arbitrary functions are being introduced as the artifice to achieve a quasi-deterministic appearance for the mathematical (differential equation) form that embodies the sense of the theory.

2. RESTATEMENT OF THE CLASSICAL THEORY

We start with the relevant mass conservation and continuity statements, such as

(2.1) $$\frac{\partial}{\partial t}(f\rho_i S_i) + \text{div}(f\rho_i S_i v_i) = 0.$$

If porosity locally does not change with time (as it might by compaction and attending rearrangement of the solid skeleton, and /or by expansion of the componenet elements with the reduction of the ambient pore fluid pressure), and if the fluids can be taken as incompressible, we have from Eq. (2.1)

(2.2) $$f\frac{\partial}{\partial t}S_i + \text{div}\, q_i = 0$$

In Eqs. (2.2), we have assumed that the macroscopic seepage flux vector, q_i, can be related to the mean value of the microscopic pore velocity, v_i, by

(2.3) $$q_i = fS_i\bar{v}_i$$

as Saffman (1959) and many others have suggested.

As the next step in the analysis (and as we shall see a doubtful one), Darcy's Law for the flow of a fluid completely filling the pore space (viz. saturated flow), is generalized so that it *appears* to apply as a description of unsaturated flow, in the form

(2.4) $$q_i = -\frac{k_i\rho_i}{\mu_i}\text{grad}\,\Phi_i \; ; \quad \Phi_i = \frac{\int dp_i}{\rho_i} + gz .$$

Hubbert (loc. cit.), for example, justifies the use of Eqs. (2.4) on the basis of the idea that the partitioned pore space occupied by each of the immiscible fluids geometrically is nothing more than another (reduced) porous medium; hence Darcy's Law should apply. The source of this idea, as it turns out, is a very old one, perhaps going back to the time of Darcy himself.

In the next section of this paper, we shall set out to show that Eqs. (2.4) should be suspected on theoretical grounds, both in their analytic form, and in the way they are commonly used to describe the unsteady displacement processes.

To use Eqs. (2.4) with Eq. (2.2), one must suppose that experiments can be done in accordance with the implied presumption of linearity between flux and force vectors, to yield the saturation-dependent relative permeability functions, or

(2.5) $$k_i/k = F_1(S_i)$$

where k is the specific permeability, that is: $k = \text{limit}_{S_i \to 1} k_i$. Then, upon introducing the notations

$$q_T = \Sigma \; q_i, \qquad P_c = p_N - p_w,$$

(2.6)

$$M = \frac{k_N}{\mu_N} \frac{\mu_m}{k_w}, \qquad \Delta\rho = \rho_w - \rho_N$$

we obtain

(2.7) $$f \frac{\partial}{\partial t} S_N + \text{div}\left[q_T \frac{M}{1 + M} \right] = \text{div} \left[\frac{-\dfrac{k_N}{\mu_N}\left(\dfrac{dp_c}{dS_w} \; \text{grad} \; S_w - \Delta\rho g \right)}{1 + M} \right]$$

$$= f \frac{\partial S_w}{\partial t} + \text{div}\left[q_T \frac{M}{1 + M} \right].$$

From Eq. (2.5), and from the third of Eqs. (2.6), we can deduce the saturation dependency for the mobility ratio parameter M, for the common case where the viscosities of the co-flowing fluids can be taken as constants

(2.8) $$M = F_2(S_i).$$

If in addition, the saturation dependence of the capillary pressure function defined by the second of Eqs. (2.6) can be established independently as

(2.9) $$P_c = F_3(S_i)$$

then Eqs. (2.7) form a complete set that can be solved, in principle, to yield solutions in the form

(2.10) $$S_i = F(x, y, z, t).$$

To verify the theory that has led to Eqs. (2.7), one performs a series of displacement experiments where the information of Eqs. (2.10), namely—saturation as a function of time and position during displacement, is observed. The comparison between what is observed and what has been predicted by obtaining the information of Eqs. (2.10) as solutions of Eqs. (2.7), gives then a measure of the consistancy of the theory. In particular, one can hope to learn if the functions, F_1, F_2 and F_3, of Eqs. (2.5), (2.8) and (2.9) uniquely exist and can be independently determined.

The displacement experiments called for in such an exercise of proof, as is well known, are extremely difficult to perform. As simplifications, we often limit attention to linear displacement where $\Delta\rho$ must be set equal to zero if flow is nonvertical, and to choose a constant rate of injection of the displacing fluid (i.e. q_T = constant). Even so, there will be many difficult problems of laboratory instrumentation to face, such as how to observe saturations simultaneously and continuously as a function of time and position (cf. Eqs. 2.10).

Indeed, it is these very difficulties of performing displacement experiments that has prompted previous workers to so laborously develop and check the theory presented in this section. The hope was that the experimental work called for in the evaluation of the saturation-dependent relative permeability and capillary pressure functions (of Eqs. 2.5, 2.8 and 2.9) would entail fewer difficulties in its execution than in the displacement experiment itself.

Thus, in the last decade, much progress has been made in the development of stable and rapidly convergent numerical solution techniques for the second-order nonlinear partial differential Eqs. (2.7). This work has yielded at least qualitatively plausible forecasts of the characteristics of immiscible displacement processes (cf. Collins, loc. cit., for some of the earlier examples of these calculations). Still no definitive laboratory work has been reported in the literature by which the quantitative significance of the theory can be judged.

And thereby hangs the tail tale! Until someone can do the experimental work necessary for the construction of a verified theory, we can only scrutinize the theory itself for insights, identifying the possible imperfections and their possible effects in the ordinary applications. This then is what we set out to do in the remaining pages of this paper.

3. THE SENSE (AND NONSENSE) OF THE F_1, F_2 AND F_3 FUNCTIONS

When the first terms of the left and right hand sides of Eqs. (2.7) are zero, there is no local displacement occuring. and The second term of the left and right hand sides of Eqs. (2.7) are in general identically zero, when q_T itself is zero (which can happen in certain counterflow situations, as well as when q_W and q_N themselves are zero). Finally the middle member of Eqs. (2.7) is equal to zero, when the capillary and gravity forces are in balance (i.e. grad Pc identically equal to $\Delta\rho g$). Thus, if the density difference between immiscible fluid is made zero, the capillary forces cease to act as soon as a constant-curvature fluid/fluid interfacial condition is achieved.

The point is that in displacement processes where wetting fluids replace non-wetting fluids (or vice versa), capillary forces cannot be ignored. This is shown, for example, by the fact first reported by Buckley and Leverett (loc. cit.) that physically implausible, multivalued solutions are obtained from Eqs. (2.7) for the saturation functions of time and position during the course of displacement, whenever the middle member of Eqs. (2.7) arbitrarily is not taken into account.

With those prefatory remarks made, we are now ready to get at the heart of the matter. We look first to see what can be said without equivocation about the relative permeability and the derived mobility ratio functions, F_1 and F_2. The capillary pressure function, F_3 ,then will be discussed separately, though for a long time formal analogies between capillary pressure and relative permeability theory have been recognized and utilized (cf. Rose and Bruce, 1949; also Rose, 1949).

The primary need for defining a relative permeability function is to facilitate the description of immiscible displacement processes that are the subject of this paper. Thus, if equations such as (2.7) are to be solvable, the functions, F_1 and F_2, given by Eqs. (2.5) and (2.8) must be independently established. This question is separate and distinct from the equally troublesome one of deciding how the capillary pressure function, F_3, is itself to be obtained (cf. for example, the further remarks of Rose, 1963).

Two main ways have been proposed for obtaining the F_1 and F_2 functions. One is by undertaking a so-called relative permeability experiment as prescribed by Eqs. (2.4), and involving well-known difficulties in laboratory procedure (cf. for example, Rose, 1951). The other is by doing the displacement experiment as prescribed by Eqs. (2.7), involving the already mentioned even greater laboratory difficulties, so that by inputing the data observed in the form of Eqs. (2.10), and upon assuming a form for F_3, the F_1 and F_2 can be calculated as the unknowns of Eqs. (2.7). In either case, some kind of presumption is being made of the idea embodied in Eqs. (2.4) that a Darcian-like law of proportionality between flux and force vectors exists, even when dealing with unsteady unsaturated flow (that is, where the S_i's are less than unity, and locally changing with time).

The proof of the reasonableness of Eqs. (2.4), must come from carefully undertaken experiments where a *uniform* saturation is maintained both in time and space over the test specimen. Then, the constant of proportionality being sought, namely the K_i's, are to be shown as explicit functions of the S_i's alone (and also implicitly of the saturation configurations because of the hysteretic effects); hence a given non-arbitrary saturation and saturation configuration condition must be maintained throughout the test specimen during the entire period of measurement. That is to rule out (if possible) that factors not appearing in Eqs. (2.4) have no independant effect on the multiphase flow process.

We may note that unlike the single phase flow where Darcy's Law sometimes does seem to apply, and therefore unlike the situation where the flowing fluid is bounded entirely by solid phase interstitial walls (where the v_1's are zero because of the microscopiaclly evident no-slip boundary condition), the immiscible fluids in undersaturated flow in part are bounded by fluid/fluid interfaces (where the v_1's are *not* zero, again because of a no-slip boundary condition). On these grounds alone one is forced to anticipate that the correct law of flow, unlike Eqs. (2.4), will show that the viscosity ratio between contiguous fluids as well as the areal extent of the fluid/fluid interface, will be factors of importance.

The difficulties being faced, however, do not only have to do with the reasonableness of Eqs. (2.4) as a proper description of *steady* multiphase flow phenomena. On the contrary, even if the possibility for transfer of momentum at fluids interfaces is ignored, the experiment to yield the relative permeability function, F_1, of Eq. (2.5) necessarily must be one where the saturations do not change with time, and where the velocity vectors of the contiguous immiscible flowing fluids are collinear. How then, we may ask, will the information obtained in such a steady-state experiment have anything to do with a displacement process where saturations not only locally change with time but exhibit spatial gradients (viz. the main features of such processes), and where in general the flux-force vectors for the adjacent fluids will not be collinear?

Indeed, it is these latter questions that have led to the so-called "Welge integration" of Eqs. (2.7), as described for example by Collins (loc. cit.). Here it is proposed, that a displacement experiment be undertaken yielding the data of Eq. (2.10), so that with an assumption about the capillary pressure function (F_3) made, so-called *dynamic* relative permeability functions, F_1 and F_2, can be deduced by solving Eqs. (2.7), (2.9) and (2.10) simultaneously. Because of the already mentioned instrumentation difficulty in scanning displacement systems in both space and time simultaneously to give the input data of Eq. (2.10), the arbitrary artifice usually employed (cf. Collin's again, loc. cit., for a description) is one where the entering fluid is chosen to have a considerably lower viscosity than the replaced fluid. What one gains thereby is a "spreading out" of the cummulative recovery versus pore volumes of injected fluid data curve, as needed to facilitate the computations (i.e. increase the resolution) of the above mentioned "Welge integration" of Eqs. (2.7).

The result of this artifice, however, is to introduce a new factor in the displacement experiment, namely the viscous fingering that characterizes the situation where the entering fluid is more mobile (e.g. considerably less viscous) than the replaced fluid. A viscous fingering, it will be recalled, describes that situation where large islands of the in situ fluid are bypassed (cf. Collins, loc. cit., for a description), in way, that clearly are not included in the development of the general theory leading, as shown in the previous section of this paper, to Eqs. (2.7).

As a rough analogy, we would say that the general theory leading to Eq. (2.7) has nothing more to do with viscous fingering phenomena, than the Navier-Stokes equations for laminar flow have to do with turbulent flow phenomena. In both cases, all one can say is that they do (at least in principle at the microscopic level), but they don't so far as usuable macroscopic representations are concerned.

The conclusion to be drawn from the above analysis is that there are overwhelming reasons to doubt the meaningfulness of Eqs. (2.4). Even if flux and force are assumed linearly proportional for *steady* unsaturated flow situations, still one does not know how or what unsteady experiment to perform in order to obtain the functions, F_1 and F_2 needed from independent measurement if the displacement Eqs. (2.7) are to be rendered determinate. From those who are not prepared to accept this conclusion and its implied indictment, we may ask for an exhibition of the laboratory data that conclusively show that the Functions F_1 and F_2 are independent of the viscosity ratio and degree of interfacial contact between contiguous immiscible fluids, independent of the degree of collinearity between the flux and force vectors of the contiguous fluids (giving rise in extreme cases to cross-flow and counter-flow displacement), and independent of the magnitude of the local saturation changes during displacement. All these factors are not even embodied in the sense of Eqs. (2.4).

But even at this negative point, the story cannot be ended, since we still must deal with the meaningfulness of the saturation-dependent capillary pressure function F_3 as given by Eq. (2.9). This last topic is an intriguing one, that reaffirms, amongst other things, that nothing substantial is gained by calling dissimilar things by the same name.

It is common practice to use for the information required from Eq. (2.9), the capillary pressure versus saturation function obtained in a laboratory experiment which measures at static equilibrium the balance between the oppositely directed gravity and capillary forces. This kind of balance (cf. Rose, 1964) is reflected by the fact that the pressure difference at the local interfaces of contact between the immiscible wetting and nonwetting fluids — after flow *ceases* — can be related both to the height of these interfaces above the reference datum of zero capillarity (say, the so-called water table), as well as to the local interfacial (geometrical) curvature, by

$$(3.1) \qquad P_c^* \equiv \lim_{q \to 0} \ (p_N - p_W) = \Delta\rho g h = \gamma \left(\frac{1}{R_1} + \frac{1}{R_2} \right) \equiv \frac{\gamma \cos \theta}{R}.$$

Accordingly, we see that at static equilibrium various fluid elements held by capillary forces above levels of minimum potential energy, internally are in tension, giving rise in turn to a stressing (hence a nonzero curvature) of the fluid/fluid interfaces. The necessary consequence is the appearance of a local capillary pressure difference across these interfaces for which, with pointed caution, we adopt the special notation, Pc^*, so that it will be distinguished from the capillary pressure term, Pc, defined by Eqs. (2.4) for the dynamic situations through the third of Eqs. (2.6).

Now, it is an undeniable fact that experiments can be done to observe the end-points of those spontaneous processes, such as an imbibition and/or drainage of wetting fluids in porous solids, where finally the sum of the free surface energy (this minimized as the solid surfaces become covered by the wetting fluid), plus the potential energy (this minimized as the denser of the immiscible fluids falls to the lowest possible levels in the system) itself is minimized, according to

$$(3.2) \qquad\qquad dE \equiv \gamma[\cos\theta \, dA_{sw} + dA_{wN}] - \Delta\rho h f dS_w \equiv 0.$$

In arriving at Eq. (3.2), use has been made of Young's well-known equilibrium statement, that equates the difference between the surface energy of solids when covered by a nonwetting fluid, and that of the same solids when covered by the conjugate wetting fluid to the interfacial tension between wetting and non-wetting fluids, γ, multiplied by the cosine of the angle of contact made by the fluid/fluid interfaces at the various solid phase junction points. Combining Eqs. (3.1) and (3.2), yields an interesting result (the first stated loosely by Leverett, 1941)

$$(3.3) \qquad \gamma[\cos \theta_{SW} A_{SW} + A_{WN}] = f \int P_c^* dS_w.$$

Thus we see that the result of the so-called capillary pressure experiment described above, is the establishment of an interlocked saturation dependency between Pc^*, the interfacial curvature, the associated areal extent of the interfacial curvature, the associated areal extent of the interfacial contacts, and the height above the reference datum where particular values of S_i's are observed. In implicit form this relationship may be expressed by:

$$(3.4) \qquad P_c^* = F_4(S_w).$$

Now it is clear, that in view of the multiplicity in hysteric possibilities, the function, F_4, may take a variety of particular forms, each reflecting not only the geometrical properties of the partitioned pore space, the fluid properties, and the surface energy properties of the fluid-solid interactions, but also reflecting the past history of imposed and constrained saturation changes.

To suppose, however, that the information of Eq. (3.4) as obtained by a capillary pressure experiment, is *equivalent* to that expressed by Eq. (2.9), would only be done as a convenience, or by confusion (that is, because Pc incorrectly was being identified as equivalent to Pc^*)! But this is what is done, by taking for the unknown but needed function F_3, the known function F_4.

But the Pc terms appearing in Eqs. (2.6), (2.7) and (2.9), unlike Pc^* itself, are not defined in terms of actual local pressure differences across actual interfaces, and of local curvatures of particular interfaces. On the contrary, the pressure difference implied by the definition of Pc, involves regional (extrapolated limit) volumetric averages of the pressures in the contiguous phases. In the dynamic situations that occur during displacement, the local fluid/fluid interfacial curvatures necessarily depend upon hydrodynamic, as well as gravity and capillary forces (cf. Rose, 1961; also Rose and Heins, 1962 for illustrations of quantitative details). Moreover, the influence of past events clearly will give rise to the establishment of *different* hysteretic aspects in the F_3 as compared to the F_4 functions.

Indeed, when one considers the extreme cases of cross and counterflow of the immiscible fluids possible during ordinary displacement processes, it is all the more hard to visualize the existance of a unique F_3 function (and harder still to propose how it should be measured!), even though when displacement ceases, a well-defined F_4 function always can be established and measured.

In fact, one might conclude that the use of F_4 for F_3 only is justified — and that in principle — at the later stages of the displacement process, the q_i's and/or the attending saturation changes aproach zero (hence, only when the associated hydrodynamic forces become negligible compared to the static interaction of the prevailing capillary and gravity forces). Even then, one would have considerable difficulty in choosing the relevant dynamic processes by which the static equilibrium F_4 function should be obtained. Even if this was properly achieved. one would still only have the possibility of accounting for the last (and therefore uninteresting) part of the displacement process being studied by Eqs. (2.7), being forced *inter aiia* to leave the initial and critical stages of displacement undescribed.

4. CONCLUSION

To recapitulate, the classical theory embodied in Eqs. (2.7), takes the form (when q_T is the imposed constant boundary condition) of

$$(4.1) \quad f\frac{\partial S_w}{\partial t} + q_T \operatorname{div}\left[\frac{F_2}{1+F_2}\right] = \operatorname{div}\left[\frac{\left(\frac{F_1}{\mu_N}\right)\{F_3\operatorname{grad}S_N - \Delta\rho g\}}{1+F_2}\right]$$

where, except for the constants μ_N, $\Delta\rho g$, q_T, the dependent variable, S_N, and the functions, F_1, F_2 and F_3 of S_N, are dependent on the variables, x,y,z,t. Thus, the implication of the classical theory is that Eq. (4.1) has available integral solutions (given the initial and boundary conditions) in the form of Eq. (2.10), as needed for the usual engineering calculations that forecast the performance of particular displacement processes.

In this paper we have set out to show that the application of the classical theory of displacement processes is beset with many difficulties, specifically we have pointed to the need of having the functions, F_1, F_2 and F_3, independently available by experimental means. These functions, as has been shown, appear in the theoretical displacement equations such as (2.7) and (4.1), as the result of the arbitrary assumption that the generalized Darcian statement given by Eqs. (2.4) is appropriate for the description of the law of force acting during displacement processes.

Further, it has been pointed out that even if Eqs. (2.4) are a sensibly correct way to describe *steady* unsaturated flow, the relative permeability experiments prescribed thereby can hardly have anything to do with establishing the form of the functions, F_1 and F_2, needed in Eqs. (2.7) and (4.1) to describe *unsteady* displacement. Likewise, we see no justification for using the function, F_4, as equivalent to the function F_3, needed in Eqs. (2.7) and (4.1), since the former reflects the local static balance of capillary and gravity forces, while the latter additionally (perhaps predominantly) reflects the averaged influence of the hydrodynamic forces as well.

Accordingly, we see that the classical theory is based on mixed incompatible parts, which expose the inherent ambiguities. Obviously, other, more general, laws of force are needed in place of Eqs. (2.4), to be combined with the valid mass conservation statements of Eqs. (2.2). Only in this way, can determination be achieved by a revised and perfected theory.

From the general ideas of coupled transport processes (cf. Rose, 1965, for an abbreviated statement in the present contexts; cf. also the basic references such as DeGroot and Mazur, 1962, etc.), we would propose

$$q_W = D_{WW}\,\text{grad}\,\Phi_W + D_{WN}\text{grad}\,\Phi_N$$

(4.2) $$q_N = D_{NW}\,\text{grad}\,\Phi_W + D_{WN}\,\text{grad}\,\Phi_N$$

$$D_{WN} \equiv D_{WN}$$

where the three relevant transport coefficients, D_{WW}, D_{WN} and D_{NN}, must be independently established as functions of saturation under the unsteady flow conditions of the displacement processes of interest. At least Eqs. (4.2) through the interaction coefficient, D_{WN}, are consistent with the idea that the motion of one of the fluids will influence the motion of the other through the transfer of momentum across the fluid/fluid interfaces.

The evaluation of the saturation dependency of the transport coefficients of Eqs. (4.2) involves undertaking three independent experiments, instead of the two called for in connection with Eqs. (2.4). But since the entropy production for dissipative (i.e. irreversible) processes must be positive-definite, it is easy to show that the absolute magnitude of the interaction coefficient, $D_{WN} = D_{NW}$, is limited by the following inequality, namely

(4.3) $$D_{WN}D_{NW} \leqq D_{WW}D_{NN}\,.$$

Furthermore, it is likely that one can assume that the entropy production rate assumes a minimum value (cf. DeGroot and Mazur, loc. cit.). Thus, from Eqs. (4.2), we may write as the rate of entropy production

(4.4) $$\dot{\sigma} = D_{WW}(\text{grad}\,\Phi_W)^2 + 2D_{WN}(\text{grad}\,\Phi_W)(\text{grad}\,\Phi_N) + D_{NN}(\text{grad}\,\Phi_N)^2 \geqq 0$$

from which the inequality of Eq. (4.3) immediately follows. By setting the variation of the entropy production rate with change in saturation equal to zero, by the calculus of variations one obtains in principle an independent condition by which the three transport coefficients, D_{WW}, D_{WN} and D_{NN}, are interrelated, through the aid of Eqs. (4.3) and (4.4). One still needs to construct an appropriate energy conservation statement with respect to the Φ_i, so that this together with the mass conservation statement, Eqs. (2.2), and with the Onsager type of law of force, Eqs. (4.2), finally will yield a determinate set of equations by which the displacement processes referred to in this paper will be fully described. Rose (1965, loc. cit.) has dealt with this latter problem tentatively (cf. Rose, 1969).

Such are the difficult tasks facing future workers, first to develop a complete and self-consistent theory of displacement processes as a substitute for the unsatisfactory classical theory, and then to devise the sensible laboratory methods by which the transport coefficients (i.e. the constitutive assumption) can be put in explicit form. The contribution intended for this paper has been to focus attention on what remains to be done at least in form if not in detail. But more to the point, we have seen here another example of how no theory is complete unless it is complemented by a compatible experimental part.

5. ACKNOWLEDGEMENT

Over the years the writer has had the good fortune of being taught by many inquisitive students, mainly in the United States, but also in Germany, France, Mexico, Israel, and now Turkey. Except for their provocations he long ago surely would have embraced and accepted the classical theory rejected by this paper, even as most other contemporary workers have found it convenient to do.

REFERENCES

1. BUCKLEY, S. E. AND LEVERETT, M. C., (1942), *Trans. A.I.M.E. 146*, 117.
2. COLLINS, R. E., (1961), *Flow of Fluids through Porous Materials*, Reinhold Publishing Co.
3. DeGROOT, S. R. AND MAZUR, P., (1962), *Nonequilibrium Thermodynamics*, North-Holland Publishing Co.
4. HUBBERT, M. K., (1956), *Trans. A.I.M.E. 207*, 222.
5. LEVERETT, M. C., *Trans. A.I.M.E. 142*, 151.
6. RICHARDS, L. A., (1931), *Physics 7*, 318.
7. ROSE, W. (1949), *Trans. A.I.M.E., 186*, 111.
8. ROSE, W., (1951) *Proc. 3rd World Petroleum Congress* Sec. II, 446.
9. ROSE, W. (1961) *Nature, 191*, No. 4785, 242.
10. ROSE, W., (1963) *Revue Institut Française du Pétrole, 18*, 1571.
11. ROSE, W. (1964) *Revue Institut Française du Pétrole, 19*, 1148.
12. ROSE, W. (1965), Penn State Conference on Petroleum Production, Prof. Donahue, Chairman.
13. ROSE, W. (1969), *METU Journal of Pure and Applied Sciences, 2*, 117–132.
14. ROSE, W., AND BRUCE (1949), *Trans. A.I.M.E, 186*, 127.
15. ROSE, W., AND HEINS (1962), *Journal Colloid Science, 17*, 39.
16. SAFFMAN, A. G. (1959), *Jour. Fluid Mechanics, 6*, 321.

W. D. ROSE, VISITING PROFESSOR.
MIDDLE EAST TECHNICAL UNIVERSITY,
ANKARA; PRESENT ADDRESS, ABADAN INSTITUTE OF TECHNOLOGY, ABADAN

PERMISELECTIVE PROPERTIES
OF POROUS MATERIALS AS CALCULATED
FROM DIFFUSE DOUBLE LAYER THEORY

P. H. Groenevelt and G. H. Bolt

INTRODUCTION

A porous material is termed permiselective with respect to the components of a solution if the transport resistance exerted by the medium differs for these components. An obvious example of this is the transport of a solution, consisting of large size solute molecules dissolved in a solvent of small molecular size, through a porous medium of which the pores have the same order of magnitude as the solute molecules. Application of a pressure gradient to the solution will then favor the transport of the smaller solvent molecules in comparison to that of the larger solute molecules, which results in "holding back" the solute with respect to the solvent. Using the name "diffusive flux," j^D, for this apparent (backward) movement of the solute with respect to the (forward) movement of the solution through the porous medium, one may conclude that in a permiselective medium an applied pressure gradient gives rise to two types of fluxes, i.e. a volume flux, j^V, and j^D. This is in contrast to the non-selective medium where an applied pressure gradient induces solely a volume flux, j^V.

A similar reasoning holds for the application of a concentration gradient to the solution phase. If the solution is present in a non-selective medium (very wide pores) the diffusive forces on the solvent and the solute molecules present in one cm^3 of solution — being equal and opposite because of the Gibbs-Duhem relation — will induce equal and opposite volume fluxes of solvent and solute. The volume flux of the solution as a whole, j^V, then equals zero and one finds solely a diffusion flux, j^D, of the solute with respect to the solvent. In the permiselective medium, however, the absolute value of the volume flux of the smaller solvent molecules will exceed that of the larger solute molecules and one finds a net volume flux, j^V, in addition to a diffusion flux j^D. Permiselectivity thus leads to coupling of the two phenomena described above, i.e. the flux of the solution as a whole and the diffusive flux of solute with respect to solvent.

The above example, i.e. permiselectivity caused by geometric constraint, obviously applies only to systems with pore sizes in the Å range. If, however, the pore walls carry a surface charge, another form of permiselectivity may arise viz. permiselectivity caused by exclusion of charged solutes (i.e. ions). Because of the relatively large range of penetration of the electric field associated with the pore-wall charge, this phenomenon may be noticeable even in pores of several hundreds of Å units. Naturally, the exclusion of one type of ions (co-ions) is then always accompanied by an accumulation of the ions of opposite charge (counter-ions). Considering the total ionic composition inside the pore (with charged walls) as consisting of neutral salts plus an excess countercharge one thus finds that such a system is permiselective with respect to the passage of neutral salts as compared to the passage of the aqueous solvent. An applied pressure gradient then induces flow of a liquid phase with a varying salt concentration which decreases from its local equilibrium value far away from the charged wall (e.g. in the axis of the pore) to much smaller values close to the wall. The result is that such a medium acts as a partial "salt-sieve," the solution at the exit-side having a lower concentration than the one entering it. As in the previous case this effect implies that the applied pressure gradient induces an apparent diffusive flux, j^D, in addition to the volume flux, j^V. At the same time, the application of a concentration gradient to such a medium induces a net flow of the solution as the total diffusive force on the solute present in a unit volume falls short of the diffusive force on the solvent present in this volume, with an amount proportional to the local deficit of salt. Thus again the volume flux, j^V, and the diffusion flux, j^D, appear to be coupled in the described system.

A complication arises furthermore because of the presence of the excess countercharge. Application of a pressure gradient will also induce a flux of this countercharge — known as the streaming current — if a provision is made for a return flow of electrons e.g. by inserting shorted electrodes in the system. In the absence of such a device a "streaming potential" will build up to such an amount that the net current flow becomes zero. Conversely the application of an electric potential gradient will induce a net driving force on the solution phase whenever an excess charge is present, thus giving rise to a volume flux, j^V, in addition to the current flow, I. This phenomenon is known as electro-osmosis.

The formal connection between all these fluxes and forces has been summarized (for the region of linear flux-force relations) in the existing literature on the thermodynamics of irreversible processes (cf. Katchalsky [4], de Groot and Mazur [3]). For isothermal systems, containing a single salt, a convenient set of equations is (cf. Groenevelt and Bolt [2]):

(1a) $j^V = L_V \Delta(-P^\dagger)/\Delta x + L_{VD} \Delta(-\pi)/\Delta x + L_{VE} \Delta(-E)/\Delta x$

(1b) $j^D = L_{DV} \Delta(-P^\dagger)/\Delta x + L_D \Delta(-\pi)/\Delta x + L_{DE} \Delta(-E)/\Delta x$

(1c) $I = L_{EV}\Delta(-P^{\dagger})/\Delta x + L_{ED}\Delta(-\pi)/\Delta x + L_E\Delta(-E)/\Delta x$

in which $j^D \equiv ((j_S/\rho_S) - j^V)$, the apparent linear diffusion flux of the salt with respect to the water (found by integrating the "point-fluxes" over the cross-secton of the pore and expressed in cm^3 per cm^2 porous medium, per second), I = electric current (also integrated over the pore, expressed in $C/cm^2 sec$); $\Delta(-P^{\dagger})$, $\Delta(-\pi)$, $\Delta(-E)$ = difference between the hydraulic head expressed in pressure units (bar), the osmotic pressure (bar) and the electric potential (Volt) respectively, at the exit and entrance side of the system, having a length in the flow direction of Δx cm;

j_S = flux of salt $(gram/cm^2 sec)$

ρ_S = concentration of salt in the corresponding equilibrium solution $(gram/cm^3$ of solution).

GENERALIZED SCHEME OF CALCULATION OF THE TRANSPORT COEFFICIENTS

In a previous publication (cf. [2]) a calculation scheme was developed, based on the following approach.

a.   Introducing a position parameter ξ, indicating the distance to the charged pore wall, the concentrations of the different components may be expressed as $c = c_e \cdot u(\xi)$, in which c_e is the local equilibrium concentration in appropriate units (i.e. the concentration at large distances from the wall), and $u(\xi)$ is a "Boltzmann factor" relating the existing concentrations to their local equilibrium value.

b.   Because of the relatively wide pores, the linear velocity, v, of all components will be dependent upon ξ, so all macroscopic fluxes must be expressed in terms of integrals of the type

(2) $$j \equiv \frac{\theta}{b} c_e \int_0^b v(\xi) \cdot u(\xi) d\xi,$$

in which θ = liquid filled fractional porosity, b = thickness of the liquid films moving along the charged pore walls.

c.   The linear velocities of the components, $v(\xi)$, may all be expressed as the velocity of the solution as a whole, $v_1(\xi)$, plus "excess" velocity components (for the solutes) due to diffusion and conduction.

d.   The solution velocity, v_1, may be related to the driving force acting on the solution, F_1, via the equation for momentum balance. For flat liquid films this becomes

(3) $$\frac{d\eta(dv_1/d\xi)}{d\xi} = F_1,$$

whereas for cylindrical pores one finds

(3c)
$$\frac{dr\eta(dv_1/dr)}{rdr} = F_1.$$

The driving force on the solution comprises the terms $\Delta(-P^\dagger)/\Delta x$ (cf. Navier-Stokes equation), the local salt concentration deficit times $\Delta\mu_S/\Delta x$ (via the Gibbs-Duhem equation) and the local volume charge density times $\Delta(-E)/\Delta x$.

d_2. The "apparent diffusion velocity" of a salt with respect to the liquid phase may be expressed as the sum of a drag term — specified as the liquid velocity times the salt *deficit* — plus a diffusion term proper, plus a contribution due to co-ion transport as a result of the gradient of the electric potential.

d_3. The electric current consists of a drag term — specified as the liquid velocity times the volume charge — plus a diffusion current, plus a conductance term.

Applying the above to a system comprising an aqueous solution of a single mono-monovalent salt, with a concentration of c_S eq. per liter, moving through flat films of thickness b cm, the coefficients specified in equations (1) are found as presented in Table I, where θ/λ indicates the macroscopic reduction factor corresponding to a liquid content θ cm^3/cm^3 and a tortuosity λ. If the parameters Λ^+, Λ^-, u^+, u^-, η appearing in these expressions are given as a function of ξ, the coefficients may be calculated. Also the coefficients for systems containing a mixture of salts may be computed in this manner, although then the "macroscopic gradient" of the osmotic pressure must be replaced by separate gradients of the type $c_j\Delta(-\mu_j)/\Delta x$.

TABLE I

(4a) $\displaystyle L_V = \frac{\theta}{\lambda b} \int_0^b d\xi \int_0^\xi \frac{d\xi}{\eta} \int_\xi^b d\xi$

(b) $\displaystyle L_{VD} = \frac{-\theta}{\lambda b} \int_0^b d\xi \int_0^\xi \frac{d\xi}{\eta} \int_\xi^b (1-u^-)d\xi$

(c) $\displaystyle L_{VE} = \frac{\theta}{\lambda b} \cdot Fc_S \int_0^b d\xi \int_0^\xi \frac{d\xi}{\eta} \int_\xi^b (u^+ - u^-)d\xi$

(d) $\displaystyle L_{DV} = -\frac{\theta}{\lambda b} \int_0^b d\xi(1-u^-) \int_0^\xi \frac{d\xi}{\eta} \int_\xi^b d\xi$

(e) $\displaystyle L_D = \frac{\theta}{\lambda b}\left\{ \int_0^b d\xi(1-u^-) \int_0^\xi \frac{d\xi}{\eta} \int_\xi^b (1-u^-)d\xi + \frac{1}{c_S F} \int_0^b \Lambda^- u^- d\xi\right\}$

(f) $\displaystyle L_{DE} = -\frac{\theta}{\lambda b}\left\{ Fc_S \int_0^b d\xi(1-u^-) \int_0^\xi \frac{d\xi}{\eta} \int_\xi^b (u^+ - u^-)d\xi + \int_0^b \Lambda^- u^- d\xi\right\}$

(g) $\displaystyle L_{EV} = \frac{\theta}{\lambda b} Fc_S \int_0^b d\xi(u^+ - u^-) \int_0^\xi \frac{d\xi}{\eta} \int_\xi^b d\xi$

(h) $\quad L_{ED} = -\dfrac{\theta}{\lambda b}\left\{ Fc_S \displaystyle\int_0^b d\xi(u^+ - u^-) \displaystyle\int_0^\xi \dfrac{d\xi}{\eta} \displaystyle\int_\xi^b (1 - u^-)d\xi + \displaystyle\int_0^b \Lambda^- u^- d\xi\right\}$

(i) $\quad L_E = \dfrac{\theta}{\lambda b} Fc_S\left\{ Fc_S \displaystyle\int_0^b d\xi(u^+ - u^-) \displaystyle\int_0^\xi \dfrac{d\xi}{\eta} \displaystyle\int_\xi^b (u^+ - u^-)d\xi \right.$

$$\left. + \displaystyle\int_0^b (\Lambda^+ u^+ + \Lambda^- u^-)d\xi\right\}$$

THE SALT-SIEVING EFFECT

For a given value of the volume flux passing through a charged porous medium, the salt-sieving effect, Δs (gram/cm^{2}sec), may be defined as the deficit in salt transport through the medium in comparison to the product of the volume flux and the salt concentration of the solution entering the medium. Thus

$$\Delta s \equiv j^V \cdot \rho_{S1} - j_S$$

in which ρ_{S1} = the salt concentration at the entrance side. Expressed as a fractional deficit, this becomes

$$f_S \equiv \Delta s / j^V \cdot \rho_{S1} = (j^V - j_S/\rho_{S1})/j^V \text{ or}$$

(5)

$$f_S = -(j^D/j^V).$$

The actual value of f_S follows from Eq (1) as

(6) $\qquad f_S = -\dfrac{L_{DV}\Delta(-P^\dagger) + L_D\Delta(-\pi) + L_{DE}\Delta(-E)}{L_V\Delta(-P\dagger) + L_{VD}\Delta(-\pi) + L_{VE}\Delta(-E)},$

and thus depends on the values of $\Delta(-\pi)$ and $\Delta(-E)$ relative to $\Delta(-P^\dagger)$ in addition to its dependence on the transport coefficients L.

In the special case that $\Delta(-E)$ is made equal to zero (by the insertion of shorted electrodes at entrance and exit of the porous medium) and $\Delta(-\pi)$ is maintained at zero (e.g. by flushing entrance and exit boundaries with solutions of identical composition) f_S reverts to the reflection coefficient at zero potential gradient, σ, according to:

(6–E–π) $\qquad f_S\{\Delta(-E) = \Delta(-\pi) = 0\} \equiv \sigma = -L_{DV}/L_V.$

In the absence of electrodes the maintenance of electroneutrality of the system requires that $I = 0$, so $\Delta(-E)$ may be expressed in terms of $\Delta(-P^\dagger)$ and $\Delta(-\pi)$ with the help of Eq. (1c). This gives

(6–I) $\quad f_S(I = 0) = -\dfrac{(L_{DV} - L_{DE}L_{EV}/L_E)\Delta(-P^\dagger) + (L_D - L_{DE}L_{ED}/L_E)\Delta(-\pi)}{(L_V - L_{VE}L_{EV}/L_E)\Delta(-P^\dagger) + (L_{VD} - L_{VE}L_{ED}/L_E)\Delta(-\pi)},$

which is the case of greatest practical interest, as it corresponds to the salt sieving effect when passing an electrolyte solution through the medium without any condition. It may be noted that if in addition $\Delta(-\pi)$ is maintained at zero level by flushing exit and entrance boundaries, $f_S(I = 0)$ reverts to the reflection coefficient at zero current, σ' (cf. Kedem and Katchalsky [5]), according to

$$(6\text{-}I\text{-}\pi) \qquad f_S\{I = 0, \Delta(-\pi) = 0\} \equiv \sigma' = -\frac{(L_{DV} - L_{DE}L_{EV}/L_E)}{(L_V - L_{VE}L_{EV}/L_E)}.$$

Obviously, both σ and σ' are properties of the system only, i.e. they are independent of the magnitude of the applied pressure gradient, $\Delta(-P^\dagger)$. In the following section the numerical value of $f_S(I = 0)$ will be estimated with the help of diffuse double layer theory.

SAMPLE-CALCULATION OF THE SALT-SIEVING

EFFECT WITH DIFFUSE DOUBLE LAYER THEORY

Once the parameters u^+, u^-, $\Lambda^\pm$ and η are specified as functions of ξ, all coefficients appearing in Table I may be calculated. In most cases such a calculation will require the application of numerical integration methods necessitating the use of a computer. As an example of such a calculation the following conditions are chosen:

a.   The system contains only monovalent cations and anions, distributed according to the Gouy theory of the diffuse double layer, having identical and position independent electric mobilities Λ.

b.   The viscosity, η, of the liquid phase is constant up to a certain distance from the charged pore wall, where it changes suddenly to very high values, giving rise to a well-defined shear plane, to be indicated with the subscript s.

Indicating the thickness of the *mobile part* of the liquid film with b', the calculations are thus limited to the range $\xi = \xi_s \to b$, with $b - \zeta_s = b'$, for concentrations ranging from $u_s c_S \to u_b c_S$, and constants values of Λ and η. As the immobile part, for $\xi < \xi_s$, does not contribute to the transport phenomena, the system behaves like one with a surface charge density — corresponding to the total mobile volume charge — situated beyond $\xi = \xi_s$. Finally, for the chosen system containing one mono-monovalent salt, $u^+(\xi) = 1/u^-(\xi) = u(\xi)$.

From the Gouy theory the relationship between u and ξ is found as

$$\kappa\xi/2 = \{F(1/u_b, \arcsin \sqrt{(u_b/u)}) - F(1/u_b, \arcsin \sqrt{(u_b/u_s)})\}/\sqrt{(u_b)}$$

with

$$(7\text{-}i)$$

$$\kappa b/2 = \{F(1/u_b, \pi/2) - F(1/u_b, \arcsin \sqrt{(u_b/u_s)})\}/\sqrt{(u_b)},$$

in which $F(k, \psi)$ indicates the elliptic integral of the first kind and $\kappa = \sqrt{(\beta c_S)}$ cm^{-1}, with $\beta \cong 1.06 \times 10^{15}$ cm/me. The Equations (7–i), valid for the truncated diffuse double layer, provide an implicit relationship between u and ξ, for given values of $u_s \equiv \exp(-e\zeta/kT)$, with ζ = electric potential at the plane of shear at $\xi = \xi_s$, and of u_b, to be calculated from a specified value of b, using the second line of Equation (7–i).

As was given elsewhere (cf.[1]), similar — though slightly more involved — relationships between u and ξ exist for systems containing a mixture of electrolytes.

If the liquid film becomes sufficiently thick (relative to κb) the value of u_b approaches unity, and the implicit relationships (7–i) degrade to a simple explicit relationship of the form

$$(7\text{–}0) \qquad \kappa \xi = \ln \frac{(\sqrt{u} + 1)}{(\sqrt{u} - 1)} - \ln \frac{(\sqrt{u_s} + 1)}{(\sqrt{u_s} - 1)}.$$

In view of the rather elaborate computations necessary to obtain the values for the different transport coefficients when using the equations for the truncated double layer, only the limiting value of $f_S(I = 0)$ at $\Delta(-\pi) = 0$ was calculated for this case. As was described elsewhere by Bolt and Römkens (cf.[1]), the coefficients L_{DV}, L_{DE}, L_{EV}, L_E and L_V were computed for different values of c_S, u_s (or ξ_s), b and Λ by numerical integration of the corresponding equations of Table 1. A typical example of the results is shown in Figure 1

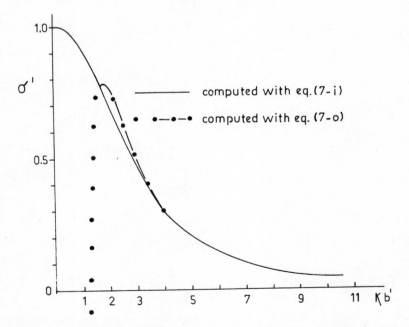

Fig. 1. The reflection coefficient, σ', as a function of $\kappa b'$, for $c_S = 0.001\ N$

(where σ' is plotted as a function of the relative film-thickness, $\kappa b'$), to be compared with the values obtained when using the much simpler equation (7–0), valid for relatively thick films. The general conclusion from this figure which was substantiated for other values of c_S, u_s and Λ (cf. [1]) — is that the use of (7–0) is fully warranted for systems with $\kappa b' > 3$, and approximately valid for $2 < \kappa b' < 3$.

The introduction of (7–0) into all equations of Table I leads to expressions which may be solved analytically. As could be expected from the form of the equations in Table I, the twin coefficients L_{VD} and L_{DV}, L_{VE} and L_{EV}, L_{DE} and L_{ED} become identical, which shows the consistency of the chosen model with the Onsager reciprocal relations. The resulting expressions are given in Table II.

TABLE II

(8a) $\qquad L_V = \dfrac{\theta b'^2}{3\lambda\eta}$

(8bd)
$$L_{VD} = L_{DV} = \frac{-4\theta}{\lambda b'\eta\kappa^3}\left\{\kappa b'\ln\frac{2\sqrt{(u_s)}}{(\sqrt{u_s}+1)} - \left\{\left(\frac{(\sqrt{u_s}-1)}{(\sqrt{u_s}+1)}\right)\right.\right.$$
$$\left.\left. - \left(\frac{(\sqrt{u_s}-1)}{(\sqrt{u_s}+1)}\right)^2\!\!\Big/4 + \left(\frac{(\sqrt{u_s}-1)}{(\sqrt{u_s}+1)}\right)^3\!\!\Big/9 - \cdots\right\}\right\}$$

(8cg)
$$L_{VE} = L_{EV} = \frac{4\theta Fc_S}{\lambda b'\eta\kappa^3}\left\{\kappa b'\ln\sqrt{(u_s)} + \ln\sqrt{(u_s)}\ln\frac{(\sqrt{u_s}+1)}{(\sqrt{u_s}-1)} - (\pi^2/4)\right.$$
$$\left. + 2(1/\sqrt{(u_s)} + 1/u_s\sqrt{(u_s)} + \cdots)\right\}$$

(8e)
$$L_D = \frac{8\theta}{\lambda\eta b'\kappa^3}\{2\ln 2 - 2\ln(1 + 1/\sqrt{(u_s)}) + (1/\sqrt{(u_s)} - 1))\}$$
$$+ \frac{\theta\Lambda}{\lambda Fc_S\kappa b'}(\kappa b' - 2 + 2/\sqrt{(u_s)})$$

(8fh)
$$L_{DE} = L_{ED} = \frac{-8\theta Fc_S}{\lambda\eta b'\kappa^3}(-1 + \ln\sqrt{(u_s)} + 1/\sqrt{(u_s)})$$
$$- \frac{\theta\Lambda}{\lambda b'\kappa}(-2 + 2/\sqrt{(u_s)} + \kappa b')$$

(8i) $\quad L_E = \dfrac{8\theta(Fc_S)^2}{\lambda\eta b'\kappa^3}(\sqrt{(u_s)} - 2 + 1/\sqrt{(u_s)}) + \dfrac{2\theta Fc_S\Lambda}{\lambda b'\kappa}(\sqrt{(u_s)} + 1/\sqrt{(u_s)} - 2 + \kappa b').$

In order to gain an impression of the magnitude of the different coefficients, a particular set of values for the system parameters was chosen, viz:

$$\eta \;=\; 10^{-2} \mathrm{g\,cm^{-1}\,sec^{-1}}$$

$$c_S \;=\; 10^{-3}\,\mathrm{me\,cm^{-3}}$$

$$\Lambda \;=\; 0.17\,\mathrm{cm\,sec^{-1}/e.s.u.\,cm^{-1}}$$

$$\sqrt{(u_s)} \;=\; 70.3$$

with $Fc_S = 2.91 \times 10^8$ e.s.u./cm^3

These values are identical with those used for the previous computations of σ' from (7–i); the particular value of $\sqrt{(u_s)}$ chosen corresponds to a mobile charge in the film of $1.37 \times 10^{-7}\,\mathrm{me/cm^2}$, which may be considered as a maximum value to be expected in clay systems. The results are presented in Table III.

TABLE III.

(9a) $$\frac{\lambda}{\theta} L_V \;=\; 3.14 \times 10^{-11} (\kappa b')^2$$

(9bd) $$\frac{\lambda}{\theta} L_{VD} \;=\; -3.77 \times 10^{-10}(0.679\kappa b' - 0.804)/\kappa b'$$

(9cg)) $$\frac{\lambda}{\theta} L_{VE} \;=\; 0.110(4.25\kappa b' - 2.32)/\kappa b'$$

(9e) $$\frac{\lambda}{\theta} L_D \;=\; -8.71 \times 10^{-10}(\kappa b')^{-1} + 5.85 \times 10^{-10}$$

(9fh) $$\frac{\lambda}{\theta} L_{DE} \;=\; -0.380(\kappa b')^{-1} - 0.170$$

(9i) $$\frac{\lambda}{\theta} L_E \;=\; 1.11 \times 10^{10}(\kappa b')^{-1} + 9.89 \times 10^7$$

The composite coefficients of Equation (6–I), after some modification, are then

$$A \equiv (\lambda/\theta)^2 L_E (L_{DV} - L_{DE}L_{EV}/L_E) = (\lambda/\theta)^2 L_E (L_{VD} - L_{VE}L_{ED}/L_E) =$$
$$= A' \cdot (\lambda/\theta)^2 L_E$$

$$A' = \{3.26 - 2.67\kappa b' + 0.0540(\kappa b')^2\}/(\kappa b')^2$$

$$B \equiv (\lambda/\theta)^2 L_E (L_D - L_{DE}L_{ED}/L_E) = B' \cdot (\lambda/\theta)^2 L_E$$

$$B' = \{-9.79 + 6.25\kappa b' + 0.0289(\kappa b')^2\}/(\kappa b')^2$$

$$C \equiv (\lambda/\theta)^2 L_E (L_V - L_{VE}L_{EV}/L_E) = C' \cdot (\lambda/\theta)^2 L_E$$

$$C' = \{-0.0538 + 0.239\kappa b' - 0.218(\kappa b')^2 + 0.348(\kappa b')^3$$

$$+ 0.00310(\kappa b')^4\}/(\kappa b')^2.$$

From this table one may obtain the value of σ' $(= - A/C)$ as a function of $\kappa b'$, valid in case of relatively thick liquid films (cf. Fig. 1).

Using the approximation for thick films the effect of the diffusion terms on the magnitude of $f_S(I = 0)$ was investigated. As follows from Equation (6–I), f_S depends — for a given value of $\kappa b'$ — on the relative magnitude of $\Delta(-\pi)$ and $\Delta(-P^\dagger)$. In Fig. 2 the generalized relationship between $f_S(I = 0)$ and $\Delta(-P^\dagger)/\Delta(-\pi)$ is shown, indicating several aspects of the salt-sieving phenomenon as characterized by $f_S(I = 0)$. Visualizing that by appropriate removal and addition of salt at entrance and exit side $\Delta(-\pi)$ is maintained at a constant level, one finds:

a.   For $\Delta(-P^\dagger)/\Delta(-\pi) \to \infty$, the diffusion becomes obviously negligible. This situation is approached by forcing the liquid through the medium under high pressure gradients, or alternatively by rapid stirring in large reservoirs with equal concentration at entrance and exit side thus maintaining $\Delta(-\pi)$ at zero value. Then the salt-sieving effect approaches the reflection coefficient, σ' (for the present case $\sigma' = 0.212$).

b.   For decreasing values of $\Delta(-P^\dagger)/\Delta(-\pi)$ the diffusion reduces the salt-sieving effect (as a "leakage" term), until for a particular ratio of $\Delta(-P^\dagger)$ and $\Delta(-\pi)$ (2.54 in the present case) this leakage just offsets the permiselectivity of the medium for the passage of salt solutions.

c.   Upon further decrease of $\Delta(-P^\dagger)$ relative to $\Delta(-\pi)$ the diffusion leakage will exceed the deficit in salt transport accompanying the volume flux and a negative value of f_S results.

d.   At a particular small positive value of $\Delta(-P^\dagger)$ in conjuction with a rather large positive value of $\Delta(-\pi)$ the volume flux itself will become zero (the counter capillary osmosis term of j^V then offsets the pressure gradient induced term) and by its definition $f_S(I = 0)$ goes through infinity. In the present case this occurs at $\Delta(-P^\dagger)/\Delta(-\pi) = 0.212$, which is obviously the same value as which was found for σ'.

e.   For a further decrease of $\Delta(-P^\dagger)/\Delta(-\pi)$ to its zero value, one finds j^V opposite to $\Delta(-P^\dagger)$ due to the dominance of capillary osmosis, while the net salt-stream remains in the direction of $\Delta(-\pi)$, giving rise to very large positive values of f_S. It should be noted that this effect could hardly be called "salt-sieving" in the proper sense as it simply corresponds to "forward" salt transport by diffusion under influence of a concentration gradient together with a small "backward" volume flux due to capillary osmosis in response to the same concentration gradient.

f.   If finally the pressure gradient is reverted, the combined volume flux increases rapidly in the "backward" direction (i.e. "uphill" with respect to the concentration gradient), still accompanied by salt diffusion in the "forward" direction. The deficit in salt transport as compared to $j^V c_S$, thus decreases

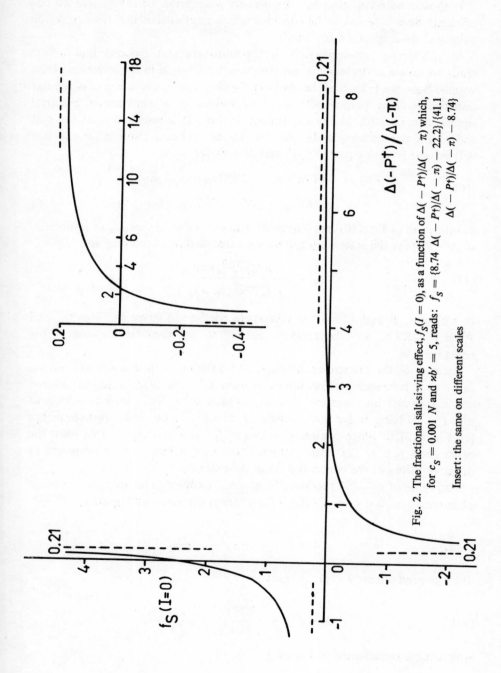

Fig. 2. The fractional salt-sieving effect, $f_S(I = 0)$, as a function of $\Delta(-P\dagger)/\Delta(-\pi)$ which, for $c_S = 0.001\ N$ and $\varkappa b' = 5$, reads: $f_S = \{8.74\ \Delta(-P\dagger)/\Delta(-\pi) - 22.2\}/\{41.1\ \Delta(-P\dagger)/\Delta(-\pi) - 8.74\}$

Insert: the same on different scales

untill at very high backward gradient of $\Delta(-P^\dagger)$ the diffusion effect becomes negligible and the limiting value σ' is reached again.

It should be noted that for the present description of the system all co-efficients were assumed to be constant which implies that the absolute value of $\Delta(-\pi)$ must remain fairly small.

As salt-sieving accompanying the pressure gradient induced liquid flow tends to create a decrease in electrolyte concentration in the direction of the volume flow, the branch of the above curve discussed in a and b is of particular importance. This branch will now be followed for a condition of practical significance, i.e. without any adjustment of the salt concentration at the exit end of the porous sample. In that case the salt concentration at the exit side will assume a value equal to j_S/j^V and $\Delta(-\pi)$ becomes

$$\Delta(-\pi) = 2RT\Delta(-c_S) = 2RT(c_{S1} - c_{S1}j_S/\rho_{S1}j^V)$$

(10)

$$= -2RTc_{S1} \cdot j^D/j^V = 2RTc_{S1} \cdot f_S.$$

Substitution of Eq. (10) into Eq. (6–I) allows one to express f_S as a function of $\Delta(-P^\dagger)$ for the stationary system with free outflow, according to

(11)
$$f_S = \frac{-A\Delta(\overline{-P^\dagger}) - Bf_S}{C\Delta(\overline{-P^\dagger}) + Af_S}$$

in which A, B and C are the composite coefficients given in Table III, and $\bar{P}^\dagger \equiv P^\dagger/2RTc_S$, i.e. the pressure relative to the osmotic pressure at the entrance side.

Because of the chosen condition, i.e. a solution with a concentration c_S, *entering a permiselective* medium and leaving it with a self-adjusted reduced concentration, this equation has significance only for positive values of $\Delta(\overline{-P^\dagger})$. Solving it for such values of $\Delta(-P^\dagger)$, one finds two branches (cf. Figure 2), of which one shows values of f_S between 0 and $-A/C$, while the other gives $f_S > 1$. The latter branch is no interest here as it corresponds to negative values of the concentration at the exit side.

The value of $\Delta(\overline{-P^\dagger})$ necessary to attain a specified value of f_S in the range of interest (viz. $0 < f_S < -A/C$) is found from the inverted equation

(11a)
$$\Delta(\overline{-P^\dagger}) = \frac{-Af_S^2 - Bf_S}{Cf_S + A}.$$

For the present case Eq. (11a) reads:

(11b)
$$\Delta(\overline{-P^\dagger}) = \frac{8.74f_S^2 - 22.2f_S}{41.13f_S - 8.74},$$

which curve is presented in Figure 3.

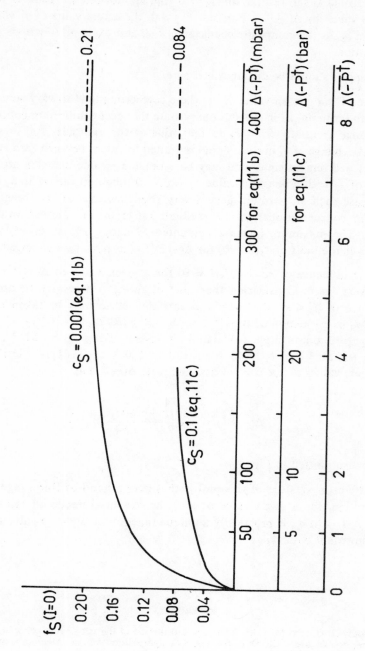

Fig. 3. The salt-sieving effect, $f_S(I = 0)$, as a function of $\Delta(-P_\dagger)$, for the condition of free outflow

Equation (11a) shows that for $f_S = 0$ the equilibrium situation is found, with zero value for $\Delta(-P^\dagger)$. For $\Delta(\overline{-P^\dagger}) > 0$, the actual value of f_S will still depend on c_S, as the composite coefficients A, B and C are all functions of the concentration.

Two aspects should be distinguished here.

a. Even for a chosen value of c_S, these coefficients tend to vary somewhat through the column, as in the stationary state the local equilibrium concentrations change gradually from c_S to the value at the exit side. As, however, the relative change of c_S in the column is limited to values between zero and f_S, this effect will remain small and may be estimated by one iteration after calculation of f_S with a constant value $c_S = c_{S1}$, throughout the column. In the present case such an iteration gave a very slight increase of f_S (because of decreasing concentrations) in the medium range of the plotted values of $\Delta(-P^\dagger)$. The maximum increase amounted to about 3% at $\Delta(-P^\dagger) = 50$ mbar, while the limit of $f_S(I = 0)$ for $\Delta(-P^\dagger) \to \infty$ remains unchanged.

b. The influence of c_{S1} on $f_S(I = 0)$ for a given value of $\Delta(-P^\dagger)$ is, of course, very large. Considering the value of $\Delta(-P^\dagger)$ necessary to obtain a given value of $f_S < \sigma'$, the main concentration effect may be taken out by plotting f_S as a function of $\Delta(\overline{-P^\dagger}) = (\Delta(-P^\dagger)/2RTc_{S1})$.

In Figure 3 one sample calculation is given, taking $c_{S1} = 0.1\ N$, while retaining $\kappa b'$ at 5 (i.e. $b' = 50$ Å instead of 500 Å at 0.001 N). Using the appropriate values of the coefficients then gives $\Delta(\overline{-P^\dagger})$ as:

$$(11c) \qquad\qquad \Delta(\overline{-P^\dagger}) = \frac{1.50 f_S^2 - 8.80 f_S}{17.9 f_S - 1.50},$$

which is also plotted in Fig. 3.

The interesting feature of this plot is that over a hundred fold range of the concentration, for a given value of $\kappa b'$, the "reduced fractional salt-sieving effect", $f_S(I = 0)/\sigma'$, is practically a unique function of $\Delta(\overline{-P^\dagger})$ only, it being almost invariant with c_{S1}.

REFERENCES

1. Bolt, G. H. and M. J. M. Römkens, Calculation of the salt-sieving effect in systems with planar double layers. Ch. 6b of the A.A.P.G. Symposium volume: Membrane and mass transport phenomena within geologic environments, Berkeley, in press.

2. Groenevelt, P. H. and G. H. Bolt (1969), Non-equilibrium thermodynamics of the soil-water system, *Journal of Hydrology*, 7, 358–388.

3. GROOT, S. R. DE AND P. MAZUR (1962), *Non-equilibrium Thermodynamics*, North-Holland Publ. Co., Amsterdam.

4. KATCHALSKY, A. AND P. F. CURRAN (1965), *Non-equilibrium thermodynamics in Biophysics*, Harvard University Press, Cambridge.

5. KEDEM, O.AND A. KATCHALSKY (1963), Permeability of composite membranes, *Trans. Faraday Society, 59*, 1918.

LABORATORY OF SOILS AND FERTILIZERS,
STATE AGRICULTURAL UNIVERSITY,
WAGENINGEN, NETHERLANDS

4

Hydrodynamic dispersion in porous media

THE TENSOR CHARACTER OF THE DISPERSION COEFFICIENT IN ANISOTROPIC POROUS MEDIA

G. De Josselin de Jong

INTRODUCTION

In their attempt to describe dispersion phenomena in mathematical form several authors arrived at the conclusion, that a porous medium possesses a coefficient of dispersion, which has the character of a tensor. This tensor is formulated in such a manner, that it represents the geometrical aspects of the porous medium responsible for the scatter of tracer particles, when carried by a fluid flowing through it. It is therefore a property of the porous medium alone, and can be considered to materialize the tortuosity of the particle trails caused by the random arrangement and the interconnectivity of the channels constituing the pore space.

Nikolaevskii (1959) originated the tensor. He constructed it with a dimension of length and predicted the magnitude of this length to be in the order of particle size. He obtained a tensor of the fourth rank for an isotropic porous medium by postulating, that the random modification of the velocity vector, from mean velocity to local velocity, is a tensor of second rank.

This postulate requires that the velocity vectors considered possess the property of linear superposition. It will be shown subsequently that this requirement is not always fulfilled, and that, for instance, Nikolaevskii's result for the isotropic medium cannot be extended to the general case of an anisotropic medium. Nikolaevskii states that the rank of the tensor must be even, but includes the possibility of any number for that rank.

Bear (1961) started with the results of the dispersion computations proposed by the author (De Josselin de Jong, 1958). These computations were executed for an isotropic model material exhibiting a particular manner of scattering tracer particles. The basic scattering mechanism adopted was apparently realistic enough to reproduce mathematically the dispersivity phenomena observed in tests and especially the difference in longitudinal and transverse dispersion.

The specific problem treated was the determination of concentration distribution developing from a point injection of tracer particles if these were carried away by the fluid flow through the porous medium. The resulting concentration distribution turns out to be almost normal (Gaussian), in all directions, with the points of standard deviation lying on an ellipsoid.

This concentration distribution can be uniquely defined by its mean point and the variances or second central moments with respect to the three space coordinates. By superposition of point injections and rotation of coordinates, Bear (1961) showed that the variances can be considered as the components of a second rank tensor, and he developed the relation between this tensor and Nikolaevskii's fourth rank tensor.

Inspired by the work of Nikolaevskii and Bear, Scheidegger (1961) suggested that the dispersion constant is also a fourth rank tensor in the general anisotropic case. This assertion was not substantiated by considering the physics of the scattering mechanism underlying the dispersion phenomenon. Therefore his suggestion is not imperative.

It is the purpose of this paper to reestablish the dispersion tensor for the anisotropic medium starting from the basic scattering mechanism. Since the mechanism used by the author, (De Josselin de Jong, 1958) proved realistic enough for the isotropic case it was considered to be. acceptable for the anisotropic case as well.

SCATTERING MECHANISM

The scattering mechanism adopted previously consists of assigning a choice to every tracer particle when arriving at a junction point in the pore channel system. It was suggested that the probability for a particle to choose a certain direction is proportional to the discharge in that direction considered as a fraction of the total discharge through all junction points and in all directions.

Once a particular channel has been chosen in junction A the particle travels a certain distance in a certain direction to the next junction point B. This distance is described by its space coordinates x, y, z with origin in A. The

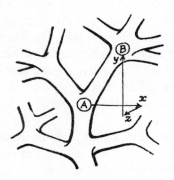

Fig. 1.

travel time required for the particle to travel from A to B is called the residence time, t.

The components x, y, z, t are stochastic variables with a probability distribution determined by the adopted flow mechanism in the channels and in the junction points. In this work use will be made of the mean values of these stochastic variables and their squares. These mean values are indicated by a bar and combinations by a circonflex.

$$
\begin{aligned}
\bar{x} &= \text{mean value of } x \\
\overline{x^2} &= \text{mean value of } x^2 \\
\overline{xy} &= \text{mean value of } x \cdot y \\
\hat{xx} = \overline{x^2} - \bar{x}^2 &= \text{mean value of square minus square of mean value} \\
\hat{xy} = \overline{xy} - \bar{x} \cdot \bar{y}. &
\end{aligned}
$$

(1)

The summation of the stochastic components x, y, z, t of each individual channel over the many channels covered by a tracer particle form its total travel path and travel time. The probability for a particle to arrive at a certain point X, Y, Z, after time T_0 is the combined probability for that particle to travel through a certain combination of channels. The probability distribution $P(X, Y, Z, T_0)$ in space of the arrival points is the tracer concentration distribution at time T_0 after the moment that an amount of particles was injected in the origin of the X, Y, Z coordinate system.

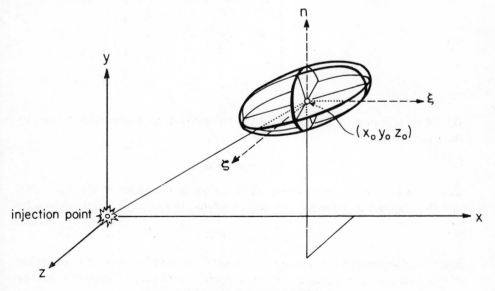

Fig. 2. Ellipsoid of standard deviation

By use of Chandrasekhar's analysis of Markoff processes and a special form of saddle point analysis this probability turns out to be given in good approximation by

(2) $$P(X, Y, Z, T_0) = \frac{\sqrt{|f|}}{(2\pi)^{3/2}} \exp\left[-\tfrac{1}{2} f_{ij}\xi_i\xi_j\right].$$

In this expression ξ_i and ξ_j are space coordinates with their origin in the point

$$X_0 = T_0\frac{\bar{x}}{\bar{t}}; \quad Y_0 = T_0\frac{\bar{y}}{\bar{t}}; \quad Z_0 = T_0\frac{\bar{z}}{\bar{t}} \quad \text{(see Fig. 2).}$$

This is the point of average displacement of the tracer particles during travel time T_0. Throughout this paper subscripts can have three values 1, 2, 3 corresponding respectively to x, y, z, if they are Roman, and if they are Greek they run over 4 values corresponding to x, y, z and t.

$|f|$ is the determinant of a matrix f with the components f_{ij}. These 9 components f_{ij} are related to the mean squares and square means of the stochastic variables, by the following expressions given here without proof.

(3) $$f_{ij} = \frac{\bar{t}}{T_0} \frac{(b_{ij}b_{\mu\nu} - b_{i\mu}b_{j\nu})\bar{x}_\mu\bar{x}_\nu}{b_{\mu\nu}\bar{x}_\mu\bar{x}_\nu}.$$

In this formula $b_{\mu\nu}$ are components of matrix b, which is the inverse of a matrix α given by

(4) $$b^{-1} = \alpha = \begin{pmatrix} \widehat{xx} & \widehat{xy} & \widehat{xz} & \widehat{xt} \\ \widehat{xy} & \widehat{yy} & \widehat{yz} & \widehat{yt} \\ \widehat{xz} & \widehat{yz} & \widehat{zz} & \widehat{zt} \\ \widehat{xt} & \widehat{yt} & \widehat{zt} & \widehat{tt} \end{pmatrix}.$$

By the inverse of α, it is meant that the components of b and α are related by the expressions

(5) $$b_{\mu\nu}\alpha_{\nu\rho} = \delta_{\mu\rho}.$$

The form (2) for the probability distribution is Gaussian in all directions, and the ellipsoid of standard deviation centered in the mean point is given by

(6) $$f_{ij}\xi_i\xi_j = 1.$$

Since the distribution is Gaussian it is uniquely defined by the second moments of the concentration around the mean point. These second moments can be combined as the components of a matrix a such that

$$
a_{pq} = \frac{\displaystyle\int\int\int_{-\infty}^{+\infty} \xi_p\xi_q P(X,Y,Z,T_0)d\xi_1 d\xi_2 d\xi_3}{\displaystyle\int\int\int_{-\infty}^{+\infty} P(X,Y,Z,T_0)d\xi_1 d\xi_2 d\xi_3}.
$$

(7)

By taking the partial derivative of (2) with respect to f_{pq} it is possible to establish, that the matrix a is the inverse of the matrix f, so that their components are related by

(8)
$$
a_{pq}f_{qr} = \delta_{pr}.
$$

By use of (3) and (4) elaboration of a_{pq} gives the following expression

$$
a_{pq} = \frac{T_0}{\hat{t}^3}\left[\hat{tt}\tilde{x}_p\tilde{x}_q - \hat{x_p t}\tilde{x}_q\hat{t} - \hat{x_q t}\tilde{x}_p\hat{t} + \hat{x_p x_q}\hat{t}^2\right].
$$

By introduction of (1) this can be reduced to

(9)
$$
a_{pq} = \frac{T_0}{\bar{t}^3}\left[\overline{t^2}\tilde{x}_p\tilde{x}_q - \overline{x_p t}\tilde{x}_q\,\bar{t} - \overline{x_q t}\tilde{x}_p\,\bar{t} + \overline{x_p x_q}\,\bar{t}^2\right].
$$

If these components a_{pq} are the components of a second rank tensor, then Bear's analysis indicates that the dispersion constant D_{ijkl} is a fourth rank tensor, because this constant relates the second rank tensor representative for the flow displacement to a_{ij}.

In order to verify the tensor character of a_{pq} in the anisotropic case it is necessary to elaborate the mean values of the stochastic variables as occurring in expression (9). For the two-dimensional anisotropic case this will be done in the next section with $\bar{x}$, $\bar{y}$, $\bar{t}$ and $\overline{x^2}$, providing the possibility to consider the form of a_{11}.

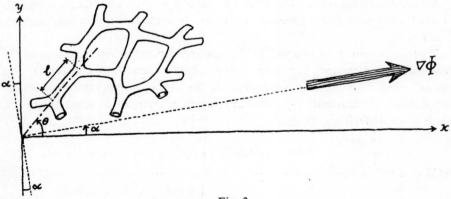

Fig. 3.

ELABORATION OF MEAN VALUES

Let $n(\theta)d\theta$ be the amount of channels having an angle between θ and $\theta + d\theta$ with the positive x-direction, such that the total amount of channels considered: $N = \int_{-1/2\pi}^{+1/2\pi} n(\theta)d\theta$ is a large number. The integral only covers half the circle, because every channel has two opposite directions.

Let $\lambda(\theta)$ be the conductivity of the channels $n(\theta)d\theta$. Anisotropy is obtained, if $n(\theta)\lambda(\theta)$ varies with θ.

Let $l(\theta)$ be the length of the channels $n(\theta)d\theta$. We will take for $l(\theta)$ a constant length l here, because it simplifies the computations without effecting the result essentially.

Let $c(\theta)$ be the cross sectional area of the channels $n(\theta)d\theta$.

In the fluid a gradient of head, $\nabla\Phi$, is assumed to exist which is everywhere equal and makes an angle α with the positive x-direction. This means that the gradient in each of the channels is equal to $\left|\nabla\Phi\right| \cdot \cos(\theta - \alpha)$. The discharge of such a channel is then

$$(10) \qquad q(\theta) = \lambda(\theta)\left|\nabla\Phi\right| \cdot \cos(\theta - \alpha).$$

According to the assumption, that the probability for a particle to choose a certain direction is proportional to the amount of fluid flowing in that direction, that probability is for a direction between θ and $\theta + d\theta$ equal to

$$(11) \qquad P(\theta \to \theta + d\theta) = \frac{n(\theta)q(\theta)d\theta}{Q}.$$

In this expression Q is the total amount of fluid passing through all the N-channels. Using (10) this becomes

$$(12) \quad Q = \int_{-1/2\pi+\alpha}^{+1/2\pi+\alpha} n(\theta)q(\theta) \cdot d\theta = \left|\nabla\Phi\right| \int_{-1/2\pi+\alpha}^{+1/2\pi+\alpha} n(\theta)\lambda(\theta)\cos(\theta - \alpha)d\theta.$$

Integration is performed over the range $-\tfrac{1}{2}\pi + \alpha < \theta < +\tfrac{1}{2}\pi + \alpha$, because only channels in those directions carry water away from the junction points. That only discharge departing from the junction points is counted, is a requirement for the probability calculus. The choice is made forward in time and not backward.

The x-component of one step in the Markoff process is the projection of the channel in x-direction. For each of the $n(\theta)d\theta$ channels, this is $l\cos\theta$. The mean value $\bar{x}$ of these x-components is this value $l\cos\theta$ multiplied by its probability of occurrence $P(\theta \to \theta + d\theta)$ integrated over all possible directions. This gives with (10), (11), (12)

$$(13) \qquad \bar{x} = \frac{\int_{-1/2\pi+\alpha}^{1/2\pi+\alpha} n(\theta)l\cos\theta\, q(\theta)d\theta}{Q} = \frac{l\int n(\theta)\lambda(\theta)\cos\theta\cos(\theta - \alpha)d\theta}{\int n(\theta)\lambda(\theta)\cos(\theta - \alpha)d\theta}.$$

For the y-component, which is $l \sin \theta$, this gives

$$(14) \quad \bar{y} = \frac{\displaystyle\int_{-1/2\pi+\alpha}^{1/2\pi+\alpha} n(\theta)l \sin \theta \, q(\theta)d\theta}{Q} = \frac{\displaystyle l \int n(\theta)\lambda(\theta) \sin \theta \cos (\theta - \alpha)d\theta}{\displaystyle\int n(\theta)\lambda(\theta) \cos (\theta - \alpha)d\theta}.$$

The residence time t of one step in the Markoff process, is the time a particle stays in one channel. If the discharge is $q(\theta)$ and the volume of the channel is $l \cdot c(\theta)$ then this residence time is $t = (l \cdot c(\theta)/q(\theta))$. The mean value $\bar{t}$ of the residence time is this value multiplied by its probability of occurrence, $P(\theta \rightarrow \theta + d\theta)$, integrated over all possible directions. This gives with (10) (11), (12),

$$(15) \quad \bar{t} = \frac{\displaystyle\int_{-1/2\pi+\alpha}^{1/2\pi+\alpha} n(\theta)l \cdot c(\theta)d\theta}{Q} = \frac{\displaystyle l \int n(\theta)c(\theta)d\theta}{\displaystyle |\nabla\Phi| \int n(\theta)\lambda(\theta) \cos (\theta - \alpha)d\theta}.$$

In the same manner we find

$$(16) \quad \overline{x^2} = \frac{\displaystyle l^2 \int n(\theta)\lambda(\theta) \cos^2\theta \cos (\theta - \alpha)d\theta}{\displaystyle \int n(\theta)\lambda(\theta) \cos (\theta - \alpha)d\theta}.$$

The general case of anisotropy is obtained by assigning to the $n(\theta)d(\theta$ channels an arbitrary distribution with respect to θ of the combined conductivities $n(\theta)\lambda(\theta)d\theta$ and the combined cross sectional areas $n(\theta)c(\theta)d\theta$. Every possible distribution can be expressed as a Fourier series. Let these be

$$(17) \quad \begin{cases} \dfrac{n(\theta)}{N} \lambda(\theta) = \displaystyle\sum_{n=0}^{\infty} A_{2n} \cos 2n(\theta - \alpha_{2n}) \\[4mm] \dfrac{n(\theta)}{N} c(\theta) = \displaystyle\sum_{n=0}^{\infty} B_{2n} \cos 2n (\theta - \beta_{2n}). \end{cases}$$

Only the even terms appear, because every channel figures in two oppposite directions.

Execution of the integrals in (13), (14), (15) and (16), all between the limits $-\frac{1}{2}\pi + \alpha$ and $\frac{1}{2}\pi + \alpha$ gives finally

$$(18) \quad \begin{aligned} \bar{x} &= \frac{l}{R} \frac{\pi}{2} [A_0 \cos \alpha + \tfrac{1}{2} A_2 \cos (2\alpha_2 - \alpha)] \\[3mm] \bar{y} &= \frac{l}{R} \frac{\pi}{2} [A_0 \sin \alpha + \tfrac{1}{2} A_2 \sin (2\alpha_2 - \alpha)] \\[3mm] \bar{t} &= \frac{l}{R} \frac{\pi}{|\nabla\Phi|} B_0 \end{aligned}$$

$$\text{with } R = \int_{-1/2\pi+\alpha}^{1/2\pi+\alpha} \frac{n(\theta)}{N} \lambda(\theta) \cos (\theta - \alpha)d\theta.$$

The real mean velocity v has x- and y-components given by $v_x = \bar{x}/\bar{t}$ and $v_y = \bar{y}/\bar{t}$. With (18) this is

$$
(19) \quad
\begin{cases}
v_x = \dfrac{\bar{x}}{\bar{t}} = \left[\dfrac{A_0}{2B_0} \cos\alpha + \dfrac{A_2}{4B_0} \cos(2\alpha_2 - \alpha) \right] \cdot |\nabla\Phi| \\[3mm]
v_y = \dfrac{\bar{y}}{\bar{t}} = \left[\dfrac{A_0}{2B_0} \sin\alpha + \dfrac{A_2}{4B_0} \sin(2\alpha_2 - \alpha) \right] \cdot |\nabla\Phi| .
\end{cases}
$$

Since the gradient of Φ has the direction α, the partial derivatives of Φ can be introduced as $\partial\Phi/\partial x = |\nabla\Phi| \cdot \cos\alpha$, $\partial\Phi/\partial y = |\nabla\Phi| \sin\alpha$. Then (19) becomes

$$
(20) \quad
\begin{cases}
v_x = \left(\dfrac{A_0}{2B_0} + \dfrac{A_2}{4B_0} \cos 2\alpha_2 \right) \dfrac{\partial\Phi}{\partial x} + \dfrac{A_2}{4B_0} \sin 2\alpha_2 \, \dfrac{\partial\Phi}{\partial y} \\[3mm]
v_y = \dfrac{A_2}{4B_0} \sin 2\alpha_2 \, \dfrac{\partial\Phi}{\partial x} + \left(\dfrac{A_0}{2B_0} + \dfrac{A_2}{4B_0} \cos 2\alpha_2 \right) \dfrac{\partial\Phi}{\partial y} .
\end{cases}
$$

These expressions form the relation between the real mean velocity vector, v and the gradient vector, $\nabla\Phi$. Because the vector components are obtained by linear superposition of the four coefficients at the right of (20), these coefficients form the components of a second rank tensor.

This tensor is the anisotropic permeability tensor divided by the porosity, because specific discharge is equal to real mean velocity multiplied by the porosity. This result is in agreement with the concept developed by Ferrandon (1948).

In the permeability tensor expressed by (20) only A_0, B_0 and A_2 appear, showing that the other terms of the infinite series (17) are of no importance to the permeability. In contrast to this simple result the dispersion phenomenon cannot be described without using all the terms in the infinite series. This will be seen by elaborating a_{pq} for the case that the conductivity distribution is expressed in an infinite series as given by (17).

Introduction of v_x and v_y in (9) gives for the coefficient $a_{11} = a_{xx}$, corresponding to $x_p = x_q = x$, the form

$$
(21) \quad a_{11} = a_{xx} = \frac{T_0}{\bar{t}} \left[\overline{t^2 v_x^2} - 2\overline{x t\, v_x} + \overline{x^2} \right] .
$$

Consider as an example the last term between brackets. Elaboration of the integral (16) with (17) gives the value of $\overline{x^2}$. Because of $T_0/\bar{t}$ before brackets, this has to be multiplied by $(RT_0 |\nabla\Phi|/\pi l B_0)$, to give finally

$$
(22) \quad \frac{T_0}{\bar{t}} \overline{x^2} = \frac{T_0 l}{\pi B_0} |\nabla\Phi| \cdot
$$

$$
\cdot \sum_{n=0}^{\infty} (-1)^n A_{2n} \frac{[(4n^2 + 3)\cos 2\alpha + (8n^3 - 10n)\sin 2\alpha - 4n^2 + 9]}{(4n^2 - 9)(4n^2 - 1)} .
$$

$$
\cdot \cos 2n(\alpha - \alpha_{2n})
$$

Considered as a polynome in $(\cos \alpha)$ and $(\sin \alpha)$, this term of a_{xx} contains $(\cos \alpha)^{2n+2}$ and $(\sin \alpha)^{2n+2}$. The contribution of the other two terms in (21) does not reduce the order of the polynome.

This means, that a_{pq} can be considered as the components of an even tensor of rank $(2n + 2)$. Since n can run up to infinity it is an even tensor of infinite rank.

CONCLUSION

Since the variance of dispersion for the general anisotropic case is a tensor of infinite rank the dispersion constant is also a tensor of infinite rank. As a consequence of this fact the differential equation for dispersion developed for the isotropic case can only be generalised to the anisotropic case by introducing a dispersion coefficient, whose dependence on the direction of flow is expressed by a series of infinite terms.

REFERENCES

1. BEAR, J. (1961), On the tensor form of dispersion in porous media, *J. Geoph. Res.*, *66*, 1185–1197.

2. DE JOSSELIN DE JONG, G. (1958), Longitudinal and transverse diffusion in granular deposits, *Trans. Amer. Geoph. Union*, *39*, 67–74.

3. NIKOLAEVSKII, V. N. (1959), Convective diffusion in porous media, *Prikl. Mat. Mekh.*, *23*, 1042–1050.

3. SCHEIDEGGER, A. E. (1961), General theory of dispersion in porous media, *J. Geoph. Res.*, *66*, 3273–3278.

5. FERRANDON, J. (1948), Les lois de l'écoulement de filtration, *Le Génie Civil*, *125*, 24–28.

CIVIL ENGINEERING DEPARTMENT,
DELFT UNIVERSITY OF TECHNOLOGY,
OOSTPLANTSOEN 25, DELFT, NETHERLANDS

ON THE DERIVATION OF A CONVECTIVE-DISPERSION EQUATION BY SPATIAL AVERAGING

Ralph R. Rumer, Jr.

ABSTRACT

Dispersion coefficients or transport coefficients for porous media flow are defined utilizing the concept of a macroscopic control volume and the spatial averaging of quantitities within the macroscopic control volume. The resulting mass balance equation is still compatible with the continuum approach normally employed in the analyses of fluid flow. For linear laminar flow the dispersion of conservative substances in flow through porous media is a result of the spatial velocity variations due to the complicated geometry of the pore system coupled with the molecular diffusion of the substance across streamlines. The derivation of the mass balance equation is extended to include decay or disappearance of the substance.

After the introduction of characteristic reference quantities, the mass balance equation (in dimensionless form) gives rise to two dimensionless groupings; namely the molecular Peclet number and the ratio of the dispersion coefficient to the effective molecular diffusion coefficient for porous media flow. The basic equation should be extended to include turbulent flow.

Beyond the linear laminar regime, the influence of viscosity becomes important in the dynamics of the flow field. Since the transport of a substance within the pore system is highly dependent upon the flow field, there results an implied dependence of dispersion coefficients upon the molecular viscosity of the host fluid.

INTRODUCTION

The process by which a soluble or miscible contaminant is transported by fluid flow through the void spaces of a porous medium can be highly complex. In addition to the convective transport by the host fluid, there is molecular diffusion of the contaminating substance across streamlines and in some cases, adsorption of the contaminating substance onto the solid boundaries of the pore system. In addition, certain contaminating substances may undergo chemical reaction during the transport process. If the flow is turbulent, increased mixing by the turbulent eddies will considerably amplify the molecular diffusion occurring within the individual pore spaces.

It has become the custom to use the term "dispersion" when this transport process is associated mainly with spatially averaged velocity variations [8]. The purpose of this paper is to develop a conservation of dispersing mass equation for this transport process based on the concept of spatial averaging of velocity variations. A "Fickian-type" assumption is made to define mass transport coefficients that account for the increased mixing due to this dispersion process. It is felt that the resulting equation gives physical insight

268

into the dispersion process. Extension of this approach to include turbulent flow would require time averaging of velocity variations as well as spatial averaging.

DESCRIPTION OF CONTROL VOLUME

It is necessary to distinguish between a microscopic control volume and a macroscopic control volume in the subsequent development of the equation for the conservation of dispersing mass. The microscopic control volume includes within it only fluid and is consistent with the continuum model normally employed in analyses of fluid flow. The macroscopic control volume contains within it both the fluid and the solid particles that make up the porous medium [6]. The continuum assumption is still made for the macroscopic volume and should be valid so long as the details of the interstitial motion can be neglected. The rationality of this approach seems all the more plausible when one considers the dimensions of most naturally occurring porous media. Normally, measurements made of velocity, pressure, or solution concentration characterize macroscopic quantities rather than microscopic quantities. The size of the macroscopic control volume must be chosen sufficiently large so that its porosity is representative of the medium as a whole.

ANALYTICAL DEVELOPMENT

Neglecting molecular diffusion relative to the mechanical mixing of dispersion due to the intertwining of flow paths, the flux of dispersing mass or substance into the microscopic control volume in the x-direction is $(\rho s_a u_a) dy\, dz$, where

$\mathbf{q}_a$ = fluid velocity vector = $\mathbf{i} u_a + \mathbf{j} v_a + \mathbf{k} w_a$ (the actual fluid particle velocity)
ρ = density of the host fluid
s_a = concentration of dispersing substance .

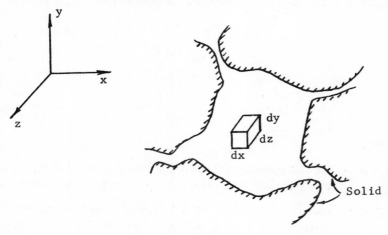

Fig. 1. Microscopic control volume

The conservation equation for the mass of the dispersing substance is obtained by determining the difference between the mass flux into and out of the microscopic control volume. This net change in mass flux must be equal to the time rate of change of the mass of the substance within the element. Disappearance of the substance due to adsorption or decay will not be considered here. The resulting equation after performing this material balance is

$$(1) \qquad \frac{\partial(\rho s_a)}{\partial t} + \frac{\partial(\rho s_a u_a)}{\partial x} + \frac{\partial(\rho s_a v_a)}{\partial y} + \frac{\partial(\rho s_a w_a)}{\partial z} = 0.$$

Expanding triple derivatives and rewriting, Eq. (1) becomes

$$(2) \qquad \frac{D(\rho s_a)}{Dt} + \rho s_a \, \mathrm{div} \, \boldsymbol{q}_a = 0.$$

For incompressible fluids, the volumetric dilation is zero [4] and Eq. (2) becomes

$$(3) \qquad \frac{D(\rho s_a)}{Dt} = 0 = \rho \frac{D s_a}{Dt} + s_a \frac{D\rho}{Dt}.$$

The continuity equation for the host or bulk fluid which must also be satisfied [4] is

$$(4) \qquad \frac{D\rho}{Dt} + \rho \, \mathrm{div} \, \boldsymbol{q}_a = 0.$$

Since $\mathrm{div} \, \boldsymbol{q}_a = 0$, $D\rho/Dt = 0$, and Eq. (3) becomes

$$(5) \qquad \frac{D s_a}{Dt} = 0 = \frac{\partial s_a}{\partial t} + u_a \frac{\partial s_a}{\partial x} + v_a \frac{\partial s_a}{\partial y} + w_a \frac{\partial s_a}{\partial z}.$$

The actual fluid particle velocities are defined as [5]

$$u_a = u + \mathring{u}$$
$$(6) \qquad v_a = v + \mathring{v}$$
$$w_a = w + \mathring{w}$$

or

$$(7) \qquad \boldsymbol{q}_a = \boldsymbol{q} + \mathring{\boldsymbol{q}}$$

where

$u, v, w = $ the components of the average fluid velocity in the neighborhood of a microscopic point (i.e. the macroscopic or seepage velocity)

$\mathring{u}, \mathring{v}, \mathring{w} = $ spatial variations of velocity components at some point in the macroscopic control volume from the local average velocity components or seepage velocity.

The substance concentration is defined similarly as

(8)
$$s_a = s + \overset{\circ}{s}$$

where

s = the average concentration of the dispersing mass in the neighborhood of a point (i.e., the macroscopic concentration),

$\overset{\circ}{s}$ = the spatial variation of the concentration at some point in the macroscopic control volume from the macroscopic concentration.

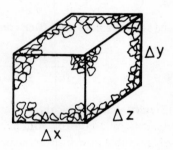

Fig. 2. Macroscopic control volume

If the velocity q_a is integrated through the pore system of a macroscopic control volume ΔV, and averaged by dividing by the pore volume, $n\Delta V$, where n is the porosity of the medium, we have

(9)
$$\bar{q}_a = \frac{1}{n\Delta V} \iiint_{\Delta V} q_a dV .$$

By definition, the macroscopic velocity is the seepage velocity and

(10)
$$\bar{q}_a \equiv q$$

The average of the spatial velocity variations becomes

(11)
$$\overline{\overset{\circ}{q}} = \frac{1}{n\Delta V} \iiint_{\Delta V} \overset{\circ}{q} dV \equiv 0$$

and

(12)
$$\bar{s}_a = \frac{1}{n\Delta V} \iiint_{\Delta V} s_a dV \equiv s$$

$\mathring{s}_a$ represents the average concentration in the macroscopic control volume, ΔV, and by definition must be equal to s, so that

(13)
$$\bar{\mathring{s}} = \frac{1}{n\Delta V} \iiint_{\Delta V} \mathring{s}\, dV \equiv 0 .$$

If Eqs. (6) and (8) are substitited into (5) and each term in the equation is averaged by integration through the macroscopic control volume, there results (after dropping the zero terms that result from the averaging procedure)

(14)
$$\frac{\partial s}{\partial t} + u\frac{\partial s}{\partial x} + v\frac{\partial s}{\partial y} + w\frac{\partial s}{\partial z} + \mathring{u}\frac{\overline{\partial \mathring{s}}}{\partial x} + \mathring{v}\frac{\overline{\partial \mathring{s}}}{\partial y} + \mathring{w}\frac{\overline{\partial \mathring{s}}}{\partial z} = 0 .$$

Adding the zero term $\quad \overline{\mathring{s}\left(\dfrac{\partial \mathring{u}}{\partial x} + \dfrac{\partial \mathring{v}}{\partial y} + \dfrac{\partial \mathring{w}}{\partial z}\right)}$

to Eq. (14) does not change its value, and after combining double derivatives, there results

(15)
$$\frac{\partial s}{\partial t} + u\frac{\partial s}{\partial x} + v\frac{\partial s}{\partial y} + w\frac{\partial s}{\partial z} + \frac{\partial}{\partial x}(\overline{\mathring{u}\mathring{s}}) + \frac{\partial}{\partial y}(\overline{\mathring{v}\mathring{s}}) + \frac{\partial}{\partial z}(\overline{\mathring{w}\mathring{s}}) = 0 .$$

The first term in Eq. (15) represents the local time rate of change of the average concentration in the macroscopic control volume. The next three terms represent the transport of the substance due to convection by the seepage velocity. The last three terms represent the additional mass flux due to the process called dispersion. Perhaps the most fundamental assumption for the additional mass flux terms is that each component at a "macroscopic" point be a function of the local seepage velocity, the local concentration gradient, and the pore system geometry at the "macroscopic" point.

It is instructive to simplify Eq. (15) by considering only unidirectional flow parallel to the x-axis. For this case Eq. (15) becomes

(16)
$$\frac{\partial s}{\partial t} + u\frac{\partial s}{\partial x} = - \frac{\partial}{\partial x}(\overline{\mathring{u}\mathring{s}}) - \frac{\partial}{\partial y}(\overline{\mathring{v}\mathring{s}}) - \frac{\partial}{\partial z}(\overline{\mathring{w}\mathring{s}}) .$$

Assuming the porous medium to be both isotropic and homogeneous, dispersion coefficients can be defined in a manner analogous to the Fickian diffusion relation, i.e. [5]

(17)
$$\overline{\mathring{u}\mathring{s}} = -D_1\frac{\partial s}{\partial x}$$
$$\overline{\mathring{v}\mathring{s}} = -D_2\frac{\partial s}{\partial y}$$
$$\overline{\mathring{w}\mathring{s}} = -D_2\frac{\partial s}{\partial z}$$

where

D_1 = coefficient of dispersion in the direction of seepage flow or longitudinal direction

D_2 = coefficient of dispersion in a direction normal to the direction of seepage flow or transverse to the flow.

Substituting Eqs. (17) into (16) results in

$$(18) \qquad \frac{\partial s}{\partial t} + u \frac{\partial s}{\partial x} = \frac{\partial}{\partial x}\left(D_1 \frac{\partial s}{\partial x}\right) + \frac{\partial}{\partial y}\left(D_2 \frac{\partial s}{\partial y}\right) + \frac{\partial}{\partial z}\left(D_2 \frac{\partial s}{\partial z}\right).$$

For this simplified situation it is readily understood how the terms $\overline{vs}$ and $\overline{ws}$ do not vanish even though the seepage components v and w are zero. Previous investigators [1, 3, 7] have shown that for an isotropic porous medium with unidirectional flow parallel to one of the major axes, only these two coefficients of dispersion (D_1 and D_2) need to be considered. For the more general case when v and w are not zero, Eqs. (17) should appear as,

$$\overline{us} = -D_{11}\frac{\partial s}{\partial x} - D_{12}\frac{\partial s}{\partial y} - D_{13}\frac{\partial s}{\partial z}$$

$$(19) \qquad \overline{vs} = -D_{21}\frac{\partial s}{\partial x} - D_{22}\frac{\partial s}{\partial y} - D_{23}\frac{\partial s}{\partial z}$$

$$\overline{ws} = -D_{31}\frac{\partial s}{\partial x} - D_{32}\frac{\partial s}{\partial y} - D_{33}\frac{\partial s}{\partial z} \;.$$

For the previous case expressed by Eq. (18), $D_{11} = D_1$ and $D_{22} = D_{33} = D_2$.

CORRELATION OF THE DISPERSION COEFFICIENTS

By the definition of Eq. (17), the longitudinal dispersion coefficient is given by

$$(20) \qquad D_1 = \frac{-(\overline{us})}{\partial s/\partial x}$$

Because of the difficulty in measuring the quantity $\overline{us}$, experimenters have tended to correlate D_1 with more easily measured quantities (e.g. the seepage velocity and particle diameter). In fact, one is led logically to this correlation since the dimensional units of D_1 are given by the product of a length multiplied by a velocity.

The assumption of gradient dependent mixing made in Eqs. (17) has quite correctly been questioned [2]. The reasoning supporting this assumption is based on an analogy made with other mixing processes, namely molecular diffusion (Brownian motion) and turbulent diffusion. Both of these latter mixing phenomena are characterized by a collision process. One can visualize a tagged particle of fluid undergoing a similar type of complicated motion on its journey through a porous medium. The combination of the changing pore velocity (both magnitude and direction) coupled with the molecular diffusion within the pores gives rise to a mixing process that apparently has the essential characteristics of the Fickian model.

The contribution of molecular diffusion to the macroscopic mass balance can be accounted for by including additional mass flux quantities in the derivation, such as $-D_m(\partial \rho s_a/\partial x)$, etc. where D_m is the molecular diffusivity for the substance in solution. The resulting equation for the simplified case of unidirectional flow parallel to the x-axis becomes (after integration through the macroscopic control volume and assuming the density to be constant),

$$(21) \quad \frac{\partial s}{\partial t} + u\,\frac{\partial s}{\partial x} = \frac{\partial}{\partial x}(\bar{D}_m + D_1)\,\frac{\partial s}{\partial x} + \frac{\partial}{\partial y}(\bar{D}_m + D_2)\,\frac{\partial s}{\partial y} + \frac{\partial}{\partial z}(\bar{D}_m + D_2)\,\frac{\partial s}{\partial z}$$

where $\bar{D}_m$ now becomes the effective molecular diffusivity in porous media [9]. It is seen that as the seepage velocity approaches zero, the disperion coefficients D_1 and D_2 as defined in Eq. (17) should approach zero since the additional mass flux quantities $\overline{u\overset{\circ\circ}{s}}$, $\overline{v\overset{\circ\circ}{s}}$, $\overline{w\overset{\circ\circ}{s}}$ would become vanishingly small. For the case when $u = 0$, Eq. (21) reduces to

$$(22) \quad \frac{\partial s}{\partial t} = \bar{D}_m\left(\frac{\partial^2 s}{\partial x^2} + \frac{\partial^2 s}{\partial y^2} + \frac{\partial^2 s}{\partial z^2}\right).$$

Eq. (21) seems appropriate for the assumed model dealt with here. As the seepage velocity increases, the quantities $\overline{u\overset{\circ\circ}{s}}$, $\overline{v\overset{\circ\circ}{s}}$, and $\overline{w\overset{\circ\circ}{s}}$ become large and the contribution of $\bar{D}_m$ can be neglected in Eq. (21). It should be noted that even when $\bar{D}_m$ can be neglected in Eq. (21), the dispersion coefficients D_1 and D_2 still retain a dependence upon D_m [2]. It is believed that this dependence arises from the lateral diffusion of mass across streamlines. Thus, even for laminar flow, the contribution of this transverse molecular diffusion at the microscopic level can significantly add to the macroscopic mixing process of dispersion.

Eq. (21) can be written in dimensionless form by introducing the following teference quantities:

$$\sqrt{k} = \text{characteristic length}$$
$$s_0 = \text{characteristic concentration}$$
$$U = \text{characteristic velocity}$$
$$\sqrt{k}/U = \text{characteristic time}$$

k denotes the intrinsic permeability of a porous medium and has the units of length squared. Since k depends upon the geometry of the pore system of a porous medium [6], it is felt that it is a more representative characteristic length than the average grain size. The resulting normalized equation becomes (primed quantities indicate dimensionless variables, e.g. $s' = s/s_0$, $x' = x/\sqrt{k}$, etc.),

$$(23) \quad \frac{\partial s'}{\partial t'} + u' \frac{\partial s'}{\partial x'}$$

$$= \frac{1}{P^*} \left[\frac{\partial}{\partial x'}(1 + D_1{}^*)\frac{\partial s'}{\partial x'} + \frac{\partial}{\partial y'}(1 + D_2{}^*)\frac{\partial s'}{\partial y'} + \frac{\partial}{\partial z'}(1 + D_2{}^*)\frac{\partial s'}{\partial z'} \right]$$

where
$$P^* = \text{molecular Peclet No., } U\sqrt{k}/\bar{D}_m$$
$$D_1{}^* = D_1/\bar{D}_m$$
$$D_2{}^* = D_2/\bar{D}_m .$$

The dependence of the dispersion process upon the fluid viscosity is not evident since only P^* and D^* emerged from Eq. (23). As has been noted [2], when the flow in the individual pores becomes nonlinear laminar or unstable and finally turbulent, the effect of viscosity upon the flow should become evident. This dependence would emerge from a separate equation of motion in which the dependence of the seepage velocity upon the molecular viscosity appears. Thus, the dependence of the dispersion coefficients upon the molecular viscosity would be through the dynamics of the flow field which is dependent upon molecular viscosity.

REFERENCES

1. BEAR, JACOB (1961), On the tensor form of dispersion in porous media, *J. Geophys. Res.*, *66* 1185–1197.

2. LIST, E. T. AND N. H. BROOKS (1967), Lateral dispersion in saturated porous media, *J. Geophys. Res.*, *72*, 2531–2541.

3. NIKOLAEVSKII, V. N. (1959), Convective diffusion in porous media, *Prikl. Mat. Mekh.*, *23*, 1042–1050.

4. ROUSE, HUNTER (1959), *Advanced Fluid Mechanics*, John Wiley and Sons, pp. 36 and 203.

5. RUMER, R. R. (1962), Longitudinal dispersion in steady and unsteady flow, *Proc. ASCE, Journal of the Hydraulics Division*, *88*, 147–172.

6. RUMER, R. R. AND P. A. DRINKER, (1966) Resistance to laminar flow through porous media, *Proc. ASCE J. Hydraulics Division*, *92*, September, 155–163.

7. SCHEIDEGGER, A. (1961), General theory of dispersion in porous media, *J. Geophys. Res.*, *66*, 3273–3278.

8. TAYLOR, G. I. (1954) The dispersion of matter in turbulent flow through a pipe, *Proc. Roy. Soc.*, A, *223*, 446–468.

9. WOODING, R. A. (1959), The stability of a viscous liquid in a vertical tube containing porous material, *Proc. Roy. Soc.*, A, *252*, 120–134.

FACULTY OF ENGINEERING AND APPLIED SCIENCES,
STATE UNIVERSITY OF NEW YORK AT BUFFALO,
BUFFALO, NEW YORK, 14214, U.S.A.

SUR LE DÉPLACEMENT BIDIMENSIONNEL
DES FLUIDES MISCIBLES DANS LES MILIEUX POREUX

T. Oroveanu et I. Spulber

1. introduction

Les considérations théoriques concernant le déplacement d'un fluide saturant un milieu poreux par un autre miscible avec lui aboutissent à l'équation

$$(1.1) \qquad \frac{\partial C}{\partial t} + g^{ij}\bar{w}_j\frac{\partial C}{\partial \zeta^i} = \frac{1}{\sqrt{g}}\frac{\partial}{\partial \zeta^i}\left(\sqrt{g}\,g^{ij}K_{jk}g^{kl}\frac{\partial C}{\partial \zeta^i}\right)$$

écrite dans un système de coordonnées curvilignes ζ^i, en utilisant les notations du calcul tensoriel. Dans cette équation, C représente la concentration, $\bar{w}_j$ les composantes covariantes de la vitesse moyenne réelle que l'on obtient en divisant la vitesse de filtration par la porosité et K_{jk} les composantes covariantes du tenseur de dispersion. De même, nous avons noté par g la valeur absolue du déterminant du tenseur métrique g_{ij} et par g^{ij} les composantes du tenseur contrevariant associé au tenseur métrique. Dans un système de coordonnées cartésiennes orthogonales x_i l'équation (1.1) prend la forme

$$(1.2) \qquad \frac{\partial C}{\partial t} + \bar{w}_i\frac{\partial C}{\partial x_i} = \frac{\partial}{\partial x_i}\left(K_{ij}\frac{\partial C}{\partial x_j}\right)$$

ou nous avons utilisé de nouveau la convention de l'indice de sommation.

L'existence d'une anisotropie en ce qui concerne le phénomène de dispersion dans un milieu poreux, même lorsque celui-ci est isotrope, a été mise en évidence par plusieurs chercheurs [1], [2], [3], [7], [11], [12]. Il faut aussi préciser que pour arriver à l'équation (1.1) ou (1.2) on doit supposer que les deux fluides sont incompressibles.

Dans ce qui suit, nous allons nous borner à l'étude de l'écoulement plan dans un milieu poreux homogène et isotrope, problème qui, à notre connaissance, a fait l'objet d'un nombre assez restreint de recherches [3], [4], [5], [8], [13]. Si nous adoptons les conclusions de [2], [7] et [12], il y a dans ce cas ($\bar{w}_3 = 0$) seulement cinq coefficients de dispersion qui ne soient pas nuls à savoir K_{11}, K_{22}, K_{33}, K_{12} et K_{21}. De plus, dans un système de coordonnées orthogonales il n'y a que quatre coefficients de dispersion différents entre eux, car dans ce cas $K_{12} = K_{21}$. Selon certaines de ces considérations théoriques, K_{33} est assez petit par rapport à K_{11} et K_{22}, ce qui nous conduit à le négliger afin d'avoir à étudier un problème essentiellement bidimensionnel. Notre but n'est pas de résoudre certains problèmes particuliers mais d'indiquer des méthodes de résolution qui pourraient être appliquées à cet effet.

La vitesse moyenne réelle n'est soumise, en principe, à aucune restriction en ce qui concerne sa grandeur. Lorsque l'écoulement est assez lent pour que la loi de Darcy soit valable, on peut déterminer assez facilement le champ des vitesses. Le seul inconvénient majeur qui se présente dans ce cas est dû au fait que la viscosité est fonction de la concentration. On peut tourner cette difficulté en supposant que les deux fluides ont la même viscosité, ce qui a été fait d'ailleurs dans beaucoup d'études théoriques concernant la dispersion dans les milieux poreux.

Les recherches, tant théoriques qu'expérimentales, ces dernières ayant trait surtout aux écoulements unidimensionnels, ont montré que les coefficients de dispersion dépendent de la vitesse. Mais il faut remarquer que les relations empiriques ne sont pas toujours en très bon accord avec les formules théoriques, sauf en ce qui concerne le régime de la diffusion mécanique pure, où la diffusion moléculaire ne joue plus aucun role [10]. Cette dernière apparaît dans les formules théoriques soit comme un terme additif [7], soit d'une manière plus compliquée [11].

Dans ce qui suit nous allons considérer d'abord le phénomène de dispersion indépendant du temps, pour passer ensuite au cas plus général qui correspond à une variation de la concentration dans le temps. De même, nous supposerons d'abord que l'écoulement n'obéit pas à la loi de Darcy; l'ultilisation de cette loi, qui facilite les calculs, sera prise en considération ensuite.

2. LA DISPERSION STATIONNAIRE

Selon les hypothèses que nous avons faites, lorsque la concentration ne dépend pas du temps elle est une solution de l'équation

(2.1)
$$K_{11}\frac{\partial^2 C}{\partial x_1^2} + 2K_{12}\frac{\partial^2 C}{\partial x_1 \partial x_2} + K_{22}\frac{\partial^2 C}{\partial x_2^2} + \left(\frac{\partial K_{11}}{\partial x_1} + \frac{\partial K_{12}}{\partial x_2} - \bar{w}_1\right)\frac{\partial C}{\partial x_1}$$
$$+ \left(\frac{\partial K_{22}}{\partial x_2} + \frac{\partial K_{12}}{\partial x_1} - \bar{w}_2\right)\frac{\partial C}{\partial x_2} = 0.$$

Cette équation est du type elliptique parce que nous avons

$$K_{12}^2 - K_{11}K_{22} = K_{11}K_{22}\left(\frac{K_{12}^2}{K_{11}K_{22}} - 1\right) < 0$$

étant donné que le coefficient de dispersion transversale est toujours plus petit que les coefficients de dispersion longitudinale. On peut ramener l'équation (2.1) à la forme canonique, le degré de difficulté des calculs étant lié à la possibilité de trouver les solutions du système d'équations

$$(2.2) \qquad \frac{dx_2}{dx_1} = \frac{K_{12} \pm i \sqrt{K_{12}^2 - K_{11}K_{22}}}{K_{11}}.$$

Une fois cette opération effectuée, nous avons

$$(2.3) \qquad \frac{\partial^2 C}{\partial \xi_2^1} + \frac{\partial^2 C}{\partial \xi_2^2} + a_1(\xi_1, \xi_2)\frac{\partial C}{\partial \xi_1} + a_2(\xi_1, \xi_2)\frac{\partial C}{\partial \xi_2} = 0$$

les nouvelles variables ξ_1, ξ_2 étant exprimées de la manière connue à l'aide des solutions du système (2.2). Les fonctions $a_1(\xi_1, \xi_2)$ et $a_2(\xi_2, \xi_2)$ étant connues, on peut envisager pour l'équation (2.3) différents problèmes aux limites. S'il agit, par exemple du problème de Dirichlet pour un domaine D simplement connexe (la concentration prend des valeurs données sur la frontière de ce domaine), la solution est unique et résulte de l'équation intégro-différentielle

$$C(\xi_1, \xi_2) = F(\xi_1, \xi_2) + \frac{1}{2\pi} \iint_D G(\xi_1, \xi_2; X_1, X_2)\left[a_1(X_1, X_2\frac{\partial C}{\partial X_1}\right.$$

$$(2.4)$$

$$\left. + \ a_2(X_1, X_2)\frac{\partial C}{\partial X_2}\right] dX_1 dX_2$$

$G(\xi_1, \xi_2; X_1, X_2)$ étant la fonction de Green pour l'équation de Laplace pour le domaine D et $F(\xi_1, \xi_2)$ une fonction harmonique qui satisfait à la condition aux limites du problème. Un procédé d'approximations successives nous permet d'écrire la solution sous la forme

$$C = \sum_{n=0}^{\infty} C_n$$

avec

$$C_0 = F(\xi_1, \xi_2)$$

et

$$C_n(\xi_1, \xi_2) = \frac{1}{2\pi} \iint_D G(\xi_1, \xi_2; X_1, X_2)\left[a_1(X_1, X_2)\frac{\partial C_{n-1}}{\partial X_1}\right.$$

$$\left. + \ a_2(X_1, X_2)\frac{\partial C_{n-1}}{\partial X_2}\right] dX_1 dX_2.$$

Il y a aussi la possibilité de ramener (2.4) à une équation intégrale; après un calcul simple dans lequel nous faisons intervenir le fait que la fonction de Green est nulle sur le contour de D, nous obtenons

$$(2.5) \quad C(\xi_1, \xi_2) = F(\xi_1, \xi_2) - \frac{1}{2\pi} \iint_D \left[\frac{\partial(a_1 G)}{\partial X_1} + \frac{\partial(a_2 G)}{\partial X_2}\right] C(X_1, X_2)dX_1 dX_2$$

equation qu'on peut aussi résoudre à l'aide d'un procédé d'approximations successives.

Si l'écoulement obéit à la loi de Darcy, nous avons en coordonnées cartésiennes orthogonales

$$\bar{w}_i = \frac{v_i}{m} = \frac{\partial \phi}{\partial x_i} \ (i = 1, 2), \quad \phi = -\frac{k}{m\mu}(p + U)$$

v_i représentant les composantes de la vitesse de filtration, m la porosité, k la perméabilité, μ la viscosité dynamique de fluide, p la pression et U le potentiel des forces extérieures. La fonction ϕ et la fonction de courant ψ sont harmoniques et conjuguées. Avec ϕ et ψ on peut définir donc un système de coordonées orthogonales, en obtenant ainsi l'équation [2]

$$(2.6) \qquad \frac{\partial}{\partial \phi}\left(K_1 \frac{\partial C}{\partial \phi}\right) + \frac{\partial}{\partial \psi}\left(K_2 \frac{\partial C}{\partial \psi}\right) - \frac{\partial C}{\partial \phi} = 0$$

K_1 étant le coefficient de dispersion longitudinale et K_2 le coefficient de dispersion transversale.

En prenant pour ces coefficients les expressions [2], [12]

$$K_1 = a_{\mathrm{I}}\bar{w}, \quad K_2 = a_{\mathrm{II}}\bar{w}$$

et en introduisant les nouvelles variables

$$\Phi = \frac{1}{\sqrt{a_{\mathrm{I}}}}\phi, \quad \Psi = \frac{1}{\sqrt{a_{\mathrm{II}}}}\psi$$

nous trouvons au lieu de (2.6) l'équation

$$(2.7) \qquad \frac{\partial^2 C}{\partial \Phi^2} + \frac{\partial^2 C}{\partial \Psi^2} + \frac{1}{\bar{w}}\left(\frac{\partial \bar{w}}{\partial \Phi} - \frac{1}{\sqrt{a_{\mathrm{I}}}}\right)\frac{\partial C}{\partial \Phi} + \frac{1}{\bar{w}}\frac{\partial \bar{w}}{\partial \Psi}\frac{\partial C}{\partial \Psi} = 0.$$

Cette équation peut être regardée comme un cas particulier de (2.3) et par conséquent le problème de Dirichlet a la solution

$$(2.8) \quad C(\Phi, \Psi) = F(\Phi, \Psi) + \frac{1}{2\pi} \iint_D G(\Phi, \Psi; \Phi, \Psi)\left[\frac{1}{\bar{w}}\left(\frac{\partial \bar{w}}{\partial \Phi} - \frac{1}{\sqrt{a_i}}\right)\frac{\partial C}{\partial \Phi}\right.$$

$$\left. + \frac{1}{\bar{w}}\frac{\partial \bar{w}}{\partial \Psi}\frac{\partial C}{\partial \Psi}\right] d\Phi d\Psi$$

déduite de (2.4). On peut appliquer un procédé d'approximations successives soit à (2.8) soit à l'équation intégrale qui en résulte, analogue à (2.5).

Nous avons considéré de plus près le problème de Dirichlet parce qu'il est le plus simple. Dans les éventuelles applications, ce problème apparaît assez particulier, les conditions aux limites étant souvent mixtes. Nous signalons seulement le fait que les études théoriques concernant les problèmes aux limites pour l'équation du type elliptique dont nous nous occupons sont assez poussées et permettent d'envisager aussi la résolution d'autres problèmes que celui de Dirichlet. Mais ces considérations demanderaient trop d'espace pour être exposées ici.

3. LA DISPERSION NON STATIONNAIRE

Lorsque la concentration dépend aussi du temps il faut considérer l'équation

$$(3.1) \quad \frac{\partial C}{\partial t} = \bar{w}^3 \left[\frac{\partial^2 C}{\partial \Phi^2} + \frac{\partial^2 C}{\partial \Psi^2} + \frac{1}{\bar{w}} \left(\frac{\partial \bar{w}}{\partial \Phi} - \frac{1}{\sqrt{a_I}} \right) \frac{\partial C}{\partial \Phi} + \frac{1}{\bar{w}} \frac{\partial \bar{w}}{\partial \Psi} \frac{\partial C}{\partial \Psi} \right]$$

si nous faisons dès le commencement l'hypothèse que l'écoulement obéit à la loi de Darcy et si nous utilisons les mêmes formules que plus haut pour K_1 et K_2.

L'équation (3.1) étant linéaire, la méthode de la séparation des variables appliquable. On peut facilement trouver la fonction qui dépend seulement du temps et en ce qui concerne l'équation qui dépend de Φ et de Ψ, il est facile de constater que celle-ci est toujours du type elliptique. Il est donc possible de développer des considérations analogues à celles que nous avons exposées plus haut.

Nous signalons enfin le fait que si la vitesse de l'écoulement dépend elle aussi du temps, le problème devient extrêmement compliqué.

4. CONCLUSION

Comme nous l'avons déjà dit, cette étude a eu pour but seulement de montrer quelles sont les possibilités de trouver les solutions des problèmes concernant le déplacement bidimensionnel des fluides miscibles. Les résultats ne sont pas particulièrement encourageants parce que la résolutions d'une équation du type que nous avons rencontré présente des difficultés de calcul assez grandes. Les méthodes qui ramènent ce problème à une équation intégro-différentielle ou intégrale ont toutefois l'avantage de permettre l'éclaircissement des questions concernat l'existence et l'unicité des solutions, parfois laissées de côté dans les applications.

La possibilité de trouver des solutions approximatives, en faisant certaines simplifications dans l'équation différentielle peut être aussi envisagée. Ces simplifications étant liées aux cas particuliers qu'on se propose d'étudier, ne rentrent pas dans le cadre de ces considérations générales.

RÉFÉRENCES

1. BEAR, J., (1961), On the tensor form of dispersion in porous media, *J. Geophys. Res.,* *66*, 4.

2. BEAR AND I. BACHMAT (1964), The general equations of hydrodynamic dispersion in homogeneous, isotropic, porous mediums, *J. Geophys. Res., 69,* 12.

3. CARRIER, G. F. (1958), The mixing of ground water and sea water in permeable subsoils, *J. Fluid Mechanics*, *4*, 5.

4. GARDER, A. O. Jr., D. W. PEACEMAN AND A. L. POZZI Jr. (1964), Numerical calculation of multidimensional miscible displacement by the method of characteristics, *Trans. AIME*, *251*.

5. HOOPES, J. A. AND D. R. F. HARLEMAN (1967), Wastewater recharge and dispersion in porous media, Proceedings of the American Society of Civil Engineers, *J. Hydraulics Division*, *93*, HY5.

6. JOSSELIN DE JONG, G. de (1958), Longitudinal and transverse diffusion in granular deposits, *Trans. Amer. Geophysical Union*, *39*, 1.

7. NIKOLAEVSKII, V. N. (1959), Konvektivnaja diffuzija v poristykh sredakh, *Prikladnaja matematika i mekhanika*, *23*, 6.

8. PEACEMAN, D. W. AND H. H. RACHFORD Jr. (1962), Numerical calculations of multidimensional miscible displacement, *Trans. AIME*, *225*.

9. PEACEMAN, D. W. (1966), Improved treatment of dispersion in numerical calculation of multidimensional miscible displacement, *Trans AIME*, *237*.

10. PFANNKUCH, H. O. (1963), Contributions à l'étude des déplacements de fluides miscibles dans un milieu poreux, *Revue de l'Institute Français du Pétrole et Annales des combustibles liquides*, *18*, 2.

11. SAFFMAN, P. G. (1959), A theory of dispersion in a porous medium, *J. Fluid Mechanics*, *6*, 3.

12. SCHEIDEGGER, A. E. (1961), General theory of dispersion in porous media, *J. Geophys. Res.*, *66*, 10.

13. STONE, H. L. AND P. L. T. BRIAN (1963), Numerical solution of convective transport problems, *AIChE J.*, *9*, 5.

INSTITUTUL DE PETROL, GAZE SI GEOLOGIE,
STR. TRAIAN VUIA 6, BUCURESTI, ROUMANIA

5

Problems of permeability, matrix deformability, consolidation, anisotropy and heterogeneity

VERTICAL AND HORIZONTAL LABORATORY PERMEABILITY MEASUREMENTS IN CLAY SOILS

W. B. Wilkinson and E. L. Shipley

1. introduction

The soils engineer, in attempting to predict the rate of dissipation of pore water pressure and settlement in clay soils subject to foundation loads, requires the values of both the vertical, c_v, and horizontal coefficients of consolidation, c_h, of the soil. These soil properties are frequently determined from laboratory tests on small samples, the results being analysed by simple Terzaghi consolidation theory. However, the consolidation behaviour of a clay may be influenced by the effect of decreasing permeability — compressibility and structural viscosity (creep properties) during a test. A number of theories to take account of these variable soil properties for both vertical (Taylor and Merchant, (1940) Barden, (1965) and horizontal consolidation (Schiffman, (1958); Berry and Wilkinson, (1969) show that c_v and c_h are not constants but functions of the permeability and compressibility changes during a test and the creep properties of the soil.

A varying permeability and compressibility will only affect the use of the laboratory c_v and c_h values in field predictions if the permeability change and and loading increment is large (Berry and Wilkinson, (1968). In such a case, in order to use the available theoretical solutions, it is necessary to measure the *actual* permeability change in the soil due to the consolidation process.

Structural viscosity effects are particularly marked at the laboratory scale, much less so in the field. The laboratory c_v and c_h values (based on volume change measurements) may appreciably underestimate the field values. These field values in soils which show marked structural viscosity effects may be deduced by combining the actual measured permeability k with a coefficient of volume change m_v (over the relevant pressure increment) in accordance with the relationship

(1)
$$c = k/\gamma_w m_v$$

Experimental and theoretical studies (Rowe, (1959; 1964; 1968), Horne, (1964) on clay soils with a more pervious macro structure such as silt layering, fissures, pervious organic inclusions etc., show that the coefficient of consolidation, c, increases with sample size, ultimately approaching a maximum value that would pertain at the field scale. Thus the use of standard laboratory methods in such soils may markedly underestimate the values that would obtain at the field scale. However, it has been shown by Rowe, (1964) that field value may be related to the overall *directly* measured permeability of the soil through Eq. 1 provided the laboratory sample is sufficiently large to be representative of the soil mass and sampling disturbance (smear etc.) is small.

In almost all consolidation theory Darcy's law is assumed to be valid (Schwartzendruber, (1962). However, a limited number of derivations from Darcy's law at low hydraulic gradients for the flow of water in fully saturated clays have been reported (Hansbo, (1960) Mitchell and Younger, (1966)). Such observations are of particular relevance in the consolidation of clays at the field scale where this low gradient condition will generally be met with. These deviations (if they exist) may not be readily observed in a laboratory consolidation test due to the large gradients that exist as a result of the limited sample size.

Although, as indicated, the *direct* measurement of vertical k_v and horizontal permeability, k_h, of clay soils and the validity of Darcy's law in these materials may be of considerable importance in estimating pore water pressure dissipation, and settlement rates, few measurements with this as the specific purpose have been made. This appears to be due firstly to the widespread belief that the deduced k from a c value was in all cases closely related to the directly measured value, in which case there would be little point in performing the direct test, and secondly to the difficulties involved in carrying out the direct k tests, particularly in the measurement of the small water flow rates in clay soils where a small leak in the system may invalidate the results.

2. PREVIOUS LABORATORY MEASUREMENTS

Many attempts have been made to measure the *vertical* permeability of clay soils following the constant and falling head methods originally proposed by Terzaghi (1925). A most sophisicated recent technique used on small samples of remoulded clay is described by Olsen (1966). In spite of the fact that in many natural soils k_h may be many times greater than k_v surprisingly few measurements of *horizontal* permeability are reported (Kenney, (1964); Raymond, and Azzouz (1969); Normand, (1964); Wit, 1967)).

In view of the importance of both k_v and k_h in relation to consolidation studies, vertical and radial permeameters were designed to enable direct permeability measurements to be made at the intermediate stages between the application of pressure increments in a consolidation test.

3. DESCRIPTION OF APPARATUS

The laboratory consolidation and permeability tests were performed in a new hydraulically loaded, sealed consolidation cell. A description of the cell and its use in consolidation testing has been given by Rowe and Barden, (1966) In order to use it as a permeameter certain modifications were necessary in addition to which constant back pressure and flow rate measuring systems are required.

Consolidation-permeameter cell — Cells of diameter 3, 6, 10 and 20 in are available all having the same basic features. The cell adapted for *vertical* permeability tests is shown in Fig. 1a. The sample is contained in a brass cylinder (1) machined to give a smooth internal surface. Two porous drainage discs (2) cover the sample at its top and bottom surfaces. Loading is applied hydraulically through a convoluted rubber jack (3). The whole of the above is contained between a metal lid (4) and base (5). These are bolted to the cylinder and a watertight seal is provided by a flange on the rubber jack (6) and a rubber 'O' ring (7). Vertical settlements are measured by a dial gauge (8) which follows a brass spindle (9) attached to the rubber jack (10). A central hole up the spindle connects the top drain to an outlet cock. The bottom drain connects to a small diameter hole (11) leading to the edge of the base and closed by a cock. The jack water pressure in the range 0–140 p.s.i. is maintained by a constant head mercury system (Skempton and Bishop, (1950).

For *radial* flow tests a central sand drain (1) is installed in the sample which is surrounded circumferentially by a porous plastic drain (2), Fig. 1b. The central sand drain rests on a porous plug (3) set in the base and connected by a small diameter hole to an external cock. The peripheral drain is in contact

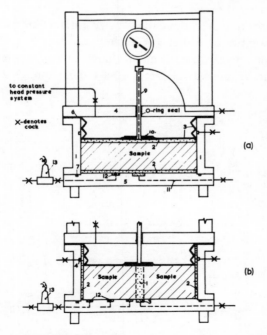

Fig. 1. Consolidation-permeability cell (a) vertical (b) radial

with a strip of porous material set in a groove (4) in the cell body this being connected by small holes to two external cocks on opposite sides of the ring.

Small ceramic sensing elements (12) set in the cell base and connected to electric pressure transducers (13) enable pore water pressure to be measured both during a consolidation test and also during steady seepage under radial flow conditions.

Flow measuring apparatus — The basic details of the flow measuring apparatus are shown in Fig. 2. A constant back pressure (range 0–30 psi) is applied to the cell drainage leads from a mercury system (1 and 2). During a consolidation tests against a back pressure where the flow rate from the sample is large the upper mercury cylinders are suspended from springs in the normal manner (Skempton and Bishop 1950). A differential mercury manometer (3) or pressure gauge set between the jack and back pressure leads records directly the final effective stress acting on the sample. During a permeability test a differential pressure causing flow across the sample is applied by adjusting one of the back pressure lines to give the required differential, the value of this is read directly on either a coloured carbon tetrachloride (4) or mercury manometer (5). When the flow rate is very low the upper mercury cylinders are transferred from the springs to screwed rods (6) so that an almost constant gradients is maintained.

One of two flow rate measuring units can be introduced into the back pressure lines by a correct manipulation of the cocks on the junction blocks (7). For very high flow rates a twin burette water/paraffin volume change apparatus with burette capacities of 50 c.c. (8) (Bishop and Henkel (1957) is used. For low flow rates the rate of travel of an air bubble in a precision bore glass tube 70 cms. long and of internal diameter 1.5 ($\pm$ 0.01) mm. (9) is recorded. The

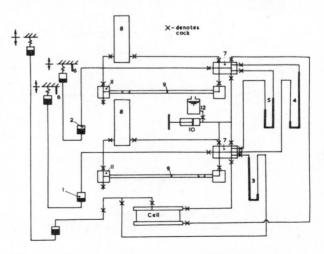

Fig. 2. Flow rate measuring apparatus

tubes are frequently changed and cleaned to keep grease and dirt contamination to a minimum (Olsen, (1966) On reaching the end of the tube the bubble is returned by the use of a small hand pump (10). This is also used to introduce a bubble into the tube through one of the junction blocks (11) and also to recharge the back pressure mercury cylinders. The whole is filled and when necessary recharged with de-aired water (12).

The flow measuring systems are duplicated so that in a permeability test both inflow and outflow at the sample may be recorded.

4. PREPARATION OF SAMPLES AND TEST PROCEDURE

Great care should be taken in the recovery and preparation of field samples in order to reduce disturbance to a minimum and to maintain full saturation. Remoulded samples are prepared by vacuuming a clay slurry in the test cell. The method of recovery, preparation and installation of samples in the cell with the required drainage boundary conditions closely follows that described by Rowe and Barden, (1966).

Once installed in the cell the samples are "bedded in" under a small effective pressure (usually 3 psi). In the tests a back pressure in the range 20 to 30 psi is generally applied. A permeability test is performed when the flow rates from the sample due to creep are relatively small with respect to the anticipated permeability stage flow rate. In general the differential pressure to cause flow is applied by increasing one of the back pressures, as a result of which the effective stress is reduced slightly and the sample swells. This gives much less of a change in the void ratio and therefore the permeability of the sample than would the consolidation resulting from a lowering of the back pressure by an equivalent amount. In addition the coefficient of swelling may be several times greater than the coefficient of consolidation (Kenney, (1964) and therefore the steady seepage state is reached much more rapidly. By following this procedure and only in usual circumstances allowing the pressure difference across the sample to exceed 15% of the applied effective stress it is felt that the effect of the imposed test conditions on the measured value of k is very small. This disturbance could possibly be overcome by clamping the top of the sample and so preventing swelling but this is difficult to accomplish in an hydraulically loaded permeameter and is in any case felt to be unnecessary. Steady state conditions are taken as established when inflow equals outflow to within 10%. Any significant leakage in the system becomes evident in the course of testing and its location and magnitude is easily determined.

In radial flow tests the seepage pressure distribution across the base of the sample is measured by the use of the ceramic pore pressure points set at different radii between the central sand drain and the peripheral drain. This distribution is compared to the theoretical as given by Eq. 2 below. Such a comparison is useful in assessing the degree of smear adjacent to the drainage boundaries as subsequently described.

Following the determination of k the sample is consolidated by increasing the jack pressure and allowing drainage. Settlement, volume change and pore pressures are recorded at suitable intervals during the test from which the c_v or c_h and m_v are determined. The use of Eq. 1 gives indirect values of k_v or k_h. At the end of each consolidation stage the permeability is measured directly as previously described.

5. DETERMINATION OF THE DIRECT k AND THE EFFECT OF SMEAR

In a vertical test k_v is simply calculated by the direct application of Darcy's law. The boundary conditions in a radial flow test are shown in Fig. 3a. The distribution of head h across the base of the sample and flow rate Q are given by the artesian well formula (Muskat, (1937). Using Eq. 3 below, k_h may be determined.

(2) $$h = h_w - [\log_e r_e/r_w(h_w - h_e)]/\log_e r_e/r_w$$

(3) $$Q = 2\pi L k_n(h_w - h_e)/\log_e r_e/r_w$$

The symbols are defined in Fig. 3.

In a clay soil with a more pervious macro structure, such as horizontal silt layering etc., smearing and therefore a reduction in k may occur at the drainage boundaries. Radial flow permeability tests would appear to be particularly

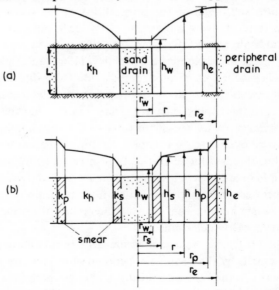

Fig. 3. Boundary conditions in a radial flow permeabilty test (a) without smear (b) with smear

susceptible to this type of disturbance. The use of Eq. 3 would, in such a case, not yield the correct value of the undisturbed k_h. Smearing effects may be given qualitative consideration by assuming a uniform layer of smear adjacent to the central sand and peripheral drains, Fig. 3b. For the boundary conditions shown in Fig. 3b the flow rate through the sample is

$$(4) \qquad Q = 2\pi L(h_e - h_w) \Big/ \left[\frac{\log_e r_e/r_p}{k_p} + \frac{\log_e r_p/r_s}{k_h} + \frac{\log_e r_s/r_w}{k_s} \right]$$

In a particular test the extent of smear is generally unknown. An erroneous measured value of permeability k_m would be obtained from Eq. 3. By equating Eq. 3 (with $k_h = k_m$) and Eq. 4 and assuming the permeability of the smeared zones at both drains is the same ($k_p = k_s$) the ratio of the undisturbed permeability k_h to the measured k_m is given by

$$(5) \qquad \frac{k_h}{k_m} = \left[\frac{k_h}{k_s} \log_e \frac{r_e}{r_p} + \log_e \frac{r_p}{r_s} + \frac{k_h}{k_s} \log_e \frac{r_s}{r_w} \right] \Big/ \log_e \frac{r_e}{r_w}$$

Neglecting smear adjacent to one drain then Eq. 5 has been plotted in Fig. 4 to show the variation of k_h/k_m with k_h/k_s (equals k_h/k_p) and r_s/r_w (equals r_e/r_p). Fig. 4 indicates that the measured permeability k_m may markedly under-estimate the undisturbed k_h if either r_s/r_w (or r_e/r_p) and k_h/k_s (or k_h/k_p) are large. Such a condition is particularly liable to occur in a soil with a more pervious macro structure.

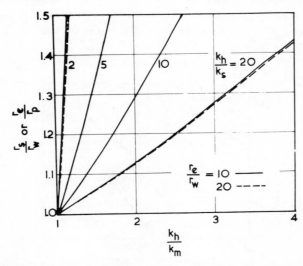

Fig. 4. Theoretical relationships between the measured permeability k_m and the undisturbed k_h in terms of the smear parameter r_s/r_w and k_h/k_s

However, the measurement of the pressure distribution across the base of the sample, as described above, enables an estimate of h_p and h_s to be made and the use of these in Eq. 3 should lead to a more correct value of k_h in a soil having a marked homogeneous anisotropy.

6. TEST RESULTS

Remoulded clays — Three saturated remoulded clays, Derwent, Kaolin and Frodsham were tested to determine their consolidation and permeability properties. Their index properties are given in Table I.

TABLE I

Clay	Description	Liquid limit	Plastic limit
Derwent	Lacustrine laminated silty clay	59	22
Kaolin	Fine china clay	72	23
Frodsham	Estuarine organic silty clay	42	20

Distilled water was used in their preparation and as the permeant. The test samples were generally 10 in. diameter and 1 to 2 in. thick. Direct k and average indirect k values (as determined from the volume change reading of the consolidation stage of the tests in accordance with Eq. 1 (Wilkinson, (1968)) are plotted to a logarithmic scale against void ratio for these three clays in Figs. 5, 6 and 7. Derwent and kaolin clays show good agreement for both vertical and horizontal tests between the direct and indirect k values. These clays have small creep properties and therefore the conventional c_h and c_v values determined from relatively small laboratory samples on such clays should predict the field behaviour.

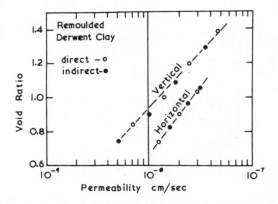

Fig. 5. Experimental void ratio — log k relationships, remoulded Derwent clay

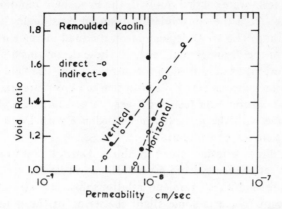

Fig. 6. Experimental void ratio — log *k* relationships, remoulded Kaolin

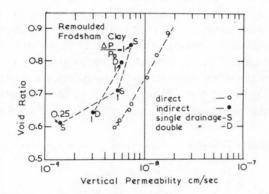

Fig. 7. Experimental void ratio — log *k* relationships, remoulded Frodsham clay

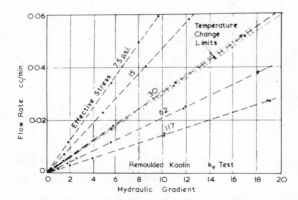

Fig. 8. Experimental flow rate — hydraulic gradient relationships, remoulded Kaolin

Only vertical tests were carried out with the Frodsham remoulded clay. The deduced k values from the consolidation tests underestimate the actual k by a factor in the range (1.5 to 4) depending on the load increment ratio ($\Delta p/p_0$) and the length of the drainage path. Fig. 7. These deviations are due to structural viscosity (creep) as described in the introduction. The field consolidation properties of homogeneous clay soils with marked creep are best obtained by performing the consolidation tests with very large samples or by combining the direct k value with the m_v value in accordance with Eq. 1 and thus deducing a coefficient of consolidation.

In all of the direct k tests described above Darcy's law was found to be valid. A typical set of results from a vertical test on kaolin are shown in Fig. 8. The flow rate:hydraulic gradient relationships are all linear and pass through the origin. There is some slight scattering of the points but this is probably explained in terms of temperature change during the test. Limit lines with respect to the 30 psi line are drawn in Fig. 8 in accordance with the temperature variation ($20\pm1°C$). Almost all points fall within these lines.

The pressure distribution in the radial flow k tests on Derwent and kaolin clays was measured and a typical kaolin result is shown in Fig. 9. The experimental points lie exactly on the theoretical line as given by Eq. 2 and indicate that in these tests the smear adjacent to the boundaries was negligible.

Natural clays — 10 and 6 in. diameter undisturbed samples of a normally consolidated estuarine silty clay containing decomposed organic rootlets were

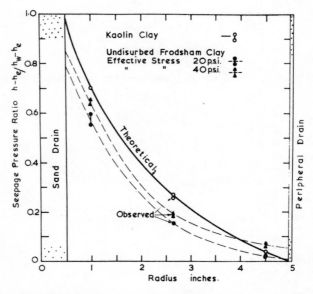

Fig. 9. Observed and theoretical seepage pressure distribution in radial flow permeability tests on kaolin and undisturbed Frodsham clay

obtained from a construction site at Frodsham (Shipley, (1967); Rowe, (1968). Vertical and radial consolidation and permeability tests were performed on these samples.

At an effective stress below 15 psi both radial and vertical samples permeated with *distilled water* gave initial direct permeability values in the range 10^{-6} to 10^{-7} cms/sec. but these reduced to values of 10^{-8} to 10^{-9} cms/sec with continued flow through the sample. The measured k was very sensitive to any change in back pressure (even though the effective stress was maintained at a constant value) or in hydraulic gradient, the tendency being for the permeability to greatly increase and then slowly decrease under the new equilibrium conditions with flow. Darcy's law was therefore invalid for these tests. In the vertical tests the flow was generally from the top to the bottom of the sample. Reversal caused an immediate increase in permeability which remained at a constant high value until the flow was again caused to be in the downward direction. A typical result for these conditions giving the relationship between permeability and flow is shown in Fig. 10.

At effective stresses greater than 15 psi the measured values of permeability for samples permeated with distilled water were independent of the amount of water that had flowed through the sample. Darcy's law was also valid at all effective stress levels for samples permeated with natural pore water collected from the site (the natural pore water contained 0.58% NaCl by weight with traces of calcium and magnesium salts).

The phenomenon described above is best explained in terms of a combined leaching and particle migration effect. Leaching due to the permeation of dis-

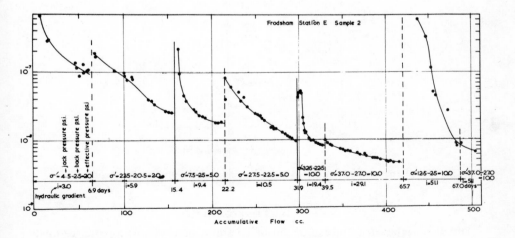

Fig. 10. Permeability — accumulative flow relationship, undisturbed Frodsham clay

tilled water would lead to a condition favourable to the formation of a mobile clay phase which may migrate under the action of the seepage forces. A decrease in permeability due to the blocking of the principal flow channels by this clay phase may result. This is probably exaggerated by the presence of organic rootlets. These are likely to form preferential flow paths provided they remain open. At an effective stress greater than 15 psi these effects are no longer observed and at this stress level the rootlets are most probably closed. These tests indicate that some consideration should be given to the choice of permeant for k tests on natural soils or misleading results may be obtained.

From consolidation-permeability tests in which Darcy's law was valid the direct and deduced permeabilities of the undisturbed clay were obtained as previously described for the remoulded materials. Fig. 11 shows a typical log k — void ratio plot for a radial consolidation-permeability test. As with the remoulded Frodsham clay structural viscosity effects are again evident, the deduced average k from the consolidation stage falling below the directly measured k value at the same void ratio.

Two typical radial steady seepage pressure distributions from such a test are shown in Fig. 9. Comparison with the theoretical distribution shows that applied seepage pressure at the boundaries is reduced due to smearing by some 20%.

Layered clays — In view of the importance of direct permeability measurements in relation to the consolidation properties of silt layered clays a series of tests on this type of material are at present being conducted at the University of Manchester and it is hoped to report the results of these at some later date.

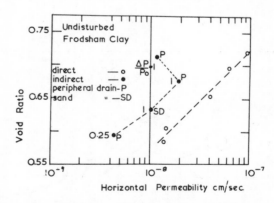

Fig. 11. Experimental void ratio — log k relationships, undisturbed Frodsham clay

7. CONCLUDING REMARKS

The direct laboratory measurement of k_h and k_v may enable a better estimate of field consolidation behaviour to be made in soils with either marked creep properties or a more pervious macro structure. This first effect is clearly demonstrated by tests on one undisturbed and three remoulded clays. Tests to demonstrate the effect of the second point are at present being made. Darcy's law was generally valid, the exception occuring in an organic clay where the permeant was not in chemical equilibrium with the pore fluid and a changing k under constant hydraulic gradient occurred, possibly as a result of particle migration. The new permeameter systems functioned well throughout the tests. A possible difficulty in the interpretation of radial flow test results due to smearing at the drains may be overcome by measuring the seepage pressure distribution across the sample base.

ACKNOWLEDGMENTS

The authors wish to thank Professor P. W. Rowe of the University of Manchester for his advice and encouragement with this work.

REFERENCES

1. BARDEN, L., (1965) Consolidations of clay with non-linear viscosity, *Géotechnique 15*: 4: 34–362.

2. BERRY, P. L. AND W. B. WILKINSON, (1969) The radial consolidation of clay soils, *Géotechnique 19*: 2: 253–284.

3. BISHOP, A. W. AND D. J. HENKEL (1957) *The Measurement of Soil Properties in the Triaxial Test*, Arnold, London.

4. HANSBO, S. (1960) Consolidation of clay with special reference to the influence of vertical sand drains, *Proc. Swedish Geotech. Inst.* No. 18.

5. HORNE, M. R. (1964) The consolidation of a stratified soil with vertical and horizontal drainage, *Int. J. Mech. Sciences 6*, 187–197.

6. KENNEY, T. C. (1964) Pore pressures and bearing capacity of layered clays, *J. Soil Mech. Fdns. Div. Am. Soc. Civ. Eng. 90*. S.M. 4, 27–55.

7. MITCHELL, J. K. AND J. S. YOUNGER, (1966) Abnormalities in hydraulic flow through fine grained soils. *Sym. Permeability and Capillarity of Soils. 69th Meet. A.S.T.M.*

8. MUSKAT, M. (1937) *The Flow of Homogeneous Fluids Through Porous Media*, McGraw-Hill, New York.

9. NORMAND, J. (1964) One-dimensional consolidation of saturated clay with large strains, M.Sc. Thesis, University of London.

10. OLSEN, H. W. (1966) Darcy's Law in saturated kaolinite, *Water Resources Research 2*, 2, 287–295.

11. RAYMOND, G. P. AND M. M. AZZOUZ (1969) Permeability determination of predicting rates of consolidation, *British Geotechnical Society Conf. on In situ Investigations Soil and Rock*.

12. ROWE, P. W. (1959) Measurement of the consolidation of lacustrine clay, *Géotechnique 9*, 107–118.

13. ROWE, P. W. (1964) The calculation of the consolidation rates of laminated, varved or layered clays with particular reference to sand drains, *Géotechnique 14*, 321–340.

14. ROWE, P. W. (1968) The influence of geological features of clay deposits on the design and performance of sand drains, *Proc. Inst. Civil Engrs. 38*, 465–466.

15. ROWE, P. W. AND L. BARDEN, (1966) A new consolidation cell, *Géotechnique 16*, 2:162–170.

16. SCHIFFMAN, R. L. (1958) Consolidation of soil under time dependent loading and and varying permeability, *Proc. Highway Res. Board*, Vol. *37*.

17. SHIPLEY, E. L. (1967) Laboratory permeability tests on undisturbed samples of an estuarine clay, M. Sc. Thesis, University of Manchester.

18. SKEMPTON, A. W. and A. W. BISHOP, (1950) The measurement of the shear strength of soils, *Géotechnique 2*, 90–108.

19. SWARTZENDRUBER, D. (1962) Modifications of Darcy's Law for the flow of water in soils, *Soil Science, 93*, 22–29.

20. TAYLOR, D. W. AND W. MERCHANT (1940) A theory of clay consolidation accounting for secondary compressions, *J. Math. Phys. 19*, 3, 167–185.

21. TERZAGHI, K. (1925) Principles of soil mechanics III — Determination of the permeability of clay, *Eng. News Record*, Vol. *95*.

22. WILKINSON, W. B. (1968) Permeability and consolidation measurements in clay soils, Ph. D. Thesis, University of Manchester.

23. WITKE, (1967) Apporatus for measuring hydraulic conductivity of undisturbed soil sampler, *Permeability and Capillarity of Soil. Special Technical Publication* No. 417 *Am. Soc. Test. Mats.*

W. B. WILKINSON, B.Sc., Ph.D. M.I.C.E.,
    LECTURER IN CIVIL ENGINEERING,
        UNIVERSITY OF MANCHESTER,
            MANCHESTER, M13 9PL, ENGLAND

E. L. SHIPLEY, B.Sc., M.Sc., ENGINEER
    PUBLIC SERVICE COMMISSION,
        ROOM 522, TOWER 'A' BUILDING, PLACE DE VILLE

THE FLOW OF AIR AND WATER
IN PARTLY SATURATED CLAY SOIL

L. Barden, A. O. Madedor and G. R. Sides

ABSTRACT

The following applies to a low clay content soil and has yet to be verified on high clay content soil which involves testing problems. The values of k_a and k_w are confirmed to be functions of the structure and saturation. Both of these are complex parameters, particularly structure which may contain various levels of micro and macrostructure; and this in turn complicates the concept of saturation, since the micropores may be saturated and the macropores unsaturated. Structure has not yet been properly defined and so cannot be measured. For the present it can only be stated that both k_a and k_w are dominated by the macropores and their degree of saturation; however, at low degrees of saturation k_w must be influenced by the microstructure. Occlusion can be caused by decreasing the suction or increasing the applied stress. In both cases k_a greatly exceeds k_w right up to occlusion, showing that pore pressure dissipation from soils with continuous air voids is governed by k_a. Following occlusion k_a is meaningless and transport of air is by a slow diffusion process. k_w now controls drainage and is influenced by the size of the air bubbles in the macropores and hence by the absolute value of u_w but the effect does not appear great.
The values of k_a decreases very abruptly as the last trace of suction is removed and it appears that occlusion finally occurs when the suction reaches zero. However, this may be merely an apparent result, since there are no methods available for measuring the local values of u_a in an occluded air bubble, particularly with such a low value of the diffusion coefficient D.

SYNOPSIS

Air and water permeabilities have been measured directly, during steady state flow conditions, on the same samples of a clay soil compacted at different moisture contents. This required the independent control of the effective stress components of applied stress and suction, which govern compression and degree of saturation, and was finally achieved in a modified triaxial cell.

It was found that both the air and water permeabilities (k_a and k_w) were functions of soil structure and degree of saturation, with macrostructure being the dominant factor. As the soil suction was decreased, either by a compression or a wetting process, the soil air voids tended to become occluded (discontinuous) and k_a, which had hitherto greatly exceeded k_w, fell by many orders towards k_w.

Following occlusion, the concept of k_a is not relevant as the transport of air is now by a true diffusion process in which the diffusion coefficient was shown to be very low due to the adsorbed nature of the pore water. The magnitude of k_w, after occlusion was shown to be influenced by the size of the air bubbles and hence by the absolute value of the pore water pressure, although the effect did not appear great.

INTRODUCTION

The consolidation process in a clay soil is dominated by the permeability of the material. In a partly saturated clay the behaviour is greatly dependent on the continuity of the air in the voids; in dry clays the air is continuous, while in a wet clay it exists in a discontinuous or occluded state. Thus in the former (dry) state there are two distinct permeabilities k_a and k_w for the separate air and water phases respectively. In the occluded state there is only a value of k_w since air transport now occurs by a true diffusion rather than a flow process.

The present investigation is part of a more general one into the consolidation and compression characteristics of partly saturated clay, with particular reference to compacted clay. In a compacted clay the division between the dry (continuous) state and the wet (occluded) state usually occurs in the region of optimum moisture content, and hence this transition zone is of considerable practical interest. Although the present tests are concerned with laboratory compacted samples, it is anticipated that the results will also be relevant to undisturbed samples of partly saturated soil.

In basic studies of k_a and k_w it is not sufficient simply to pass air or water through the samples as this will alter the state of the soil. This has been recognised by workers in Soil Science, Richards [10], Corey [4], who have independently controlled the air and water pressures and hence the degree of saturation of the material. These earlier studies were concerned mainly with changes of saturation in a sensibly unchanging pore geometry (structure). However, in studying consolidation it is clearly necessary to include the compression of the soil skeleton, and thus both changes in structure and saturation must be controlled. This can be done in terms of the effective stress [2]

$$\sigma' = (\sigma - u_a) + \chi(u_a - u_w)$$

where σ is the total stress, u_a the pore air pressure, u_w the pore water pressure and χ an empirical coefficient. The component $(\sigma - u_a)$ is termed the applied stress and $(u_a - u_w)$ the suction.

This has been recognised by M.I.T. [8] who constructed a one-dimensional compression cell, which appears to be a development from the apparatus of Corey [4], in which the vertical confining stress σ can be controlled and compressions measured. An alternative approach is to use transient flow measurements [5]. However, for the present research, it was decided to use the steady state flow method, and the first apparatus to be tested was a modified version of the M.I.T. cell. This finally led to the development of a modified triaxial cell, as described in the following section.

DEVELOPMENT OF APPARATUS

A Rowe consolidation cell [11] was modified after the manner of the M.I.T. cell [8] and is illustrated in Fig. 1.

The compacted sample was laterally confined in the 6 in diameter cell body. Continuity of water was provided by Aerox Celloton Grade VI ceramic discs (air entry value 30 psi, $k_w = 3 \times 10^{-6}$ cm/sec) sealed into the top and base platens using araldite. Continuity of air was provided by a 0.5 in square gridwork of fine grooves scribed in the front face of the ceramic. The platens at the back face of the ceramic contained a spiral groove, used to flush out any air that might collect following diffusion through the water filled ceramic and thus interrupt the continuity of the water circuit. This arrangement is successful in separating the air and water flowing through a sample, since the water is prevented from entering the coarse air grooves by the fact that $u_a > u_w$ and air is prevented from entering the fine ceramic pores by the 30 psi air entry value.

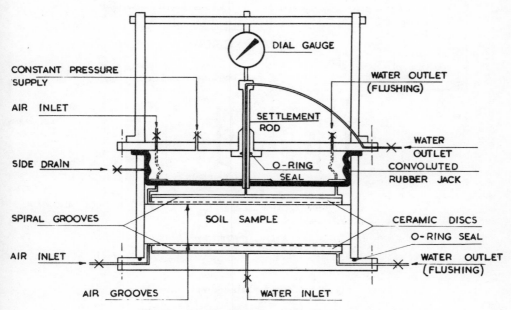

Fig. 1.   Modified Rowe consolidation cell

A vertical total stress was applied by means of a hydraulic pressure acting on the rubber jack behind the rigid top platen, as described in [11]. Vertical strain was measured by means of a dial gauge reading to 0.0001 in.

The independent air and water circuits are illustrated in Fig. 2.

The water pressures u_w were controlled from constant pressure mercury head systems [3]. Flow through the sample was caused by a small water pressure

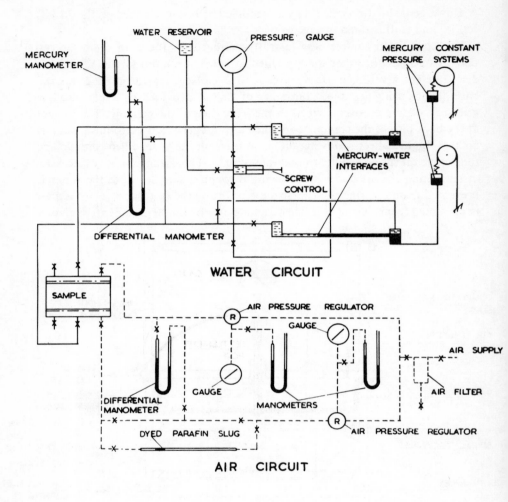

Fig. 2.   Independent air and water circuits

difference measured on a mercury differential manometer. Flow rates into and out of the sample were measured at a mercury-water interface travelling in horizontal bore glass tubes. The air pressure u_a were controlled from A.E.I. air reducing valves accurate to $\pm$ 0.1 psi. The very small pressure difference necessary to cause air flow were measured on a differential manometer containing carbon tetrachloride. The air supplied from a compressor was very damp and so no saturator was used. In this case only the rate of flow out of the sample was measured at a slug of red dyed paraffin travelling in a 5 m horizontal length of 2 mm bore polythene tube.

To conduct a permeability test the stresses σ, u_a and u_w were adjusted to their chosen values and the sample allowed to come to equilibrium for 24 hours. The amount of vertical compression and the net inflow or outflow of water were used to calculate the change in void ratio and in degree of saturation. To measure k the calue of u was raised at the sample top and lowered at the base by suitable small amounts, thus keeping the mean value unchanged and causing very little departure from the equilibrium condition. In the case of k_w the rates of flow were very low and it is necessary to check that inflow and outflow rates are equal before accepting a measurement. In the case of k_a the flow, rate was usually much higher and the steady state outflow rate was accepted once it had become constant over a period of 2 hours.

A preliminary series of tests on the apparatus of Fig. 1 showed that even when the suction term had been reduced to zero, thereby causing occlusion of the air voids, the value of k_a often remained appreciable. This indicated that the sample, although compacted directly into the cell body, was not forming a reliable seal at its perimeter. The use of various seals such as silicone grease, jointing compound, etc. failed to produce a reliable seal with the clay tested. This lack of seal is also evident in the M.I.T. test results [8], although other workers report a satisfactory seal with silicone grease [6]. The efficiency of the seal may depend on soil type.

The above sealing difficulties, which would clearly be greater if undisturbed samples were to be tested, made it necessary to revert to a triaxial type of apparatus as illustrated in Fig. 3.

The air and water circuits remained as in Fig. 2. The top and base platens were of the same type as in Fig. 1 and were of stainless steel to resist attack from mercury. The sample was 4 in diameter by 1 in thick, sealed in a conventional triaxial latex membrane, but surrounded by a mercury jacket to prevent any diffusion of air and also to permit volume change measurement if desired [3]. In the present series of tests the total stress system was isotropic, however it is simple to apply an anisotropic system using the triaxial cell axial loading plunger. The efficiency of the mercury seal was established in tests on occluded samples in which it was confirmed that the flow of air was not detectable. For a more detailed description of the apparatus and testing techniques, see Reference [7] and [12].

The apparatus in Fig. 3 has been found suitable for testing soils with a low clay content where the suctions are less than the 30 psi air entry value of the Aerox Celloton VI ceramic. For soils with high clay content and high suctions a similar apparatus has been constructed, incorporating an Aerox Celloton CC2 70 psi air entry ceramic.

SOIL PROPERTIES AND SAMPLE PREPARATION

The soil was West water clay, clay fraction (illite) 10%, Liquid limit = 20%, Plastic limit = 10%, Proctor optimum = 10.8%. The powdered clay was mixed with the required amount of distilled water and allowed to come to moisture equilibrium for 24 hours in a sealed container. It was then compacted with full Proctor effort to form a sample 4 in diameter which was trimmed to a height of 1 in taking care not to smear the top or bottom faces of the sample. The initial density and moisture content were measured to allow the calculation of the degree of saturation, and the sample set up in the apparatus of Fig. 3.

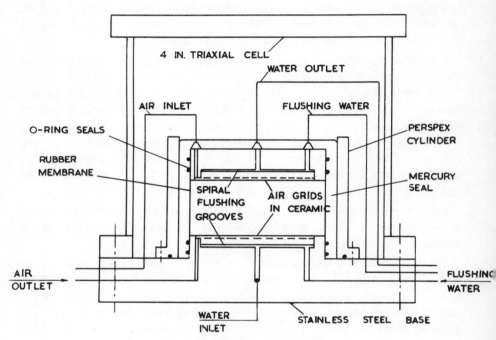

Fig. 3.   Modified triaxial apparatus

TEST PROGRAM

Series 1. A sample was compacted at an initial water content $w = 9.96\%$ which is just dry of Proctor optimum. It was placed in the triaxial cell and brought to equilibrium under $\sigma = 15$ psi, $u_a = 7$ psi and $u_w = 2$ psi. In this series σ and u_a were held constant and u_w was increased in stages up to 7 psi. Thus the applied stress $(\sigma - u_a)$ remained constant while the suction $(u_a - u_w)$ was reduced in stages from 5 psi to zero. After each change in suction the sample was allowed to come to equilibrium for 24 hours; then first the value of k_a was measured, followed by the value of k_w by consecutively applying small gradients of u_a and u_w measured at the differential manometers of Fig. 2. The respective steady state flow rates were measured as detailed earlier.

The values of k_a and k_w were calculated assuming the validity of Darcy's Law in the form $v = k \ H/L$ where

> v is the steady flow velocity, cm/sec.
> k is the permeability, cm/sec.
> H is the head difference in cm of relevant fluid.
> L is the length of the sample, cm.

The resistance to flow at the top and base platens was small enough to be neglected in these tests (ceramic $k_w = 3 \times 10^{-6}$ cm/sec.) The results of Series 1 have been plotted in Fig. 4 as k_a and k_w versus suction and degree of saturation.

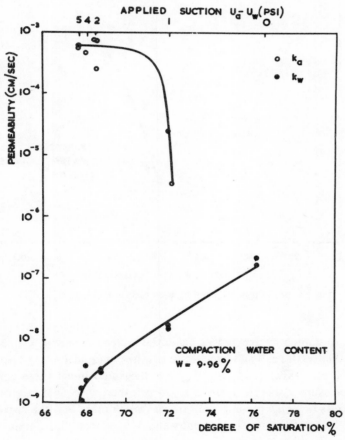

Fig. 4. Variation of k_a and k_w with suction and saturation

Series 2. A drier sample compacted at an initial water content $w = 7.45\%$ was tested as detailed in Series 1. The results are plotted in Fig. 5.

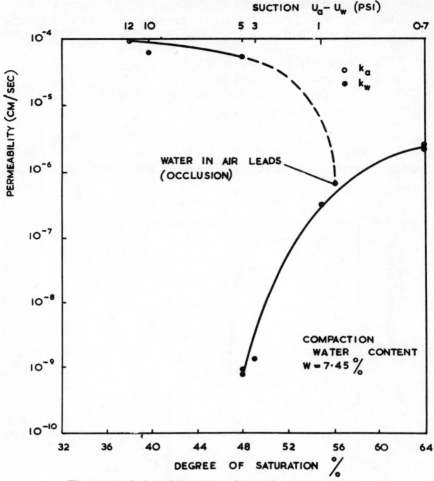

Fig. 5.   Variation of k_a and k_w with suction and saturation

Series 3. A sample compacted at an initial water content $w = 10.38\%$ was placed in the triaxial cell and brought to equilibrium under $\sigma = 12$ psi, $u_a = 8$ psi, $u_w = 2$ psi. Values of k_a and k_w were measured under these conditions. The cell pressure σ was then raised to 25 psi, the consequent increase in the applied stress $(\sigma - u_a)$ causing the sample to compress and the suction term term $(u_a - u_w)$ to decrease. To prevent water flowing from the sample the value of u_a was reduced, the value of u_w remaining constant at $u_w = 2$ psi. When equilibrium had been established for 24 hours under the new stress system, the values of k_a and k_w were measured. The cell pressure was raised in similar stages to $\sigma = 40, 60, 75$ and 90 psi and the values of k_a and k_w measured as the soil gradually approached occlusion by compression at constant water content. The results are plotted in Fig. 6 as k_a and k_w versus the effective stress components $(\sigma - u_a)$ and $(u_a - u_w)$.

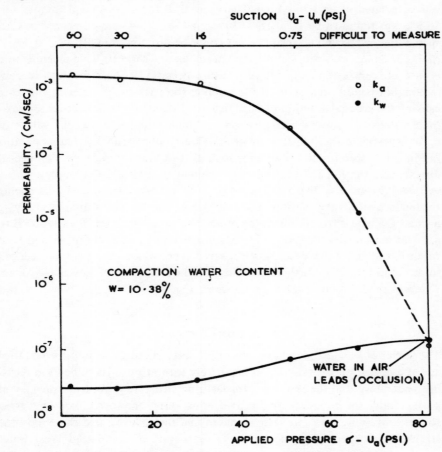

Fig. 6. Variation of k_a and k_w with suction and applied pressure

Series 4. A sample was compacted at an initial water content $w = 16.5\%$ which is wet of Proctor optimum, and was sealed in the triaxial cell. It was found that the air voids were occluded and that when an air pressure gradient was applied to the sample, water flowed into the air outflow circuit. It was clear that k_a no longer had relevance since any transport of air must now be by a true diffusion process in the water phase. To measure the amount of air diffusing through the occluded sample the base of the sample was sealed with a

0.001 in sheet of latex rubber which is virtually impermeable to water but relatively permeable to the flow of air [12]. The rate of diffusion of air was measured under carefully controlled conditions of temperature $20 \pm 0.1°C$ by submerging the entire apparatus in a water bath. The rate of flow was very low and the time to steady state approximately 4 days. The diffusion coefficient D was calculated using Ficks Law. The vaue of D for an occluded sample $w = 16.5\%$ was found to be $D = 3.8 \times 10^{-7}$ cm^2/sec. The detailed procedures for such diffusion tests on a variety of clays have been reported in [2].

Series 5. In an occluded sample the parameter k_a ceases to have meaning and even k_w becomes very difficult to measure because of the difficulties in controlling an equilibrium state in the sample, the passage of water in this case causing the gradual solution of occluded air. Despite these difficulties approximate values of k_w were measured on occluded samples at various values of u_w to investigate the influence of occluded air bubbles on k_w. The apparatus in Fig. 1 has been found suitable for such k_w tests and a sample was compacted directly into the cell at an initial water content $w = 15.0\%$. On loading with a vertical pressure $\sigma = 50$ psi a value of $u_w = 37$ psi was measured by balancing the tendency for water outflow. The value of k_w was then measured by applying a small gradient at this mean value of u_w. The value of u_w was then reduced to $u_w = 25$ psi causing the sample to consolidate and when equilibrium had been reached, the new value of k_w was measured. The value of u_w was then reduced to zero and following further consolidation the final value of k_w was measured. The results of duplicate tests are shown in Fig. 7.

DISCUSSION OF RESULTS

The values of k_a and k_w are functions of the soil structure (including void ratio) and the degree of saturation. Unfortunately saturation (S) is difficult to measure accurately and the problem of measuring structure is yet to be defined, at present only the roughest qualitative terms being employed. For any given soil, the structure exists at various macro and micro levels and these have the strongest influence on permeability. The structure of compacted West water clay formed at various compaction water contents dry and wet of optimum has been studied in the scanning electron microscope at magnifications up to X 20,000, Reference [12]. While there was little obvious difference in microstructure there were major differences in macrostructure. The peds in the dry soil were strong enough to resist distortion during compaction, leaving large interpedal macropores and channels. In the wet soil the peds were more easily distorted during compaction leaving very few interpedal spaces and producing a much more homogeneous structure. The presence of continuous macropores clearly has a most significant influence on both k_a and k_w. Compare for instance the concepts discussed in [9].

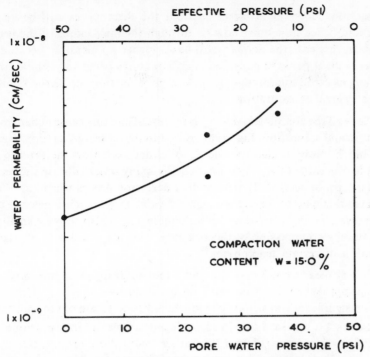

Fig. 7.   Variation of k_w with pore water pressure and effective stress

In Series 1 and 2 tests the effective stress component $(\sigma - u_a)$ is held constant and only the suction component $(u_a - u_w)$ is varied. A separate series of compression tests conducted in the cell of Fig. 1 have shown that for such samples of West water clay the compression caused by a decrease in suction *alone* is negligible, [1]. Hence, it can be assumed that during the Series 1 and 2 tests the structure remained essentially constant during a given Series, although it is clearly different in Series 1 and 2 because of the different compaction water contents. Thus in Fig. 4 and Fig. 5 the variation in k_a and k_w is due to changes in S alone.

The values of k_a remain fairly constant at a high value as S increases towards occlusion, suggesting that it is dominated by a few interpedal macropores. Thus changes in suction and hence in S cause the smaller pores to fill up first, exerting little influence on k_a. However, as the last trace of suction is removed the largest pores finally occlude and the value of k_a is seen to drop abruptly towards the value of k_w.

Similar factors appear to govern the variation in k_w. Thus at small suctions the soil in Series 2 has a higher k_w than in Series 1 presumably because its water filled macropores are larger, because of the higher strength of the peds during compaction. However at larger suctions the macropores will be emptied of water and hence flow must now be through the micropores of the peds themselves. In this case the Series 1 soil has a higher k_w, possibly because there is greater contact between peds, the microstructures being very similar under the scanning electron microscope, [12]. At a given suction the Series 1 soil has the higher degree of saturation.

In Series 3 the component $(\sigma - u_a)$ is increased and this causes changes in both structure and saturation. Fig. 6 shows that in this case the suction approaches zero much more gradually and it is easier to follow the drop in k_a right down to the value of k_w. Again k_a falls by many orders during the removal of the last 1 psi of suction. It appears that occlusion occurs when $(u_a - u_w) = 0$. However this may be merely an apparent value as we cannot measure the local internal value of u_a in an occluded air bubble. Fig. 6 also shows a small increase in k_w despite compression of the structure. This may be due to an increase in saturation of certain macropores.

It is very clear from Series 1, 2 and 3 that k_a drops most abruptly as occlusion is approached and that right up to occlusion $k_a \gg k_w$. There appears to be no significant transition zone in which k_a and k_w are of the same order.

In Series 4 it was seen that in an occluded sample the transport of air is by a true diffusion process (Ficks Law) rather than a flow process (Darcy's Law) and hence D rather than k_a is the relevant parameter. The measured value of D for an occluded sample $D = 3.8 \times 10^{-7}$ cm²/sec is much smaller than the value of $D = 2.2 \times 10^{-5}$ cm²/sec for the diffusion of air through free water. It is shown in Reference [12] that this may be accounted for by the adsorbed nature of the water in a clay soil. This low value of D confirms the difficulty of measuring or controlling the value of u_a in an occluded soil. Hence in the present tests it can only be stated that the *apparent* suction approaches zero as the soil occludes.

In Series 5 this same difficulty prevents independent control of the air and watei phases by means of the pressure u_a and u_w as in Series 1, 2 and 3. Thus in an occluded sample the passage of water progressively increases the saturation by dissolving air and transporting it out of the soil. Except at the limit of complete saturation reliable steady state flow condition cannot be obtained, and hence only approximate values of k_w can be deduced from short term 'steady state' flow observations. In the tests reported in Fig. 7 the decrease in u_w causes both compression of the soil and the expansion of the occluded air bubbles in the larger voids, and hence changes in both structure and saturation. The expansion of the air bubbles in the macropores is apparenly not a major factor in reducing k_w with this wet compacted structure, possibly because the interpedal macropores are not so dominant as in a dry compacted structure. Similar results have been found on other occluded samples [12].

REFERENCES

1. BARDEN, L., A. O. MADEDOR, AND G. R. SIDES, (1969) Volume change characteristics of partly saturated clay, *Proc. A.S.C.E.*, Vol. *95*: SM1.

2. BISHOP, A. W. (1959) The principle of effective stress *Teknisk Ukeblad*, Vol *106* : 39.

3. BISHOP, A. W. AND D. J. HENKEL, (1962) *The Measurement of Soil Properties in the Triaxial Test*, Arnold, London.

4. COREY, A. T. (1957) Measurement of water and air permeability in unsaturated soil, *Proc. Soil. Sci. Soc. Am.*, Vol. *21*.

5. GARDNER, W. R. (1956) Calculation of capillary conductivity from pressure plate outflow data, *Proc. Soil Sci. Soc. Am.*, Vol. *20*.

6. LANGFELDER, L. J., C. F. CHEN, AND J. E. JUSTICE, (1968) Air permeability of compacted cohesive soils, *Proc. A.S.C.E.*, Vol. *94* : SM4.

7. MADEDOR, A. O. (1967) Consolidation characteristics of compacted clay, *Ph. D. Thesis, University of Manchester*.

8. M. I. T. (1963) Engineering behaviour of partially saturated soils, *Soil Eng. Div. Pub.* No. 134, M. I. T, Boston.

9. OLSEN, H. W. (1962) *Hydraulic Flow Through Saturated Clays, Clay and Clay Minerals*, Vol. II, Pergamon Press.

10. RICHARDS, L. A. (1931) Capillary conduction of liquids through porous mediums, *Physics*, Vol. 1.

11. ROWE, P. W. AND L. BARDEN, (1966) A new consolidation cell, *Géotechnique*, Vol. *16* : 2.

12. SIDES, G. R. (1968) Pore pressure and volume change characteristics of compacted clay, *Ph. D. Thesis, University of Salford*.

L. BARDEN,
    SIMON ENG. LABORATORIES,
        MANCHESTER UNIVERSITY, ENGLAND

A. O. MADEDOR,
    THE FEDERAL MINISTRY OF WORKS,
        LAGOS, NIGERIA

G. R. SIDES,
    DEPT. OF CIVIL ENG., SALFORD UNIVERSITY
        MANCHESTER, ENGLAND

AN INVESTIGATION INTO THE FLOW BEHAVIOUR THROUGH COMPACTED SATURATED FINE-GRAINED SOILS WITH REGARD TO FINES CONTENT AND OVER A RANGE OF APPLIED HYDRAULIC GRADIENTS

J. S. YOUNGER AND C. I. LIM

ABSTRACT

A short review is given of research carried out to investigate the validity of Darcy's law as applied to flow through soil systems. The paper is then concerned with the conditions of flow through laboratory compacted saturated fine-grained soils and the results of tests carried out on samples with varying fines content over the range of gradients 0.2 and 12.0 are presented. Some interesting trends are revealed showing that peak deviations from Darcy's law were obtained when the samples had a fines content of about 30% when using a single size type of coarse grading. (For the purpose of comparison it was assumed that "true" Darcy behaviour was exhibited at a relatively large hydraulic gradient of 12). Using a well-graded coarse fraction the trends were more complex. The evidence suggests particle migration particularly since the clay contents, and hence clay structural effects, were small. Mostly test runs were initiated at a gradient of 1 but a few tests were commenced with this second grading at even smaller gradients, giving "true" Darcy or greater behaviour. The results suggest, where gradients less than 1 should be examined, that occasionally it may be more appropriate even critical that permeability tests be carried out in this range for certain soil types and gradients.

INTRODUCTION

The validity of Darcy's law as applied to hydraulic flow through soils has interested soils engineers for some considerable time. The relationship known as Darcy's law is expressed in the equation

$$v = ki$$

where v = velocity of flow

k = coefficient of permeability (hydraulic conductivity)

i = applied hydraulic gradient.

Evidence has been presented many times to show that deviations from flow through porous media exist. Most of the evidence, however, has been related to observations on laboratory prepared samples with one or two notable exceptions in behaviour of natural soils. In design problems involving flow behaviour it is customary, and often justifiably so, to consider the permeability of the soil mass as constant for a particular direction. If linearity is assumed any subsequent analysis is normally fairly straight-forward. It is interesting to note, however, that non-Darcy behaviour in the form of a threshold gradient has been recognised and assumed for some time [1, 2].

312

After reviewing briefly some of the work carried out on flow behaviour through porous media, including mention of the reasons for the abnormalities recorded, the paper is concerned with a study made on flow through laboratory compacted saturated soils with varying fines content. Artificial gradings of the coarse and fine fractions of the soil were made up with a view to relating any deviations from Darcy's law with the fines content of the sample at constant compaction conditions. In addition, each sample was tested over a range of relatively low to fairly high applied hydraulic gradients to investigate the relationships between any deviation recorded and the corresponding applied hydraulic gradient.

BRIEF REVIEW OF LITERATURE

Since Darcy's law was proposed many permeability investigations have been carried out over a wide range of soil types. Depending on the soil type and other important factors confirmation and disagreement with the law have both been expressed. Where disagreement has been met the deviations have been found to be either one of two main types, namely, (a) increasing hydraulic gradient causes either an increase or a decrease in flow rate; (b) a threshold gradient exists that must be exceeded before flow occurs.

Generally, where the soil involved has been largely sand then reasonable agreement with linear flow behaviour has been found [3, 4, 5]. On the other hand, where the soil has contained fine-grained particles many researchers have recorded deviations from Darcy's law. Evidence accumulated to date shows that the main factors influencing permeability in fine-grained soils are: properties of permeants, soil composition, compaction conditions, arrangement of clay particle structure, void ratio, degree of saturation, soil-water interaction, ionic concentration, thixotropic effects and bacterial activity [6]. Causes for the invalidity of the law have been expressed in terms of electroviscous drag, plugging and unplugging of voids, particle migration, quasi-crystalline water structure, the "cluster" concept [7] and experimental error particularly from contamination of the measuring tubes.

Space does not permit a wide discourse of the reasons for these explanations of anomalous behaviour except to state that the type of deviations found seem to be largely a function of the soil type used and method of sample preparation and possibly testing employed. A summary of some of the more important points involving soils with fines content is now made with particular reference to particle migration research.

Considering first the research on pure clay systems, the evidence points strongly to the existence of a threshold gradient, [8, 9] in some cases of considerable magnitude, depending largely on the main clay ion involved. The fact that no flow was evident below a certain gradient has suggested considerable clay-water interaction with non-Newtonian flow also possible [10, 11]. Lutz and Kemper [12] for water flow in pure and natural clays observed a curvi-linear velo-city-gradient relationship in some cases to considerable gradients. The per-meability of clay systems is also influenced by the permeant [13, 14].

Hansbo [15], reporting tests on undisturbed natural clay samples with a sensitive apparatus under equal and constant temperature conditions, found that, for gradients less than 5, the velocity-gradient relationship was curvi-linear but thereafter linear. He suggested that this relationship should be ex-pressed as shown in Fig. 1. In no case did Hansbo find the existence of a thre-shold gradient.

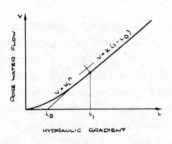

Fig. 1

It was postulated that the curvi-linear velocity-gradient relationship could be due to particle migration of movable finer clay particles in the pores of the coarser particles and load-carrying skeleton of the clay. These particles could be moved by seepage forces to plug or unplug flow channels in the skeleton pores [15, 16].

Olsen [17] concluded from experimental information that the discrepancies between measured hydraulic flow rates in saturated clays and those predicted from Darcy's law were caused mainly by unequal pore sizes in the clays. This cluster concept explained that the clay matrix was composed of unequal pore sizes, most of the flow passing through only that part of the clay porosity that was distributed among the larger pores. As the porosity decreased by compres-sion, the larger pores mainly were reduced. At very low porosities, volume changes of clay also brought changes in the porosity distributed among the smaller pores. At a given porosity, the measured flow rates varied with the chemical compositions of the clay and permeant. This was explained as a result of the dependence of the cluster mass and cluster void ratio parameters on the degree of dispersion of a clay-permeant system, where both parameters decreased with increasing dispersion. Olsen further pointed out that high vis-cosity and tortuous flow paths failed completely to account for the discrepan-cies between measured and predicted flow rates in saturated clays, and the possible errors of Darcy's law and electrokinetic coupling were insignificant.

Further work by Olsen [17] showed that experimental errors could be made with readings made in conventional capillary tubes as a result of contamination of liquid and solid surfaces from grease absorbed from the atmosphere. This causes the advancing menisci to have greater contact angles than the reading menisci with a resulting pressure drop across the sample. Therefore, all readings at very low gradients made in this manner must necessarily be considered with some reservation. Olsen concluded that contaminated capillary tubes could account for deviations from Darcy's law of types 1, 2 and 3 but not of types 4, 5 and 6 as shown in Fig, 2.

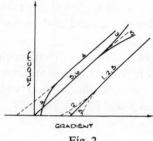

Fig. 2

In studies of the permeability characteristics of compacted clays Mitchell et al. (18) found the following main characteristics. The permeability values of compacted clay samples on the dry side of optimum of the standard dry density-water content compaction curve were many times higher than on the wet side. The reason is explained in terms of clay particle structure. Many soils compacted dry of optimum had a more random particle orientation (flocculent structure) and a larger average pore size than when compacted wet of optimum where the particles had a more parallel arrangement (dispersed structure). The larger the individual pores for any given total pore area, the greater the flow would be, since the permeability varied as a power function of pore size [19]. The permeability was also shown to be a function of compactive effort, method of compaction, saturation and aging (thixotropic characteristics) [20].

Olsen [21] found that flow through saturated kaolinite over a range of gradients from 0.2 to 40 and a wide range in porosity agreed with Darcy's law. He explained that the flow channels in this clay type were too large to be affected by the influence of quasi-crystalline water, and also possible changes in the clay fabric from seepage forces was prevented by the confinement of the sample. He further concluded that there may be many natural clays and clayey sediments, sufficiently confined and with adequately large pores, in which Darcy's law was obeyed. However discrepancies might be apparent in very fine-grained clays, in granular soils containing a small amount of clay and also where high gradients were present in shallow, unconfined sediments.

Von Engelhardt and Tunn [22] found that the velocity-gradient relationship for flow through clay bearing sandstones was curvi-linear.

Evidence for particle migration was presented by Bodman and Harradine (23) from the experimental results of six different untreated soils using distilled water as the permeant. Their results are summarised in Fig. 3 and it can be noticed that significant migration of clay particles took place in five out of the six soils tested where the clay content was 3% to 28%, but where the clay content was 60% almost no movement of particles was observed. They concluded that decrease in permeability with time appeared to be the result of pore-blocking in the columns partly by deeply migrated and dispersed, and partly by locally dispersed and locally migrated fine particles. The intensity of migrations depended upon both chemical and physical properties of the soils concerned.

Hallsworth [24] observed that in a soil with high clay content, migration of the clay was severely restricted. Little movement of kaolinite occurred at clay contents of 40%, or of montmorillonite at clay contents of 20% while some movement took place when these figures were 20% and 10% respectively as shown in Fig. 4.

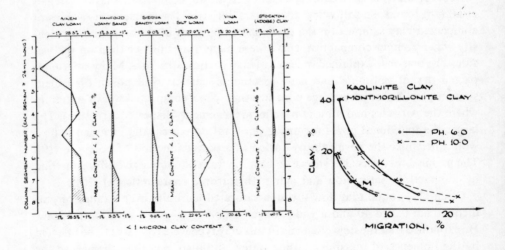

Fig. 3. Final distribution of < 1 μ clay with   Fig. 4
        respect to depth for each soil column

Mitchell and Younger [25] observed the non-Darcy behaviour of compacted saturated silty clay samples and stated that particle migration was more likely the cause in this case than abnormal water properties. Only those particles which were not involved in the load-carrying skeleton could be moved by seepage forces.

DESCRIPTION OF SOIL, APPARATUS AND TESTING PROCEDURES

The soil used in this investigation was a mountain till (angular in shape) having the following properties: liquid limit = 27, plastic limit = 17 and plasticity index = 10. All sizes above B.S.14 were discarded and the remainder was further divided by wet sieving into three main groupings —

a) coarse material — passing BS14 and retained on BS100

b) coarse material — passing BS100 and retained on BS200

c) fine material — silt and clay fraction passing BS200

Samples were prepared from these gradings by mixing the appropriate quantities of fines and coarse material.

The ranges of particle size distribution and compaction characteristics are given in Fig. 5 and Tables I and II respectively. The procedure for preparing the samples has already been described fully [18] and will not be repeated.

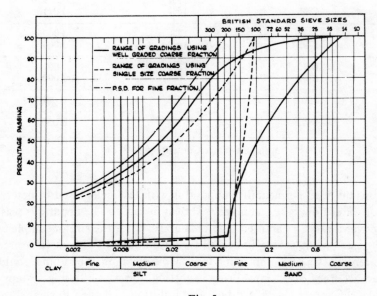

Fig. 5

TABLE I

Compaction and Permeability Characteristics of Samples Using Single Size Coarse Fraction

% Fines	Dry Density Kg/m³	Initial Water Content $m_i - \%$	"Darcy" Permeability $k - \mu m/s$
5	1730	9.6	6.66000
20	1828	12.1	0.10400
30	1834	11.8	0.14000
40	1840	11.3	0.21800
60	1822	12.5	0.00407
80	1838	12.2	0.00249

Optimum Water Content $\doteq 14\%$

TABLE II

Compaction and Permeability Characteristics of Samples Using Well-Graded Coarse Fraction

% Fines	Dry Density Kg/m³	Initial Water Content $m_i - \%$	"Darcy" Permeability $k - \mu m/s$
5	1755	8.4	6.24000
	1753	16.2	1.9030
15	1842	12.0	0.70200
	1793	16.8	0.15880
25	1840	11.8	0.11440
30	1793	16.4	0.01728
40	1844	11.4	0.07920
	1795	15.1	0.01135
46	1839	12.1	0.00348
	1795	16.7	0.00392
64	1840	12.1	0.00147
	1800	16.4	0.00050
76	1807	15.9	0.000453
88	1843	11.6	0.00078

Optimum Water Content $\doteq 14\%$

The constant head permeability apparatus with test cells used to determine the rate of water flow in the samples is shown in Fig. 6. The apparatus was devised to cover a range of applied gradients up to 12 for the size of sample used. Capillary tests on the tubing indicated that a minimum gradient to overcome capillary tension was 0.1 approximately. Movement on the tubing could be read to an accuracy of 0.5 mm. Prior to testing, the whole tubing system was thoroughly cleaned out with a dilute acid solution and then flushed through

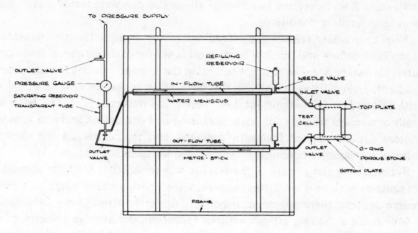

Fig. 6

and set with fresh distilled de-aired water. The porous discs in the test cells were always boiled in water for at least ten minutes to ensure saturation. Saturation, always greater than 97%, was achieved using back pressure gradually increased to a maximum of 700 KN/m^2. The permeability of compacted fine-grained soils increases with increasing saturation but little increase in permeability is realised for saturations greater than 97% [18]. The required heads to cause flow were obtained from the difference in level of the menisci. The tests were carried out under constant temperature conditions ($76°F \pm 1°F$).

It was recognised that air would be taken into the water at the air/water interface in the tubing. However, since the diameter of the tubing was small, the time of testing relatively short with continued replenishment of the water in the horizontal tubing, and since the mean of the inflow and outflow readings (little difference was experienced) was taken, it is felt that any errors from air dissolution were small and could be ignored.

Test runs were commenced by opening both cell valves simultaneously and taking periodic readings of inflow and outflow as indicated after a small time interval. In most cases, tests were started at gradients of 1 and then increased gradually in steps to 12. Thereafter the gradients were decreased in similar stages. In addition, some samples were also tested at gradients less than 1. Finally, the samples were subjected to sudden applications of a gradient of 12.

PRESENTATION AND ANALYSIS OF RESULTS

The Reynold's number, $R = dv\gamma/\mu g$, where R = Reynold's number, d = effective grain size (d_{10}), v = velocity, γ = density of water, μ = viscosity, g =

gravitational acceleration, was calculated for each sample. It was found that in all cases R was less than 1 and hence all the samples were tested under theoretical laminar flow conditions.

Velocity-gradient relationship: Readings of the distance the menisci travelled along both inflow and outflow horizontal tubes for each sample at several time intervals during a test were taken to obtain the average velocity for each applied gradient. Irregularities usually in the form of an initial decreasing of velocity with subsequent fairly constant flow, were often observed [25]. The latter constant or average velocity for both increasing and decreasing gradient runs were plotted against applied hydraulic gradient, and typical results are shown as follows for samples containing widely different fines content.

Summarising the results of the tests it was found that for tests carried out on samples with only a 5% fines content mixed with either a single or a graded coarse content, there appeared to be little deviation from Darcy behaviour in a wide sense although, almost without exception, the average velocity of flow on the second run at a particular gradient was noticeably lower than that obtained in the first run at that gradient. This effect appeared to reduce slightly at higher gradients. It is interesting to note that by projecting the results of the first runs obtained at the lower gradients that the projected velocity corresponding to a gradient of 12 would be much higher than that obtained from the testing procedure adopted. Conversely, had the test runs been continuous over a much longer period of time than available in this apparatus then a gradual decrease in the flow velocity might have been observed as is often found in standard constant head tests on soils with a similar coarse grading.

Consider next the samples tested with the single size coarse grading. Deviations in the form of a less than proportional relationship were observed at the lower range of gradients throughout the gradings investigated from a 20% fines content and upwards as shown in Fig. 7. However, a definite maximum discrepancy was observed when the fines content was in the range of 20–40%. Whenever the fines content was either less or greater than this the deviations decreased. At the higher gradients the deviations decreased with increasing seepage force. The results have been summarised in Fig. 7 and again diagrammatically in Fig. 8.

The picture obtained using the well-graded coarse material was more complicated. As above, similar trends were observed for fines contents less than about 40% as shown in Fig. 9. At 40%, however, the velocity-gradient relationship was approximately linear, or more than proportional when the fines content was only 6% greater, checked slightly at a gradient of approximately 2, and then increased more than proportionally but at a decreasing rate with increasing gradient. For the two samples tested at 64% and 88% fines, again a more than proportional relationship was obtained with gradients less than about 2 but a definite change in the slope was then observed with a subsequent

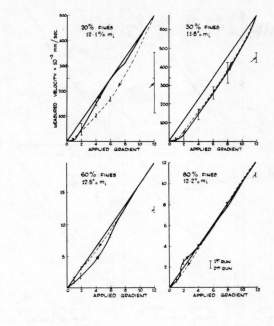

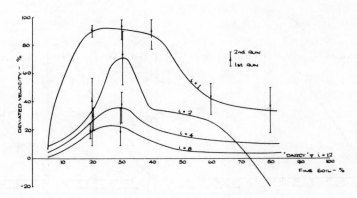

Figs. 7 and 8

less than proportional relationship. At the highest fines contents for each type of grading a certain similarity in flow behaviour is apparent at the lower gradients as might be expected since the composition in each relevant test would be similar.

For samples compacted on the wet side of optimum with the well-graded coarse material a diagrammatic summary is given in Fig. 10. Similar trends were observed compared with the samples compacted dry of optimum except that

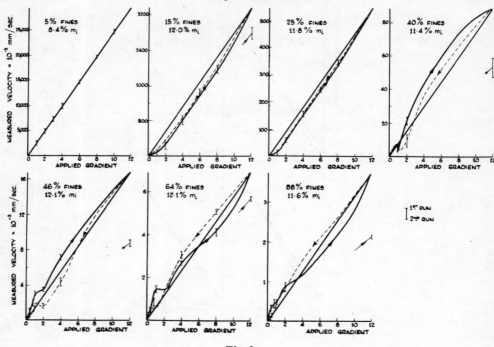

Fig. 9

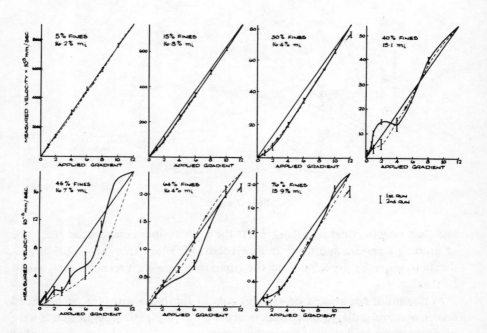

Fig. 10

a proportional rather than greater than proportional velocity-gradient relationship was observed at the lower gradients.

In no case was there any evidence of a threshold gradient.

DISCUSSION

In general, the results of all the velocity-gradient plots showed deviations from Darcy behaviour.

Where clay contents are high abnormal water properties in the form of the pore water having a "quasi-crystalline" structure with non-Newtonian flow characteristics may be expected. In these tests, however, the gradings were such that, even at the highest fines contents tried, the corresponding clay fraction, and hence particles having surface charges, was approximately 25% maximum. In addition, where the largest discrepancies were observed the clay content was considerably less. Further, with the above phenomenon, a more regular deviation pattern might be expected than was observed. Consequently a particle migration concept (i.e. an actual change in soil fabric during flow) would seem to account largely for the anomalies recorded. Some evidence of this phenomenon in a compacted silty clay soil has already been given [25]. An approximate analysis of the magnitude of seepage forces relative to inter-particle bond strengths in the solid skeleton suggests, however, that only particles not participating in the actual load-carrying structure are susceptible to movement during normal seepage conditions.

These loose particles, not rigidly fixed in position, could under application of a critical seepage force be moved either by rotating about a point on the particles, particularly with high angularity in particle shape as in this case, or by direct translation through a pore space up the pertinent flow channel until a constriction was met. Further increase in seepage force could effect further movement of some of the loose particles. The effect of particle migration in this case is to cause plugging and unplugging of flow channels in similar manner to that described by Martin [16].

In the light of this concept, interpretation of the results obtained revealed some interesting points. Firstly, with regard to the tests conducted using the single size coarse fraction, there would appear to be a definite range, 20–40%, of fines content in which peak deviations from Darcy's law could be expected at lower gradients. A similar trend is at first apparent using the well-graded coarse fraction with fines contents less than about 40%. However, a totally different behaviour pattern, particularly at low gradients, was observed in the samples with greater fines contents. In addition, in these samples, instead of an initial applied gradient of 1 as generally operated throughout the test series, gradients were applied in three incremental stages (0.2, 0.4, 0.8) up to 1 before proceeding as before. Comparison of the results, therefore, suggests that a sudden application of a gradient, even only of 1, may be sufficient to cause significant particle movement with consequent deviational behaviour. Furthermore, the value of permeability obtained at gradients of 12 or greater may be an underestimate of the true permeability of the medium particularly where applied field gradients, as is very often the case, are less than 1. Another piece of evidence in this light was given in these tests when at the end of a particular test a sudden application of a gradient of 12 was made to the sample. Almost invariably, as indicated by the separate arrow in Figs. 7, 9, 10, the consequent flow and hence permeability was less than that obtained by incremental application of gradients.

On the other hand, when comparing the series wet of optimum with the above for the well-graded coarse fraction, it can be seen that an approximately proportional relationship was obtained for fines contents of 40% and greater for the samples wet of optimum. In these tests the initial applied gradient was generally about 1. In addition, structural differences as described by Mitchell et al. [18] for fine-grained soils wet and dry of optimum are not expected to have been particularly significant in these samples excepting those with the highest fines content, the gradings being measurably coarser. Therefore, as was observed, permeability values for samples having the same fines content and of the same initial dry density but compacted wet or dry of optimum, would be expected to be similar. Hence, particle migration for samples compacted wet of optimum as described would also seem to be the main cause for the anomalies recorded. The deviation pattern, therefore, whether wet or dry of optimum, would not seem to be wholly a function of the incremental gradient increase at the critical low gradients, but also of the type of soil, in this case angular, and coarse grading employed.

SUMMARY AND CONCLUSIONS

A brief review of literature, principally of research into particle migration during permeability tests, has been made. Evidence is subsequently given of permeability tests over a range of hydraulic gradients carried out on soil samples made up of artificial gradings with varying fines contents. Very apparent deviations from a regular flow pattern were observed suggesting that fine particle movement occurred.

Using a single size coarse grading a definite pattern of non-Darcy behaviour was recognisable with peak deviations occurring when the fines content was of the order of 20–40%. When a well-graded coarse fraction was used similar deviations to that obtained using the single size coarse fraction were obtained when the fines fraction was 30% or less. For greater amounts of fines a more complex pattern was evident wherein greater than Darcy behaviour was obtained until gradients of about 2 were applied. Thereafter there was a noticeable change in the flow behaviour. Almost similar trends were evident whether the compaction conditions were wet or dry of optimum. The apparent similar trends suggested that a critical gradient of the order of 1 to 2 for these gradings and soil type was required to effect considerable migration. It was stated that clay particle structure would have little effect on the compaction conditions wet or dry of optimum except perhaps at the highest fines contents used. The velocity of flow in the tests carried out confirmed this.

It is suggested, therefore, that it may be important for certain types of soils, taking account both of grading and particle shape, that permeability tests be carried out under the gradient conditions appropriate to the problem to be analysed.

ACKNOWLEDGEMENTS

The authors gratefully appreciate the help and encouragement given to Mr. Lim by Mr. T. Dymock, Mr. W.G.N. Geddes and Mr. G. Rocke of Messrs. Babtie, Shaw & Morton, Consulting Engineers, Glasgow, with whom he was engaged whilst working on his Master of Science degree. The assistance of Mr. D. Mitchell, Mr. W. Carter and personnel of the soil mechanics technical staff who built and maintained the apparatus is also gratefully acknowledged. Thanks are also due to Miss. M. Simpson and drawing office staff who prepared the illustrations and to Miss E. Reilly who typed the manuscript.

REFERENCES

1. FLORIN, V. A. (1951) Consolidation of Earth Media and Seepage under conditions of variable porosity and consideration of the influence of bound water, *Izvestia Acad. of Sc., U.S.S.R., Section of Technical Science*, No. 11.

2. ROZA, S. A. and KOTOV, A. I. (1958) Experimental Studies of the Creep of Soil Skeletons, *Zapiski Leningradlkovo ordenor Lerrina i Trudovovo Krasnovo Znameni gornovo instituta um G. A. Plakhanova*, Vol. *34*, No. 2.

3. FANCHER, G. H. AND LEWIS, J. A. (1933) Flow of simple fluids through porous materials, *Ind. Eng. Chem.*, *25*.

4. MEINZEN, O. E. AND FISCHEL, V. C. (1934) Test of permeability with low hydraulic gradients, *Trans. Amer. Geoph. Union*, *15*.

5. FISCHEL, V. C. (1935) Further tests of permeability with low hydraulic gradients, *Trans. Amer. Geoph. Union*, *16*.

6. GUPTA, R. J. AND SWARTZENDRUBER, D. (1962) Flow-associated reduction in the hydraulic conductivity of quartz sand. *Soil Sci. Soc. Proc.*, *26*.

7. OLSEN, H. W. (1962) Hydraulic flow through saturated clays. *Clay and Clay Min.*, Vol. *9*.

8. LOW, P. F. (1960) Viscosity of water in clay systems, *8th Nat. Conf. on Clay and Clay Min.*, 170–182, Pergamon Press, New York.

9. MILLER, R. J. AND LOW, P. F. (1963) Threshold gradient for water flow in clay systems. *Proc. Soil. Sci. Soc. of Amer.*

10. LOW, P. F. (1961) Physical chemistry of clay-water interaction, *Adv. in Agronomy*, *13*.

11. SWARTZENDRUBER, D. (1962) Modification of Darcy's law for the flow of water in soils. *Soil Sci.*, Vol. *93*.

12. Lutz. N. F. and Kemper, W. D. (1959) Intrinsic permeability of clay as affected by clay water interaction. *Soil Sci.*, Vol. *88*.

13. Michaels, A. S. and Lin, C. S. (1954) The permeability of kaolinate, *Ind. Eng. Chem.* Vol *46*.

14. Fireman, M. and Bodman, G. B. (1939) The effect of saline irrigation water upon the permeability and base status of soils. *Proc. Soil Sci. Soc. of Amer.*

15. Hansbo, S. (1960) Consolidation of clay, with special reference to influence of vertical sand drains. *Swed. Geot. Inst. Proc.* No. 18.

16. Martin, R. T. Absorbed water on clay: a review, *Clays and Clay Min.*, Vol. *9*.

17. Olsen, H. W. (1963) Deviations from Darcy's law in saturated clays. *Proc. Soil Sci. Soc. of Amer.*

18. Mitchell, J. K., Hooper, D. R. and Campanella, R. G. (1965) Permeability of Compacted Clay. *Jour. S. M. and F. D., A.S.C.E.*

19. Carman, P. C. (1956) Flow of gases through porous media. *Acad. Press Inc., New York.*

20. Mitchell, J. K. (1960) Fundamental aspects of thixotropy in soils, *Jour. S. M. and F. D., A.S.C.E.*

21. Olsen, H. W. Darcy's law in saturated kaolinite. Water Res. Research, Vol. *2*, No. 2.

22. Von Engelhardt, W. and Tunn, W. L. M. (1955) The flow of fluids through sandstones — see *Illinois State Geol. Surv. Circ.* 194.

23. Bodman, G. B. and Harradine, E. F. (1938) Mean effective pore size and clay migration during water percolation in soils. *Proc. Soil Sci. Soc. of Amer.*,

24. Hallsworth, E. G. An examination of some factors affecting the movement of clay in an artificial soil. *Jour. of Soil Sci. 14.*

25. Mitchell, J. K. and Younger, J. S. (1966) Abnormalities in hydraulic flow through fine-grained soils. *ASTM Symp. on Permeability and Capillarity.*

J. S. Younger, Lecturer,
Department of Civil Engineering,
University of Strathclyde, Glasgow C.1.

C. I. Lim, Assistant Engineer,
Messrs. Babtie, Shaw & Morton,
Consulting Civil Engineers, Glasgow C.2.

NON-DARCIAN FLOW OF WATER IN SOILS — LAMINAR REGION

A Review

M. Kutílek

The flow of water in soils obeying the Darcy's equation $v = kI$ will be called the Darcian flow, while the flow not expressible by the Darcy's equation will be called the non-Darcian flow. The liquid not characterized by the relation $\eta = \tau dy/dv$ will be classified generally as non-Newtonian, since there is no directly measured rheologic parameter of soil water allowing its precise rheologic classification. In the above mentioned equations, v is the velocity of flow, k is the hydraulic conductivity, I is the hydraulic gradient, η is the viscosity, τ is the shearing stress. In the majority of cases, I will deal with the flow in soils saturated with water (saturated flow).

To keep the discussion lucid, the $v(I)$ curves as determined by various authors will be classified into twelve types denoted from A to R, (see Fig. 1). The Darcian flow is represented by a straight line passing through the origin. The types A to F demonstrate a more than proportional growth of v with I, and the deviations from the Darcian flow are usually explained by the theory on the altered viscosity of soil water and by the action of streaming and osmotic potential. The deviatons of these types are only rarely supposed to be caused by the change of the geometric conditions within the soil sample. The non-Darcian flow as characterized by the types L to O where v grows less than proportional with I is probably dominantly influenced by the shift of particles, and by the change of the tortuosity within the soil sample. The P and R types have obviously resulted from the combination of the types B–L or E–M.

EXPERIMENTAL ERRORS

The experimental errors playing a possible role in experiments will be divided into seven groups and indicated by letters (a) to (g). The first group of possible errors follow from the nature of the experimental material.

(a) Heavy soils and clays, mostly examined in the non-Darcian experiments, swell after wetting and when the sample is not perfectly closed, it incrases its porosity with time. On the other hand, under the action of rising hydraulic gradient the porosity of the unclosed sample will decrease thus resulting in one of the types L to O of the $v(I)$ curve.

327

(b) Even in a closed sample, the swelling causes a gradual closing of coarse pores thus inducing a decrease of flow velocity with time at constant gradient, and only after a longer time period after the wetting, is the steady flow reached. If the experiment is performed during swelling, the shape of the $v(I)$ curve depends upon the direction of the test. When I is raised, we get a L type, when I is decreased, we get a B type, if the flow were otherwise Darcian.

(c) When the soil is wetted under the atmospheric pressure, it will contain the closed air which is either gradualy pushed out during the flow and the velocity of flow will rise until it reaches approximately a constant value. Or, more frequently, the entrapped air will form a barrier affecting the reduction in flow. When the non-deaerated water is used in the experiment, air will be imported into the soil sample and the flow velocity will change similarly as mentioned above. When during both these processes the dependence $v(I)$ is measured, the resulting curve does not represent the real flow conditions. The errors will be suppressed if the measuring starts after reaching the steady state conditions.

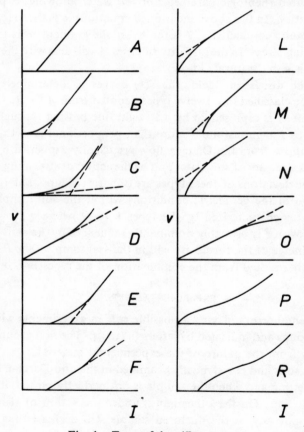

Fig. 1.   Types of the $v(I)$ curves

(d) Since all the flow experiments must be performed during long time intervals, the bacteria developing at the inflow end may cause the reduction in flow velocity with time (Gupta, Swartzendruber, (1962)).

(e) The negative effect of bacteria development will be hindered if 0.1 % phenol solution is used. However, the question remains what happens with the "quasi-crystalline" structure of water induced by the solid surface after the phenol was added to water, especially when we take into consideration the recently found influence of the organic cations of pyridinium and quinolinium upon the flow of water in clay minerals (Kutílek, Salingerová, (1966)).

(f) In majority of the flow tests, the velocity of flow is in the range of extreme low values and the rate of flow is measured by the progress or regress of the air–water meniscus in the capillary tube. When the progress of an air buble is measured, then according to the analysis done by Olsen (1965), the contamination of the wall of the capillary can give rise to the hysteresis of the wetting angle causing the losss of the hydralic head and thus the apparent deviations from the Darcy's equation.

(g) The results of experiments on unsaturated flow have to be analysed with regard to the above mentioned sources of errors. When the non-steady flow is evaluated, it is evident that the condition on constancy of geometry inside of the sample cannot be satisfied. When for example the infiltration in a dry soil is evaluated, the differences of the swelling presure at the wetting front, the thixotropy of soil water and the probable displacement of particles are factors that can cause the deviations from the Darcy's equation, too. However, there are no quantitative studies available.

CAUSES OF THE NON-DARCIAN FLOW

A hypothesis frequently used for the explanation of the non-Darcian flow in the laminar region is the theory on the altered viscosity of water in soil due to the influence of the soil solid surface. Low (1961) developed the theory on the quasi-crystalline arrangement of water molecules in the proximity of the solid surface. A complex approach about the role of different factors in the deviations from the Darcy's equation is found in papers of Swartzendruber, starting from 1962, when he analysed experiments of several authors with the conclusion on non-Newtonian flow of soil water as one of the reasons for deviations from the Darcy's equation. The school of Deryaguin and Nerpin (reviewed in Nerpin, Tchudnovskij, 1967) supposes that water in capillaries has the properties of a Bingham body. Applying this theory to the porous media and soils, the Eq. (12) was developed together with a detailed discussion on rheologic parameters. Kutílek (1964, 1967) considers the flow of water in soils to be composed of two categories of water, one being the Newtonian, the other one a non-Newtonian liquid. The second category, the non-Newtonian liquid, is not uniform in its properties and it is subdivided into two categories of different natures. Some other authors, as for example Von Engelhardt and Tunn (1955), Lutz and Kemper (1959), Hansbo (1960), Hadas (1964), Thames (1966) explain the non-Darcian flow also by the non-Newtonian properties of soil water.

Swartzendruber (1966, 1967) has discussed later the electrical streaming potential effect as one of the possible causes of the non-Darcian flow. The direction of the gradient of the electrical streaming potential as induced by the flow of water is opposite to the direction of the hydraulic gradient. The velocity of flow of water v_w will be therefore expressed, under the assumption of the linearity, by

$$(1) \qquad v_w = k_{ww} \frac{\Delta\Phi}{L} - k_{we} \frac{\Delta\phi}{L} ,$$

where k_{ww} is the transport coefficient for water when the hydraulic gradient $\Delta\Phi/L$ is a driving force, k_{we} is the linked transport coefficient for water when the flow is caused by the gradient of the streaming potential $\Delta\phi/L$. Swartzendruber (1966) arranged the Eq. (1) in

$$(2) \qquad v_w = \left[k_{ww} - k_{we} \frac{\Delta\phi}{\Delta\Phi} \right] \frac{\Delta\Phi}{L} ,$$

and has demonstrated the possibility that the term $\Delta\phi/\Delta\Phi$ is not constant, and as it decreases with the growing $\Delta\Phi$, this fact has been considered as one of the causes of the non-Darcian flow. However, with regard to the irreversible thermodynamics, using the relations (Prigogine, 1955, p. 61–64, De Groot, and Mazur, 1962, p. 438)

$$(3) \qquad \left(\frac{\Delta\phi}{\Delta\Phi} \right)_{i=0} = - \frac{k_{ew}}{k_{ee}} ,$$

or

$$(4) \qquad \left(\frac{\Delta\phi}{v_w} \right)_{i=0} = - \frac{k_{ew}}{k_{ww}k_{ee} - k_{we}k_{ew}} ,$$

we can see that if the conditions

$$(5) \qquad \left(\frac{\Delta\phi}{\Delta\Phi} \right)_{i=0} = \text{const.}$$

are not fulfilled, any of the transport coefficients can be considered as not being constant. Therefore, the non-constancy of $\Delta\phi/\Delta\Phi$ cannot be used as a proof against the hypothesis about the influence of the non-Newtonian viscosity upon the deviations from the Darcy's equation. A similar treatise can be applied for the other linked coefficients.

The non-Darcian flow is probably caused by the change of the geometric arrangement of particles inside of the soil sample, too. The simplest is the case of transport of small particles in coarse pores, where either gradual blocking of outflow channels of big pores, or unblocking of inflow channels into big pores, can occur. The combination of both with a dominance of one of them is highly probable. This phenomenon is supposed to have an important role in heterogeneous mixtures of coarse materials with colloid particles, and in coarse grained sediments containing less rigidly cemented particles.

The flowing water acts upon the compression of the sample which is opposed by the swelling pressure. Blackmore and Marshall (1965) have shown that in the suspension of bentonite the shift of particles causes a gradual decrease in spacing, the change being continuous and dependent upon the applied hydraulic gradient, or velocity. Gairon (1968) has experimented with bentonite clay paste, and according to his results the flow was less than proportional, at least at low concentrations. He explains these deviations from the proportional Darcian flow mainly by the displacement of particles within the column of the clay paste. Let us choose therefore a simple model of plates with ratio of the height h to the length L of plates being $h/L = 1/100$. The dependence of the permeability K upon the half spacing d' or upon the porosity ε is plotted in Fig. 2, when the specific surface of bentonite was taken as equal to 800 m²/g, and if a simple Kozeny's equation was used, K will be expressible by

$$(6) \qquad K = ad'^b$$

where a, b are coefficients dependent upon the type of model, $b > 0$. Let us

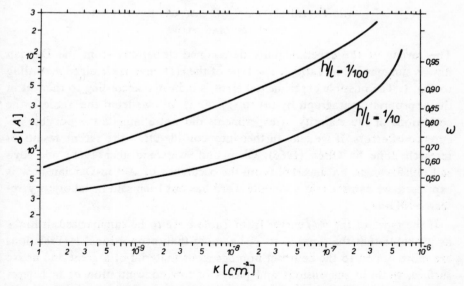

Fig. 2 The dependence of the permeability K upon the half spacing d' and upon the porosity ε as calculated for a model of plates of bentonitic type. L is the length, h the height of the plates

assume $d' = z^{-c}$ for $c > 0$, and rising with the hydraulic gradient I. Since the velocity of flow in an elementary layer of the height dz is

$$(7) \qquad\qquad v = - K(z) \frac{dh}{dz}$$

then after integration, the permeability of the total column K_T is

$$(8) \qquad\qquad K_T = \frac{a(bc + 1)}{l^{bc}}$$

where l is the height of the column of suspension. Even if the procedure bears the features of a great simplification, it demonstrates the dependence of hydraulic conductivity upon the hydraulic gradient in suspensions of platty particles.

Miller and Low (1963) however consider the rearrangement of particles during their flow experiments to be parallel to the flow path and the relation $v(I)$ was more than proportional as the toruosity decreased. Emersleben's and Iberall's analysis (acc. to Scheidegger, 1957, p. 108–110) of flow in fibrous sample gives the $v(I)$ relation less than proportional. Similarly Zaslavski (1964) gets from the analysis of flow and shift of particles the hydraulic conductivity dependent upon time and gradient.

THE EXPERIMENTALLY DETERMINED DEVIATIONS

FROM DARCIAN FLOW

The review of the experimentally determined deviations from the Darcian flow is summarised in Table I. The type of the $v(I)$ curve is classified according to Fig. 1. The possible experimental error is indicated according to the text in the appropriate paragraph by (a) to (g). As it follows from the Table I, the conditions of the majority of experiments did not eliminate the possible occurence of errors. If we take further into consideration the recent results of research done by Olsen (1966), Olson and Swartzendruber (1968), and Russell (1968) where an uncertainty on the occurence of a non-Darcian flow is expressed, we can see that a definite word has not been said in laboratory research till now.

If the types of the $v(I)$ curves from Table I are to be summarized, it looks as probable that the types A to F, i.e. more than proportional $v(I)$ relations are more bound to the compact homogeneous material of a great and active surface, while in suspensions and pastes of low concentration or in heterogeneous materials the types L to O, or even P and R will prevail, i.e. the $v(I)$ relations are less than proportional or they represent the combination of both

EMPIRICAL EQUATIONS OF NON-DARCIAN FLOW

Some authors demonstrating the existence of the non-Darcian flow have suggested equations describing the determined deviations. All these equations which are now going to be described, have the empirical character with the exception of Eq. (12) and they were not compared with equations of flow of non-Newtonian liquids of certain rheologic types in porous media. However, the empirical procedure cannot be criticized since even the rheological equation of a certain model has its rheologic parameters of more or less empirical nature.

The simplest empirical procedure is the introduction of the threshold gradient I_0 below whch no flow occurs. Puzyrevskaya (1931, acc. to Polubarinova Kotchina, 1952) was probably the first to formulate

$$(9) \qquad v = k(I - I_0) .$$

In the same year, Izbash (1931) suggested

$$(10) \qquad v = kI^{\,m}$$

where $m > 1$. A further developed type of Eq. (10) is the suggestion of Slepička (1961)

$$(11) \qquad v = \alpha \left(\frac{\eta}{\sigma} \right)^{f-1} k^f I^f$$

where α is a coefficient dependent upon the nature of the solid surface, η is the viscosity, σ is the surface tension, and the empirical exponent $f > 1$.

Nerpin (in Nerpin and Tchudnovskij, 1967) developed the theory that the soil water behaves like a Bingham body and he suggested an equation identical with the flow of Bingham liquid through a system of capillary tubes:

$$(12) \qquad v = k_0 I \left[\frac{1}{3} \left(\frac{I_0}{I} \right)^4 - \frac{4}{3} \frac{I_0}{I} + 1 \right]$$

where I_0 is the threshold gradient expressible by

$$(13) \qquad I_0 = \frac{\tau_0 S}{\gamma \varepsilon}$$

where S is the specific surface, ε is the porosity, γ is the density, τ_0 is the threshold shearing stress. The equation used by Volarovitch and Tchuraev (1964)

$$(14) \qquad v = AI + \frac{C}{I^3} - B$$

is identical with Eq. (12), A, B, C being empirical coefficients. Their physical meaning can be simply derived from Eqs. (12) and (13).

Swartzendruber (1962) suggested after analysing the $v(I)$ curves the equation

$$(15) \qquad v = B[I - J(1 - e^{-CI})]$$

---

334   *M. Kutílek*

TABLE I

REVIEW OF EXPERIMENTS

Author	Material	Flow saturated	Flow unsaturated	Type of $v(I)$ curve	Possible error
Wellitschkovski, Wollny, King (1898, acc. to Swartzendruber, 1962)	sand and sandstones	+		A	?
Izbash (1931) Puzyrevskaya (1931)	gravel	+		A, B	?
acc to Polubarinova Kotchina, 1952) von Engelhardt,	clay	+		A	?
Tunn (1955)	sandstone	+		B, E?	—
Lutz, Kemper (1959)	bentonite, halloysite, clay pastes	+		B,N, (E?)	a, (b?)
Hansbo (1960)	clay	+		B	f
Slepička (1961)	sandstones	+		B, C	(c?)d
Miller, Low (1963)	bentonite paste	+		M, A, C—F, R	f
Li (1963)	clay	±		A	f
Rawlins, Gardner (1963)	silty clay loam		+	A, B	g
Davidson et al. (1963)	silt loam		+	B, E	g
Hadas (1964)	loess, clay		+	B—C	+
Volarovitch, Tchuraev (1964)	peat	+		E	—
Kutilek (1964, 1967)	kaolinite, illite montmorillonite	±		C (E?)	a
Tchuraev, Yashtchenko (1966)	sand, humic soils	+		C	—
Thames (1966)	sandy loam, silt loam		+	B—C—E	g
Nerpin, Tchudnovskij (1967)	a review: clays	+		E	?
Olson, Swartzendruber (1968)	mitxures sand, silica flour, kaolinite		+	(L), (B)	e
Russell (1968)	mixtures dtto and bentonite	+		L, P, R	b
Gairon (1968)	bentonite paste	+		L, P	—

where B, J, C are parameters, J^2C is a characteristic of the non-Darcian flow, B is the assymptote of the $v(I)$ curve, J is the section on I axis cut by the assymptote. Kutílek (1964) developed the equation with parameters of the initial conductivity M', of conductivity M when $I \to \infty$ and of constant J

$$(16) \qquad v = M\left[\frac{1}{B}\ln(A + e^{BI}) - J\right]$$

where

$$(17) \qquad B = \frac{1}{J}\ln\frac{M}{M'}$$

$$(18) \qquad A = \frac{M}{M'} - 1 = e^{BJ} - 1.$$

FLOW OF SOME NON-NEWTONIAN LIQUIDS IN POROUS MEDIA

Even if our present knowledge on the viscosity and on rheologic parameters of soil water is rather spare, it will be useful to compare the empirical Eqs. (9) to (18) with equations developed for flow in porous media when the flowing liquid is expressed as a definite rheologic model. These rheologic equations were derived for a simple capillary model. First of all the equation of the volume flow rate in tube was calculated, if not available in basic literature. Then the velocity of flow in a bundle of capillaries was derived according to the assumption $r = 2\varepsilon/S$ where r is the radius, ε is the porosity, and S is the specific surface. The pore velocity v_0 is in relation with the flow velocity $v = v_0\varepsilon$. In the next text, the rheologic parameters are denoted by m, n, A, B, ϕ_0, ϕ_1, α, μ_0, μ_∞, μ_s, μ_0, χ, and their meaning follows from the basic rheologic equations.

For the Newtonian liquid, the physical meaning of k, usually called Kozeny's equation, is well known. The coefficient k_0 in Eq. (12) of the Bingham flow has the same physical meaning as k. Pseudoplastic liquids described by

$$(19) \qquad \tau = -m\left(\frac{dv}{dy}\right)^n$$

will have

$$(20) \qquad v = kI^{1/n}$$

where

$$(21) \qquad k = \frac{n}{3n+1}\varepsilon^{(2n+1)/n}\left(\frac{2}{s}\right)^{(n+1)/n}\left(\frac{\gamma}{2n}\right)^{1/n}.$$

The expression formally similar to Eqs. (20) and (21) will be valid for the flow of liquids defined by

$$(22) \qquad \tau = -f\left(\frac{dv}{dy}\right)$$

only with the exception that n denotes a functional relationship dependent upon the functional relationship of the term dv/dy in Eq. (22). The Eyring's model

$$\tau = A \operatorname{arc\,sinh}\left(-\frac{1}{B}\frac{dv}{dy}\right)$$

gives

$$(24) \qquad v = \frac{M}{I} - \frac{N}{I^2}\frac{1}{2}(e^{pI} - e^{-pI}) + \frac{O}{I^3}\left[\frac{1}{2}(e^{pI} + e^{-pI}) - 1\right]$$

where

$$(25) \qquad M = \frac{2AB\varepsilon}{\gamma}$$

$$(26) \qquad N = \frac{4A^2BS}{\gamma^2}$$

$$(27) \qquad O = \frac{4A^3BS^2}{\gamma^3\varepsilon}$$

$$(28) \qquad p = \frac{\gamma\varepsilon}{AS}.$$

The model of Ellis

$$(29) \qquad \frac{dv}{dy} = (\phi_0 + \phi_1\tau^{\alpha-1})\tau$$

when applied to porous media gives

$$(30) \qquad v = CI + DI^\alpha$$

where

$$(31) \qquad C = \frac{\phi_0\varepsilon^3\gamma}{2S^2}$$

$$(32) \qquad D = \frac{\phi_1(2\varepsilon)^{\alpha+2}\gamma^\alpha}{S^{\sigma+1}}.$$

The model of Reiner-Philipoff

$$(33) \qquad \frac{dv}{dy} = \left[\frac{1}{\mu_\infty + \dfrac{\mu_0 - \mu_\infty}{1 + (\tau/\tau_\bullet)^2}}\right]\tau$$

will be expressed by

(34) $$v = \frac{M}{I}[F_1(I) + F_2(I)\ln 2F_2(I) - N] + OI$$

where

(35) $$M = \frac{(\mu_0 - \mu_\infty)S^2\tau_s^2}{2\varepsilon\mu_\infty^2\gamma}$$

(36) $$N = \frac{2\varepsilon^2}{S^2}$$

(37) $$F_1(I) = N\ln f_1(I) = \frac{2\varepsilon^2}{S^2}\ln\frac{\mu_\infty\gamma^2 I^2}{\tau_s^2}$$

(38) $$F_2(I) = N + f_2(I) = \frac{2\varepsilon^2}{S^2} + \frac{2\mu_0\tau_s^2}{\mu_\infty\gamma^2 I^2}$$

(39) $$O = \frac{\varepsilon^3\gamma}{2\mu_\infty S^2}.$$

The Reiner's equation

(40) $$\frac{dv}{dy} = [\phi_\infty - (\phi_\infty - \phi_0)e^{-\tau^2/x}]\tau$$

where

(41) $$\chi = \frac{\phi_\infty - \phi}{d\phi/d\tau^2}$$

gives for the flow in a capillary model

(42) $$v = MI + \frac{N}{I^2}e^{-pI^2} - \frac{O}{I^4}$$

where

(43) $$M = \frac{\phi_\infty\varepsilon^3\gamma}{2S^2}$$

(44) $$N = \frac{\chi^2 S^2(\phi_\infty - \phi_0)(4 + \chi)}{4\varepsilon\gamma^2}$$

(45) $$p = \frac{\varepsilon^2\gamma^2}{\chi^2 S}$$

(46) $$O = \frac{\chi^3 S^2(\phi_\infty - \phi_0)}{\varepsilon\gamma^4}.$$

The Gheorghitza's (1964) vector analysis of the flow of a thixotropic liquid in porous media and Raats' (1967) application of continuum mechanics for non-Darcian flow must be mentioned at this place, too.

After comparing the above derived equations with the empirical equations, we find that the Eq. (9) does not correspond to any rheologic type of liquid. The Eqs. (12) and (14) are equations of flow of a Bingham body in porous media represented by a capillary model. The Eqs. (10) and (11) have the meaning corresponding with the flow of a pseudoplastic liquid described by Eqs. (19) and (20). The empirical Eqs. (15) and (16) are not identical with the rheologic equations. However, after expanding the terms expandable in series and after ignoring of terms of higher degree, the equation corresponding with the Ellis' model was obtained. A wider use of rheologic Eqs. (24), (34), or (40) is less probable due to the high number of coefficients. It would be useful for the time being to check and compare these equations with experimental data.

A different approach to the theoretical solution was suggested by Klausner and Kraft (1966) who derived the distribution of velocity across a capillary taking into consideration the wall forces. Then the porosity of a parallel capillary model was introduced. They managed to explain in this way different types of the $v(I)$ curves, at least of the A, B, C, E types.

REFERENCES

1. BIRD, R. B., W. E. STEWART AND E. N. LIGHTFOOT (1956), *Transport Phenomena*, John Wiley, New York.

2. BLACKMORE, A. V. AND T. J. MARSHALL (1965) Water movement through a swelling material, *Austral. J. Soil Res.*, *3*, 11–21.

3. DAVIDSON, J. M., J. W. BIGGAR AND D. R. NIELSEN (1963) Gamma radiation attenuation for measuring bulk density and transient water flow in porous materials, *J. Geophys. Res. 68*, 4777 4783.

4. DE GROOT, S. R. AND P. MAZUR (1962), *Non-Equilibrium Thermodynamics*, North-Holland Publ. Co., Amsterdam.

5. GAIRON, S. (1968), Private communication.

6. GHEORGITZA, S. I. (1964) On the non steady motion of viscoplastic liquids in porous media, *J. Fluid Mech. 20*, 273–280.

7. GRADSHTEIN, I. S. AND I. M. RYZHIK (1963) *Tablicy Integralov, Summ, Ryadov i Proizvedeniy. GIFML*, Moskva.

8. GUPTA, R. P. AND D. SWARTZENDRUBER (1962), Flow-associated reduction in the hydraulic conductivity of quartz sand, *Soil Sci. Soc. Am. Proc. 26*, 6–10.

9. HADAS, A. (1964) Derivations from Darcy's law for the flow of water in saturated soils, *Israel J. Agr. Res., 14*, 159–168.

10. HANSBO, S. (1960), Consolidation of clay with special reference to influence of vertical sand drains, *Swed. Geotech. Inst. Proc. 18*, Stockholm.

11. IZBASH, S. V. (1931), O filtracii v krupnozernistom materiale. *Izv. Nauchno-issled. Inst. Gidrotekhniki (NIIG)*, Leningrad.

12. KLAUSNER, Y. AND R. KRAFT (1966), A capillary model for non-Darcy flow though porous media, *Trans. Soc. Rheol. 10*, 603–619.

13. KUTÍLEK, M. (1964) The filtration of water in soils in the region of the laminar flow, 8th Int. Congr. ISSS, II: 45–52, Bucharest.

14. KUTÍLEK, M, (1967) Temperature and non-Darcian flow of water. Int. Soil Water Symp. 35–50, Prague.

15. KUTÍLEK, M. AND J. SALINGEROVA (1966), Flow of water in clay minerals as influenced by adsorbed quinolinium and pyridinium. *Soil Sci. 101*, 385–389.

16. LI SEUNG PING (1963), Measuring extremely low flow velocity of water in clays, *Soil Sci., 95*, 410–413.

17. LOW, P. F. (1961) Physical chemistry of clay-water interaction, *Advances Agron. 13*, 269–327.

18. LUTZ, J. F. AND W. D. KEMPER (1959), Intrinsic permeability of clay as affected by clay-water interaction, *Soil Sci. 88*, 83–90.

19. MILLER, R. J. AND P. F. LOW (1963), Threshold gradient for water flow in clay systems, *Soil Sci. Soc. Am. Proc. 27*, 605–609.

20. NERPIN, S. V. AND A. F. TCHUDNOVSKIJ (1967), *Fizika Potchvy*. Nauka, Moskva.

21. NERPIN, S. V. (1967) Influence of plastic resistance on equilibrium and water movement in soils, *Int.. Soil Water Symp.* 279–289, Prague. see also: BONDARENKO, N. (1967), Liquid movement in capillary-porous media. *Int. Soil Water Symp.*, Discussion 99–114, Prague.

22. OLSEN, H. W. (1965) Deviations from Darcy's law in saturated clays, *Soil Sci. Soc. Am. Proc. 29*, 135–140.

23. OLSEN, H. W. (1966) Darcy's law in saturated kaolinite, *Water Resources Res. 2*, 287–295.

24. OLSEN, T. C. AND SWARTZENDRUBER (1968) Velocity gradient relationships for steady state unsaturated flow of water in nonswelling artificial soils, *Soil Sci. Soc. Am. Proc. 32*, 457–462.

25. POLUBARINOVA KOTCHINA, P. J. (1952) *Teorije dvizhenija gruntovykh vod. GITTL*, Moskva.

26. PRIGOGINE, I. (1961) *Introduction to Thermodynamics of Irreversible Processes* Interscience Publ., New York.

27. PUZYREVSKAYA, T. N. (1931) Prosatchivanije vody tcherez pestchanye grunty, *Izv. NIIG, 1*, Leningrad (acc. to 25).

28. RAWLINS, S. L. AND W. H. GARDNER (1963), A test of the validity of the diffusion equation for unsaturated flow of soil water, *Soil Sci. Soc. Am. Proc., 27*, 507–511.

29. REINER, M. (1949), *Deformation and Flow*, H. K. Lewis, London.

30. RUSSELL, D. A. (1968), Velocity-gradient relationships for water-saturated porous media, Ph.D. Thesis, Purdue University.

31. SCHEIDEGGER, A. E. (1957) *The Physics of Flow Through Porous Media*, University Toronto Press.

32. SLEPICKA, F. (1961), *Filtrační zákony*. VUV Praha-Podbaba.

33. SWARTZENDRUBER, D. (1962), Modification of Darcy's law for the flow of water in soils, *Soil Sci., 93*, 22–29.

34. SWARTZENDRUBER, D. (1962), Non-Darcy flow behavior in liquid-saturated porous media, *J. Geophys. Res., 67*, 5205–5213.

35. SWARTZENDRUBER, D. (1963), Non-Darcy behavior and the flow of water in unsaturated soils, *Soil Sci Soc. Am. Proc., 27*, 491–495.

36. SWARTZENDRUBER, D. (1964), Comments on paper by J. M. Davidson, J. W. Biggar and D. R. Nielsen, "Gamma-radiation attenuation for measuring bulk density an transient water flow in porous media", 1964 *J. Geophys. Res., 69*, 1679–1677.

37. SWARTZENDRUBER, D. (1966), Soil water behavior as described by transport coefficients and functions, *Advance Agron. 18*, 327–370.

38. SWARTZENDRUBER, D. (1967), Non-Darcian movement of soil water, *Int. Soil Water Symp.*, 207–222, Prague.

39. SWARTZENDRUBER, D. (1968), The applicability of Darcy's law, *Soil Sci. Soc. Am. Proc., 32*, 11–18.

40. TCHURAEV, N. V. AND YASHTCHENKO, A. I. (1966), Experimentalnoe izutchenije processov filtracii vjazkoplastitchnych zhidkostej tcherez poristyje tela. *Koll. zhurn. 28* 302–307.

41.   Thames, J. L. (1966) Flow of water under transient conditions in unsaturated soils, Ph.D. Thesis, Univ. Arizona.

42.   Volarovitch, M. P. and N. V. Tchuraev (1964), Vlijanoe poverkhnostnykh sil na peredvizhenie vlagi v poristykh telakh. Issled. v oblasti poverkhnostnykh sil, Nauka, 234–243, Moskva.

43.   Von Engelhardt, W. and W. L. M. Tunn (1954), The flow of fluids through sandstones, *Heidelberger Beitr. Mineral. u. Petrographie 2*, 12–25.

44.   Zaslawsky, D. (1964), Saturated and unsaturated flow equation in an unstable porous medium, *Soil Sci.*, *98*, 317–321.

Soil Science Laboratory,
      Technical University,
            Karlovo Nám. 3, Prague 2, Czechoslovakia

HYDROSTATICS AND HYDRODYNAMICS
IN SWELLING MEDIA*

J. R. PHILIP

ABSTRACT

The point of departure is hydrodynamics in unsaturated non-swelling media, particularly the "diffusion analysis" developed in mathematical soil physics. Generalisations of this approach to two- and three-component horizontal systems in swelling media are described.

The generalization to vertical systems demands reconsideration of hydrostatics in swelling media. The total potential includes, in this case, an additional component, the *overburden potential*, Ω, Evaluation of Ω leads to the condition for equilibrium in the vertical. This reduces to a first-order linear diff. eqn. with singular coefficients. It follows that there are three types of equilibrium profile: *hydric* profiles with the moisture gradient $d\theta/dz < 0$; *pycnotatic* profiles with $d\theta/dz = 0$; and *xeric* profiles with $d\theta/dz > 0$. Both hydric and xeric profiles approach the pycnotatic state (of maximum apparent specific gravity) in depth. Classical concepts of groundwater hydrology (tacitly based on the behaviour of non-swelling media) fail completely for swelling media.

The steady vertical flow equation in swelling systems reduces to a second-order linear d.e. with singular coefficients. The, somewhat complicated, set of possible flows is established. These include upward flows against the moisture gradient.

The non-steady vertical flow equation is derived. It is shown how its solution leads to the theory of one-dimensional infiltration in swelling media.

1. INTRODUCTION

Eleven years ago, I concluded a review of the modern theory of hydrodynamics in unsaturated non-swelling porous media (Philip, 1958) with the recognition that the task remained of generalizing the theory to include swelling media. In the soils context this meant the extension of the methods of mathematical soil-physics to embrace the processes of volume-change which play a central rôle in soil mechanics and (as we shall see) in the hydrology of swelling soils.

Most work in mathematical soil physics since 1958 has, in fact, been concerned with exploring details in hydrodynamics in non-swelling media; and it seems that it is only recently that a coherent attack on hydrostatics and hydrodynamics in swelling media has been developed. It is my purpose to devote this article to a review of these recent and current developments.

---

* We use the term "swelling media" for media subject to bulk volume-change. Such media, of course, may shrink as well as swell.

For definiteness, I shall use the terminology of soil physics. I shall consider the usual soils application in which the medium contains, in general, water and air, with pressure differences in the air phase negligible. The soil particles may be taken as inelastic and the water as incompressible. I shall not labour the generalizations to other media and other fluids; but these will be obvious to members of this Symposium.

2. HYDRODYNAMICS IN UNSATURATED NON-SWELLING MEDIA

Our point of departure is the theory of water movement in unsaturated non-swelling porous media, and, in particular, that part of the theory known sometimes as the "diffusion analysis." The foundations of the theory were laid by Buckingham (1907) and its further development was due to Richards (1931), Childs and Collis-George (1950), Klute (1952), Philip (1954, 1955a, 1957a) and others. A critical discussion of the physical basis of the theory and a unified treatment of many of its mathematical aspects are given in a lengthy review by Philip (1968a).

For present purposes it is sufficient to remark that the classical diffusion analysis is based on the following concepts:

(i) Water movement takes place in response to a gradient of the total potential Φ, made up of the moisture potential Ψ and the gravitational potential $-z$, so that

$$(1) \qquad\qquad \Phi = \Psi - z.$$

We here define potentials per unit weight of water, so that z is the vertical coordinate, positive downward. We take cm as the convenient unit of length (and also of potential).

(ii) Darcy's law holds, so that

$$(2) \qquad\qquad v = -K\nabla\Phi.$$

Here v (cm sec^{-1}) is the vector flux density of water and K (cm sec^{-1}) is the hydraulic conductivity.

(iii) Ψ and K are unique functions of the volumetric moisture content θ (or indifferently, θ and K are unique functions of Ψ), at least for the particular flow process to be analysed.

It then follows that $\Psi(\theta)$ and $K(\theta)$ comprise the necessary and sufficient hydrodynamic specification of the medium "on the Darcy scale." Combining (1) and (2) with the continuity requirement, we obtain the non-linear Fokker-Planck equation

$$(3) \qquad\qquad \frac{\partial \theta}{\partial t} = \nabla.(D\nabla\theta) - \frac{dK}{d\theta}\frac{\partial \theta}{\partial z},$$

where t is time (sec) and the moisture diffusivity D is the function of θ defined by

$$(4) \qquad\qquad D = Kd\Psi/d\theta.$$

For horizontal* and vertical one-dimensional systems, Eq. (3) takes the forms (5) and (6) respectively:

(5)
$$\frac{\partial \theta}{\partial t} = \frac{\partial}{\partial x}\left[D\frac{\partial \theta}{\partial x}\right];$$

(6)
$$\frac{\partial \theta}{\partial t} = \frac{\partial}{\partial z}\left[D\frac{\partial \theta}{\partial z}\right] - \frac{dK}{d\theta}\frac{\partial \theta}{\partial z}.$$

In (5) x (cm) is the horizontal coordinate.

Mathematical methods for solving these non-linear Fokker-Planck and diffusion equations are available for many problems of practical and theoretical interest.

3. THE EXTENSION TO SWELLING MEDIA

An important limitation of this approach has been its inability to deal with swelling media. As we shall see, however, it may be generalized to apply to the processes of flow and volume-change (and to their interactions) in such media. Before going into details, we remark on certain overall properties of the generalization:

(i) For one-dimensional systems (which we discuss principally), the extension does not depend on any particular theory of stress-distribution or stress-strain relations in the medium.

(ii) The mathematical formalism for many problems of flow in swelling media is found to be identical with that for non-swelling media. Mathematical methods developed for classical, non-swelling, media thus apply to analogous problems in swelling media.

(iii) Certain more complicated problems of flow in swelling media lead to formalism similar to, but rather more elaborate than, that for non-swelling media. For such problems the classical mathematical apparatus provides a useful point of departure.

In *non-steady* systems exhibiting volume-change the solid particles of the medium are, in general, in motion, so that it is essential to recognise that Darcy's law applies to *flow relative to the particles*. This elementary consequence of the Navier-Stokes equation was first remarked by Gersevanov (1937), according to Tsien (1961) and Zaslavsky (1964). We therefore rewrite Eq. (2) in the more general form

(7)
$$v_r = -K\nabla\Phi,$$

---

* In this paper we describe as "horizontal" systems in which the influence of gravity is negligible. This includes not only horizontal systems but also the "small-time" behaviour of non-steady vertical systems (and of systems of arbitrary geometry in two and three dimensions). Cf. Philip (1968a).

where v_r (cm sec^{-1}) is the vector flux density of water relative to the soil particles. For one-dimensional horizontal systems this becomes

$$(8) \qquad\qquad v_r = -K \partial \Psi / \partial x.$$

The following Sections 4 and 5 treat non-steady one-dimensional horizontal systems. When the influence of gravity is significant, the matter is more complicated. In such cases, hydrostatics in swelling media becomes fundamentally different from that in non-swelling media, as we show in Section 6. In Sections 7 and 8 the methods and results of Sections 4–6 are employed to yield the theory of steady vertical flows in swelling media and the general theory of non-steady flow and volume change in vertical systems in such media.

4. ONE-DIMENSIONAL HORIZONTAL TWO-COMPONENT SYSTEMS

The simplest class of non-steady phenomena of water movement and volume-change are those in one-dimensional, horizontal, two-component (solid, water) systems. Such two-component systems exhibit *normal* volume-change in the terminology of Keen (1931) and later authors (e.g., Marshall (1959)): that is, a change in the volume of water in the medium produces an equal change in the bulk volume of the medium.

Two separate approaches to this and other non-steady one-dimensional phenomena in swelling media may be distinguished: (i) an *Eulerian* analysis in terms of the *physical space coordinate*; (ii) a *Lagrangian* analysis in terms of a *material coordinate*, in which the solid particles of the medium remains at rest.

The *Eulerian* analysis preserves a direct connection between the mathematics and the physical processes, but it has the disadvantage that the mathematics becomes very complicated. Prager (1953) used this approach in a related context. Philip (1968d) analysed the class of problem considered in this section in this way.

The *Lagrangian* analysis has the virtue that the mathematics is elegant and relatively simple; but it suffers from some loss of immediacy in the connections between the mathematics and the physics. This approach was used in related contexts by Hartley and Crank (1949), McNabb (1960), de Jager et al. (1963), Fatt and Goldstick (1965) and Gibson et al. (1967). In connection with the class of problem of this section, Smiles and Rosenthal (1968) and Philip (1968d) introduced the material coordinate m, defined by

$$(9) \qquad\qquad dm/dx = (1 + e)^{-1},$$

where e is the void ratio (i.e., the ratio of the volume of voids to the volume of solid particles). We use the Lagrangian approach in the present account, noticing that in two-component systems

$$(10) \qquad\qquad e = \theta / (1 - \theta).$$

Combining Eq. (8) and the continuity requirement for element dm, we obtain

(11)
$$\left[\frac{\partial e}{\partial t}\right]_m = \frac{\partial}{\partial m}\left[K\frac{\partial \Psi}{\partial x}\right],$$

since v_r is the flux density across the plane $m = $ constant and $e\,dm$ is the volume of water in element dm. Use of (9) and the chain rule for differentiation then yields

$$\frac{\partial e}{\partial t} = \frac{\partial}{\partial m}\left[\frac{K}{1+e}\frac{\partial \Psi}{\partial m}\right],$$

i.e.

(12)
$$\frac{\partial e}{\partial t} = \frac{\partial}{\partial m}\left[D_m\frac{\partial e}{\partial m}\right],$$

where

(13)
$$D_m = \frac{K}{1+e}\frac{d\Psi}{de} = \frac{K}{(1+e)^3}\frac{d\Psi}{d\theta} = \frac{D}{(1+e)^3} = (1-\theta)^3 D.$$

Eq. (12) is a non-linear diffusion equation of the same form as (5) and may, of course, be solved by the same methods. Raats (1965) arrived at an equation similar to (12) by means of a lengthy analysis based on continuum mechanics (Truesdell and Toupin, 1960).

Philip (1968d) made a detailed Eulerian analysis of the processes of absorption (swelling) and desorption (consolidation) in one-dimensional horizontal systems. He obtained an integro-differential equation exactly equivalent to (12; and solved this for initial and boundary conditions corresponding to a step-function change of surface moisture potential or load in an initially uniform, effectively semi-infinite system. (The same results, of course, could have been obtained from a Lagrangian analysis in which the solution of (12) is recast into space coordinates. (Cf. Philip and Smiles (1969).) A great variety of properties of the process follow immediately from the basic solution, including such details as the instantaneous distribution of flux densities of water and solid, and the displacement histories of the solid particles.

Important physical implications of the solutions include the following:

(i) The *mass flow* of water associated with the movement of the solid particles is of the same order of magnitude as the Darcy flow relative to the particles.

(ii) Water exchange during desorption tends to be faster than during the converse absorption. This behaviour is opposite to that in non-swelling media, and is attributed to mass flow effects.

(iii) The classical theory of consolidation and swelling due to Terzaghi (1923) (a) takes K and $d\Psi/de$ as constant, and (b) totally neglects the mass flow due to particle movement. The first of these deficiencies is unimportant in the limit of small strains often of interest in soil mechanics; but the second is basic and is operative for small strains (Philip, 1968d), despite the belief to the contrary expressed by Gibson *et al.* (1967). The analysis described here, of course, overcomes both limitations of the classical theory.

Although our principal concern is with one-dimensional systems, we note in passing that the Lagrangian approach may be applied also to non-steady flow and volume-change in two-component cylindrical and spherical systems. The kinematics of volume-change in such systems has interesting features. (Philip, 1968e).

5. ONE-DIMENSIONAL HORIZONTAL THREE-COMPONENT SYSTEMS

The approach may be extended readily to three-component (solid, water, air) systems (Philip and Smiles, 1969). The new element which enters the analysis is that the $e(\theta)$ relation is no longer fixed by the constraint of normal volume-change. It becomes, instead, an *independent functional characteristic* of the medium, so that $\Psi(\theta)$, $K(\theta)$ and $e(\theta)$ now make up the necessary and sufficient hydrodynamic specification of the medium.

In this case the combination of (8) and the continuity requirement for element dm yields

$$(14) \qquad \left[\frac{\partial[(1+e)\theta]}{\partial t}\right]_m = \frac{\partial}{\partial m}\left[K\frac{\partial \Psi}{\partial x}\right],$$

since here $(1+e)\theta\, dm$ is the volume of water in dm. Applying (9) and the chain rule for differentiation gives

$$(15) \qquad C\frac{\partial \theta}{\partial t} = \frac{\partial}{\partial m}\left[D_m \frac{\partial \theta}{\partial m}\right],$$

where

$$(16) \qquad C = 1 + e + \theta\, de/d\theta \quad \text{and} \quad D_m = D/(1+e).$$

The formalism is slightly simpler in terms of the *moisture ratio* ϑ, defined by

$$(17) \qquad \vartheta = (1+e)\theta.$$

Eq. (15) then reduces to

$$(18) \qquad \frac{\partial \vartheta}{\partial t} = \frac{\partial}{\partial m}\left[D_\vartheta \frac{\partial \vartheta}{\partial m}\right],$$

with

$$(19) \quad D_\vartheta = \frac{K}{1+e}\frac{d\Psi}{d\vartheta} = \frac{D}{(1+e)(1+e+\theta\, de/d\theta)} = \frac{D(1+e-\vartheta\, de/d\vartheta)}{(1+e)^3}.$$

Raats (1965) used the concepts of continuum mechanics in an elaborate analysis which yielded an equation equivalent to (18). Eq. (18) is also of the form (5), so that mathematical methods developed for non-swelling media again apply here.

Philip and Smiles (1969) solved in detail the problems of absorption (swelling) and desorption (consolidation) for initial and boundary conditions corresponding to a step-function change of surface moisture potential in a three-component initially uniform, effectively semi-infinite, horizontal, one-dimensional system. (The same mathematical methods apply to the case of a step-function change of load, and the results are analogous. Concepts to be developed in Section 6 below, however, are needed in connection with that problem.)

The physical implications of these, more general, solutions are similar to those for two-component systems. Again the *mass flow* of water is, in general, of the same order of magnitude as the Darcy flow and exerts an important influence on the phenomena. The solutions (including those for a step-function change of load) provide the generalization to three-component systems of the theory of consolidation and swelling.

A remark on the applicability of this analysis to *cracking media* is required. Philip and Smiles (1969) point out that, since the approach deals with one-dimensional volume-change, crack volume is counted as void space in the definition of e. On this understanding, the analysis applies, so long as the characteristic functions $\Psi(\theta)$, $K(\theta)$ and $e(\theta)$ can be defined; but the results of the analysis will have significance only as averages over volumes with linear dimensions no smaller than the characteristic length of the cracking pattern. A more refined analysis, allowing for local disequilibrium between cracks and macroaggregates and following the methods of Philip (1968b, c), seems possible in principle.

6. HYDROSTATICS IN SWELLING MEDIA

We have been able to treat flow and volume-change in two-component and (unloaded) three-component horizontal systems without any careful examination of the total potential Φ in swelling media because $\partial\Phi/\partial x = \partial\Psi/\partial x$ in these systems. Further progress, however, requires this examination.

Soils in nature tend to be constrained so that bulk volume-change takes place one-dimensionally and vertically. Our primary concern here is with hydrostatics in swelling media constrained in this way (though we mention briefly horizontal systems also).

Using the device of the scalar potential, first applied to problems of hydrostatics in soils by Buckingham (1907), Philip (1969a) showed that, in a one-dimensional vertical system of a swelling medium, with the particles constrained from moving only at its base, Eq. (1) is replaced by

$$(20) \qquad\qquad \Phi = \Psi - z + \Omega,$$

where Ω is the *overburden potential*. The addition of any small quantity of water to the medium demands a local increase in bulk volume, with an associated movement of the particles. Ω is the work performed, per unit weight of water added, to realize the required movement of the particles against gravity and any external load. It follows that

$$(21) \qquad\qquad \Omega = \frac{de}{d\vartheta} \cdot P(z)$$

$P(z)$ (cm) is the total vertical stress. Where there is an external load $P(0)$ on the upper surface $z = 0$, we have

$$(22) \qquad\qquad P(z) = P(0) + \int_0^z \gamma \, dz,$$

where the *apparent wet specific gravity*, γ, is defined by

$$(23) \qquad\qquad \gamma = (\vartheta + \gamma_s)/(1 + e).$$

γ_s is the particle specific gravity which is taken to be greater than 1 in the further developments. $\gamma_s \approx 2.7$ for mineral soils.

For the horizontal column $x > 0$ constrained at its remote end and subject to external pressure $P(0)$ at $x = 0$,

$$(24) \qquad\qquad \Omega = \frac{de}{d\vartheta} \cdot P(0).$$

Note that $de/d\vartheta = 1$ for normal volume change and $0 < de/d\vartheta < 1$ for residual volume change in the terminology of Keen (1931).

Coleman and Croney (1952), through a lengthy thermodynamic argument, arrived at a result essentially equivalent to (21). As Philip (1969a) stated, the experimental determination of $e(\vartheta)$ should be made under the constraint of one-dimensional volume-change. Coleman and Croney made no such stipulation and, in fact, measurements of $e(\vartheta)$ by their laboratory are explicitly for three-dimensional volume-change. It is a matter of some surprise that the implications of their work have been largely neglected. The reason may be, in part, the difficulty of their argument and, in part, the inaccessibility of their original communication.

The condition for equilibrium moisture distribution in the vertical in a swelling medium is that

$$(25) \qquad \Phi = \Psi - z + \frac{de}{d\vartheta} \left[P(0) + \int_0^z \gamma \, dz \right] = \text{constant} = -Z,$$

where we have evaluated Φ through (20), (21), (22). In hydrological contexts Z is the "water table depth" in the operational sense that the equilibrium level of free water in a (suitably encased) borehole is at distance Z below the surface.

Philip (1969a) showed that (25) reduces to the differential equation

$$(26) \qquad dz/d\vartheta = -M - Nz,$$

where the coefficients M, N are (known) functions of ϑ. The general solution of (26) is in closed form and is well-known (Ince, 1927). M and N, however, are singular for

$$(27) \qquad de/d\theta = \gamma^{-1}.$$

Philip showed that (27) is satisfied only at an extremum of γ (as a function of ϑ) and that, under realistic and relatively weak conditions on $e(\theta)$, there is *one and only one* such extremum, and that this is a *maximum*.

Eq. (26) thus has a singular solution at the ϑ-value at which the apparent wet specific gravity of the medium is maximum. This ϑ-value is denoted by ϑ_p and is designated the *pycnotatic point* [Gk. πυκνότατος = "densest"]. It follows that there are three distinct types of possible equilibrium vertical moisture profiles in a swelling medium, distinguished by their value of ϑ_0, the value of ϑ at the surface $z = 0$:

$$(28) \quad (a) \qquad \text{For } \vartheta_0 > \vartheta_p, \ \vartheta > \vartheta_p \ \text{ and } \ d\vartheta/dz < 0 \ \text{ for } \ z \geqq 0.$$

Profiles of this type are wettest at the surface and are everywhere wetter than ϑ_p. This type is conveniently designated *hydric*.

$$(29) \quad (b) \qquad \text{For } \vartheta_0 = \vartheta_p, \ \vartheta = \vartheta_p \ \text{ and } \ d\vartheta/dz = 0 \ \text{ for } \ z \geqq 0.$$

The medium is everywhere in the pycnotatic state of maximum apparent wet specific gravity. Such profiles are described as *pycnotatic*.

$$(30) \quad (c) \qquad \text{For } \vartheta_0 < \vartheta_p, \ \vartheta < \vartheta_p \ \text{ and } \ d\vartheta/dz > 0 \ \text{ for } \ z \geqq 0.$$

These profiles are driest at the surface and are everywhere drier than ϑ_p. They are converse in their properties to hydric profiles and are appropriately termed *xeric*.

For all three types of profile

$$(31) \qquad \lim_{z \to \infty} \vartheta = \vartheta_p.$$

For both xeric and hydric profiles, the approach to ϑ_p at great depths is like that of a hyperbola to its asymptote.

In mineral soils $\gamma_{max} \approx 2$, so that, in view of (27), the pycnotatic point occurs near the moisture ratio at which $de/d\theta = 0.5$. This implies that, for soils exhibiting a range of normal volume-change, ϑ_p is rather less than the moisture ratio at the dry end of the normal range.

It should be understood that gravity influences equilibrium (and movement) of water in swelling media in a manner completely different from its influence in non-swelling media. It follows from (25), in fact, that the effect of gravity in a swelling medium is *approximately* $(1 - \gamma \, de/d\vartheta)$ times that in a non-swelling one. For a mineral soil this factor is about -1 in the normal range, increases to 0 at $\vartheta = \vartheta_p$, and approaches $+1$ as $de/d\vartheta \to 0$ at small values of ϑ.

Contrary to a commonly expressed notion (Philip 1957d, Rose *et al.*, 1965; Rose, 1966) the effects of overburden potential manifest themselves right to the surface: it is not the magnitude of Ω, but that of $d\Omega/dz$, which is important.

Philip (1969a, b) solved Eq. (25) for various surface moisture conditions and for various combinations of water table depth and surface loading and used the results to illustrate the implications for *hydrology* and *soil mechanics*. Various classical concepts of groundwater hydrology (tacitly based on the behaviour of non-swelling media) fail completely for swelling media. The distributions of saturation and of hydraulic conductivity relative to the water table differ entirely fron the conventional picture, as we may infer at once from (28)–(31). Variations of surface topography affect moisture distribution in swelling soils so that, for example, appreciable moisture gradients adjoin bodies of surface water in such soils. Classically, such a distribution would be interpreted as due to impermeability of the soil; but it may, in fact, represent a true equilibrium and arise, not from a lack of hydraulic conductivity, but from a lack of difference in total potential. Soil-mechanical problems analyzed by Philip (1969b) include the variation of equilibrium soil levels with water table depth and with water depth over the soil, and the effect of surface loading on equilibrium moisture profiles and soil levels.

Philip (1969b) has indicated how the approach of this section may be used in the investigation of two- and three-dimensional problems of equilibrium moisture distribution. The extension depends on several assumptions and the use of an appropriate stress distribution theory such as that of Boussinesq (1885) or the modification due to Fröhlich (1934).

7. STEADY VERTICAL FLOWS IN SWELLING MEDIA

The analysis of steady vertical flows in swelling media follows (Philip, 1969c). Apart from its significance in steady problems, this is an essential preliminary to the theory of one-dimensional infiltration in swelling media.

Combining the one-dimensional vertical form of (7) with the expression for Φ in swelling media (first equality of (25)), we obtain

$$(32) \qquad v_r = -K \left[\left\{ \frac{d\Psi}{d\vartheta} + \frac{d^2 e}{d\vartheta^2} \left[P(0) + \int_0^z \gamma \, dz \right] \right\} \frac{\partial \vartheta}{\partial z} + \gamma \frac{de}{d\vartheta} - 1 \right].$$

For steady flows we omit the suffix r from v (since the particles are stationary) and we rewrite the partial derivative as total. Rearrangement of (32) then gives

$$(33) \qquad \frac{v}{K} + \gamma\frac{de}{d\vartheta} - 1 = -\frac{d\vartheta}{dz}\left\{\frac{d\Psi}{d\vartheta} + \frac{d^2e}{d\vartheta^2}\left[P(0) + \int_0^z \gamma\, dz\right]\right\}.$$

Eq. (33) reduces to the differential equation

$$(34) \qquad dz'/d\vartheta = -M_1 - N_1 z'$$

where the coefficients M_1, N_1 are (known) functions of ϑ, and z' denotes $dz/d\vartheta$. The general solution in the form $z'(\vartheta)$ is well-known (Ince, 1927), and an integration with respect to ϑ yields $z(\vartheta)$ in closed form. M_1 and N_1, however, are singular for

$$(35) \qquad v = K(1 - \gamma\, de/d\vartheta)\,[= \kappa(\vartheta)].$$

It follows from the singular behaviour of M_1 and N_1 that

$$(36) \qquad z \to \infty \quad \text{as } \kappa(\vartheta) \to v,$$

so that

$$(37) \qquad v = \kappa(\vartheta_\infty)$$

where we write ϑ_∞ for $\lim_{z\to\infty}\vartheta(z)$. Thus, in the limit as $z \to \infty$, v is wholly determined by ϑ_∞.

Philip (1969c) showed that, under realistic and relatively weak conditions on $e(\vartheta)$, $\kappa(\vartheta)$ has *one and only one maximum*, which is at

$$(38) \qquad \vartheta = \vartheta_k < \vartheta_p.$$

ϑ_k, like ϑ_p, has an important place in the hydrodynamics of swelling media. In view of (37) it is the value of ϑ_∞ for which the greatest possible steady downward flow occurs. We call this point the *katotatic point* [Gk. κατωτάτω = "most downward"].

The singular solutions of (33) are readily established through (35). Nonsingular solutions must simultaneously satisfy two conditions: (i) $z(\vartheta)$ must be monotonic. (ii) $\kappa(\vartheta_0) - \kappa(\vartheta_\infty)$ must be of the same sign a $\vartheta_\infty - \vartheta_0$. Identification and classification of the possible solutions is rather elaborate. A classification based primarily on the value of v, for example, involves five classes, each with up to five sub-classes depending on the value of ϑ_0.

We cannot here give a detailed exposition. Figure 1, however, gives a graphical summary. For $\vartheta_\infty > \vartheta_p$, we have steady upward flow, depending only on ϑ_∞ and independent of ϑ_0. For $\vartheta_\infty = \vartheta_p$, we have the no-flow equilibria treated in Section 6. For $\vartheta_k \leq \vartheta_\infty < \vartheta_p$, we have steady downward flow depending only on ϑ_∞, provided $\vartheta_0 > \Theta$, a function of ϑ_∞ which varies monotonically from 0 at $\vartheta_\infty = \vartheta_p$ to ϑ_k at $\vartheta_\infty = \vartheta_k$; no steady vertical flows are possible for $\vartheta_0 \leqq \Theta$. For $\vartheta_\infty < \vartheta_k$, the only possible steady vertical flows are the singular downward ones for which $\vartheta_0 = \vartheta_\infty$. Flows on the diagonal $\vartheta_0 = \vartheta_\infty$ are singular, with $d\vartheta/dz = 0$ everywhere in the profile; flows above this diagonal, i.e. for $\theta_0 > \vartheta_\infty$, have $d\theta/dz < 0$; flows below this diagonal, i.e. for $\vartheta_0 < \vartheta_\infty$, have $d\vartheta/dz > 0$.

We remark on two physical implications: (i) The diagonal line of purely singular downward flows $\vartheta_0 = \vartheta_\infty < \vartheta_k$ occupies the whole diagonal of the figure for non-swelling media. (ii) Steady upward flows occur against a moisture gradient when $\vartheta_0 > \vartheta_\infty > \vartheta_p$. Behaviour of this type during evaporation from swelling media was reported by Hallaire (1963) and by Nutter and White (1968).

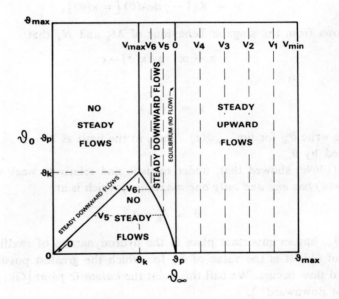

Fig. 1. Schematic representation of possible steady vertical flows in swelling media. ϑ_0 is the surface moisture ratio and ϑ_∞ that in the limit of large depths. ϑ_{max} is the moisture ratio for $\Psi = 0$. ϑ_p is the pycnotatic point and ϑ_k the katotatic point. Symbols v_{max} to v_{min} indicate values of v, the flux density. v_{max} is the maximum possible downward steady flux density; v_{min} is minus the maximum possible upward steady flux density. $v_{min} < v_1 < v_2 < v_3 < v_4 < 0 < v_5 < v_6 < v_{max}$.

8. NON-STEADY VERTICAL FLOWS IN SWELLING MEDIA. INFILTRATION.

We treat the general equation describing non-steady vertical flow and volume-change in swelling media by bringing together various elements of this exposition.

Combining the continuity requirement for element dm with Eq. (32), we obtain the required equation:

(39)

$$\frac{\partial \vartheta}{\partial t} = \frac{\partial}{\partial m}\left[\frac{K}{1+e}\left\{\frac{d\Psi}{d\vartheta} + P(0)\frac{d^2e}{d\vartheta^2}\right\}\frac{\partial \vartheta}{\partial m}\right]$$

$$- \frac{\partial}{\partial m}\left[K\left\{1 - \gamma\frac{de}{d\vartheta} - \frac{d^2e/d\vartheta^2}{1+e}\int_0^m \gamma(1+e)\,dm\frac{\partial \vartheta}{\partial m}\right\}\right].$$

The material coordinate m is now defined by

(40)
$$dm/dz = (1+e)^{-1}.$$

Eq. (39) is of the form

(41)
$$\frac{\partial \vartheta}{\partial t} = \frac{\partial}{\partial m}\left[D_* \frac{\partial \vartheta}{\partial m}\right] + \text{term of dimensions } m^{-1},$$

where D^* is the function of ϑ defined by

(42)
$$D_* = \frac{K}{1+e}\left[\frac{d\Psi}{d\vartheta} + P(0)\frac{d^2e}{d\vartheta^2}\right].$$

D_* is the generalization of D_* (Eq. (19)) which embraces the influence of surface load $P(0)$.

Evidently we may develop the theory of various non-steady processes of flow and volume-change in vertical systems by solving (39) subject to appropriate conditions. Translation of solutions in material coordinate m back to space coordinate z is a trivial matter of detail.

We limit our attention here to infiltration, but it will be understood that capillary rise, drainage and evaporation are amenable to similar treatment. Further, the theory of vertical consolidation and swelling of layers so thick that gravity cannot be neglected is established in the same way.

Infiltration into an effectively semi-infinite, homogeneous, vertical system of uniform initial moisture ratio ϑ_∞ from water at moisture potential $\Psi(\vartheta_0)$ continuously available at upper surface $m = 0$ from time $t = 0$ is described by Eq. (39) subject to conditions (43):

$$t = 0, \ m > 0, \ \vartheta = \vartheta_\infty;$$

(43)

$$t \geqq 0, \ m = 0, \ \vartheta = \vartheta_0.$$

The symbol ϑ_∞ is appropriate for the initial moisture ratio, which is $\lim_{m\to\infty}(\vartheta) = \lim_{z\to\infty}(\vartheta)$ for all $t \geqq 0$.

It is a consequence of (41) that, for small enough t, the second term on the right of (39) is negligible compared with the first, and (39) becomes, effectively,

$$(44) \qquad \frac{\partial \vartheta}{\partial t} = \frac{\partial}{\partial m} \left[D_* \frac{\partial \vartheta}{\partial m} \right]$$

The solution of (44) subject to (43) is

$$(45) \qquad m(\vartheta, t) = \phi t^{\frac{1}{2}},$$

where ϕ is a function of ϑ which is readily evaluated by numerical (Philip, 1955b) or analytical (Philip, 1960) integration of an ordinary nonlinear differential equation. ((45) is the sorption solution of Philip and Smiles (1969), slightly generalized to include surface loading $P(0)$.)

Treating (41), then, as a perturbation of (44), we may develop the solution of (39) subject to (43) in the form

$$(46) \qquad m(\vartheta, t) = \phi_1 t^{\frac{1}{2}} + \phi_2 t + \phi_3 t^{3/2} \cdots,$$

where the ϕ's are functions of ϑ which are readily evaluated. ϕ_1 is simply ϕ of (45); and ϕ_2, ϕ_3 etc. are solutions of linear equations which are established, for example, by a process of "equating coefficients." This solution is similar to that developed by Philip (1957a, b) for the non-linear Fokker-Planck equation (6), which is also of the form (41). Just as for Eq. (6), we may supplement the series solution (which converges for a range of t which is usefully large in the application) by the asymptotic solution for large t.

For Eq. (6) there are two modes of asymptotic behaviour, depending on whether the terms of the right side are of the same, or of different signs. When the signs are the same, (e.g. for infiltration in non-swelling media) the "profile at infinity" solution holds (Philip, 1957c; see also Irmay, 1956). When the signs differ (e.g. for capillary rise in non-swelling media) the asymptotic solution is a steady distribution (Philip, 1957b, 1966). For Eq. (39), however, the matter is more complicated, since, in general, the signs differ over part of the ϑ-range and agree over the rest.

Two classes of asymptotic solution of (39), (43) are found. For $\vartheta_\infty \geqq \vartheta_k$, the solution is a steady distribution satisfying Eq. (34). For $\theta_\infty < \vartheta_k$, however, the solution asymptotically approaches a steady distribution for $\vartheta_k \leqq \vartheta < \vartheta_0$ connected to a moving front (reminiscent of the "profile at infinity") for $\vartheta_\infty \leqq \vartheta < \vartheta_k$. For this latter class the final infiltration rate is v_{max}, independent of the value of ϑ_∞.

REFERENCES

BOUSSINESQ, J. (1885), *Application des Potentiels à l'Étude de l'Équilibre et du Mouve ment des Solides Élastiques*, (Gauthier-Villars: Paris.)

BUCKINGHAM, E. (1907), *U.S. Dept. Agr. Bur. Soils Bull.*, 38.

CHILDS, E. C. AND N. COLLIS-GEORGE (1950), *Proc. Roy. Soc.*, A 210, 392.

COLEMAN, J. D. AND D. CRONEY (1952), The estimation of the vertical moisture distribution with depth in unsaturated cohesive soils, *Road Res. Lab. Note* RN/1709/JDC., DC (unpublished).

FATT, I. AND T. K. GOLDSTICK (1965), *J. Colloid Sci.*, *20*, 962.

FRÖHLICH, O. K. (1934), *Druckverteilung in Baugrunde*, Springer, Berlin.

GERSEVANOV, N. M. (1937), *The Foundations of Dynamics of Soils*, 3rd. ed., Stroiizdat, Moscow-Leningrad.

GIBSON, R. E., G. L. ENGLAND AND M. J. L. HUSSEY (1967), *Géotechnique*, *17*, 261.

HALLAIRE, M. (1963), *Ann. Agron*, *14*, 393.

HARTLEY, G. S. AND J. CRANK (1949), *Trans. Faraday Soc.*, *45*, 801.

INCE, E. L. (1927), *Ordinary Differential Equations*, Longmans, London.

IRMAY, S. (1956), *Symposia Darcy*, I.A.S.H., *2*, 57.

JAGER, E. M. DE, M. VAN DEN TEMPEL AND P. DE BRUYNE (1963), *K. Ned. Akad. Wet.*, B, *66*, 17.

KEEN, B. A. (1931), *The Physical Properties of the Soil*, Longmans, London.

KLUTE, A. (1952), *Soil Sci.*, *73*, 105.

MCNABB, A. (1960), *Q. App. Math.*, *17*, 337.

MARSHALL, T. J. (1959), *Relations between Water and Soil*, Comm. Agr. Bur., Farnham, Royal, Bucks.

NUTTER, W. L. AND D. P. WHITE (1968), *Agron. Abst.*, p. 78.

PHILIP, J. R. (1954), *J. Inst. Eng. Australia*, *26*, 255.

——— (1955a), *Proc. Nat . Acad. Sci., India*, A, *24*, 93.

——— (1955b), *Trans. Faraday Soc.*, *51*, 885.

——— (1957a), *Soil Sci.*, *83*, 345.

——— (1957b), *Australian J. Phys.*, *10*, 29.

——— (1957c), *Soil Sci*, *83*, 435.

——— (1957d), *Soil Sci*, *84*, 163.

——— (1958), *Spec. Rep. Highway Res. Board Wash.*, *40*, 147.

——— (1960), *Australian J. Phys.*, *13*, 1.

——— (1966), *Proc. UNESCO-Neth. Symp., Water in the Unsaturated Zone*, p. 557.

——— (1968a), *Adv. Hydroscience*, *5*, 215.

——— (1968b), *Australian J. Soil Res.*, *6*, 1.

——— (1968c), *Australian J. Soil Res.*, *6*, 21.

——— (1968d), *Australian J. Soil Res.*, *6*, 249.

——— (1968e), Sorption and normal volume-change in cylindrical and spherical aggregates, Unpublished.

——— (1969a), *Australian J. Soil Res.*, *1*, 99.

——— (1969b), *Australian J. Soil Res.*, *1*, 121.

——— (1969c), Steady vertical flows in swelling soils. Unpublished.

PHILIP, J. R. AND D. E. SMILES (1969), *Australian J. Soil Res.*, *7*, 1.

RAATS, P. A. C. (1965), Development of equations describing transport of mass and momentum in porous media, with special reference to soils. Ph.D. Thesis, University of Illinois.

RICHARDS, L. A. (1931), *Physics*, *1*, 318.

ROSE, C. W. (1966), *Agricultural Physics*, Pergamon, London.

ROSE, C. W., W. R. STERN AND J. E. DRUMMOND (1965), *Australian J. Soil Res.*, *3*, 1.

SMILES, D. E. AND M. J. ROSENTHAL (1968), *Australian J. Soil Res.*, *6*, 237.

TERZAGHI, K. (1923), *Sitzb. Akad. Wiss. Wie*, Abt. IIa, *132*.

TRUESDELL, C. AND R. A. TOUPIN, (1960), *Handbuch der Physik*, III/1, 226, Springer, Berlin.

TSIEN, H. S. (1961), In *Problems of Continuum Mechanics*, p. 565, SIAM, Philadelphia.

ZASLAVSKY, D. (1964), *Soil Sci.*, *98*, 317.

Note added in proof.

The developments of Sections 6–8 are not exact for three-component systems under variable load. See communications by E. G. Youngs and G. D. Towner, and by J. R. Philip, in the August 1970 number of *Water Resources Research*, *6*.

CSIRO DIVISION OF PLANT INDUSTRY,
  P.O. BOX 109,
    CANBERRA CITY, A.C.T., 2601, AUSTRALIA

MODEL TESTS TO STUDY GROUNDWATER FLOWS USING RADIOISOTOPES AND DYE TRACERS

D. KLOTZ, H. MOSER AND W. RAUERT

1. DESCRIPTION OF THE MODELS

To study the behavior of groundwater flows, in particular in the vicinity of borings, two types of flow models have been built on a scale of 1:1.

Fig. 1 shows the schematic drawing of the three-dimensional models, Table I

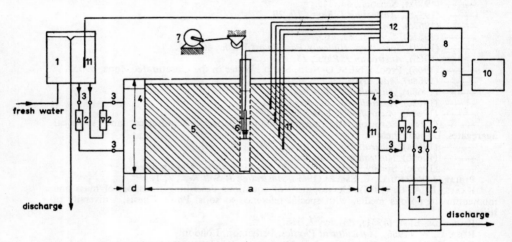

Fig. 1. Schematic plan of model setup for filtration velocity measurements
by the point dilution method (dimensions see Table I)
1 overflow, 2 flow meter, 3 control valve, 4 stilling chamber, 5 aquifer (sand or gravel), 6 filter tube with probe, 7 mixing motor, 8 scaler, 9 rate meter, 10 chart recorder,
11 resistance thermometer, 12 6-point recorder

TABLE I (see Fig. 1)

Dimensions of the three-dimensional flow models, in mm

Number of model	length a	width b of porous aquifer	height c	length of fore chambers
1	5 200	2 500	1 800	400
2	2 480	1 500	1 300	300
3	1 200	1 200	800	150
4	2 200	1 200	1 000	150
5	1 500	1 000	4 000	150

containing the essential dimensions. Up to the present, the aquifer has been simulated by quartz sands and gravels with grain sizes between 0.5 and 1.5 mm, 2 and 4 mm, and 4 and 6 mm. Intake and discharge of water are symmetrical through inlets and outlets distributed evenly over the end walls, and across stilling chambers. Throughflow is regulated by overflow tanks and valves, and measured by float flow meters. So far, filtration velocities have been set between 1.5 mm/day and 60 m/d, the upper limit being governed mainly by the water supply and the head that can be adjusted. As insulation against temperature changes, the models have been enclosed by Styrocell panels. Tracer experiments were performed to measure the velocity distribution over the cross section of the models; potential distribution can be followed continuously by pressure measuring points on the bottom and on the walls. Both measurements revealed a constant velocity over the model cross section.

Fig. 2 shows the diagram of a quasi-two-dimensional model on which any

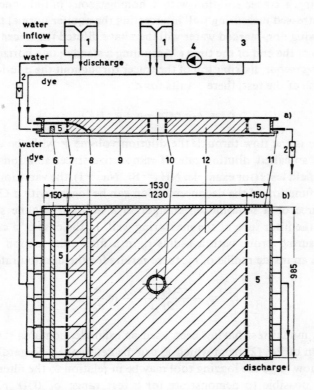

Fig. 2. Quasi-two-dimensional Plexiglass model for flow tests using dye tracers. (a) cross section of model in horizontal position; (b) model seen from above; (dimensions: mm) 1 overflow, 2 flow meter, 3 dye reservoir, 4 pump, 5 stilling chamber, 6 baffle, 7 perforated metal, 8 dye injectors, 9 Plexiglass plate, 10 well screen, 11 U-iron frame, 12 gravel

section through the aquifer can be represented. The model consists of a steel frame supporting Plexiglass panes sandwiching the aquifer. Inflow and outflow, as well as control and measurement of the water flowing through, are similar to the three-dimensional models. The tracer solution (dye or radioactive tracer) can be introduced into the aqifer through nozzles to produce flow lines, or injected at any desired point (for instance, in a simulated boring).

The models described have been used in developing single-well tracer methods to determine filtration velocity (point-dilution method) and direction of flow in the unpumped aquifer. Results have been good, as will be shown by examples in the following.

2. MODEL TESTS RELATING TO THE POINT-DILUTION METHOD TO DETERMINE FILTRATION VELOCITY

Principle of the method (compare [1, 2, 3, 4, 6, 7, 9,10, 11]):

Introducing a tracer solution with a homogeneous initial concentration c_0 into an unstressed measuring well intersecting the aquifer normal to the stream lines, inflowing non-marked water will then have diluted the tracer to a concentration of c at the end of the time t. Assuming a steady-state horizontal current and a homogeneous distribution of the tracer throughout the dilution volume V in the course of the test, there results for c:

$$(1) \qquad c = c_0 \cdot \exp(-v_g \cdot q \cdot t/V),$$

where $v_g \cdot q$ is the flow through the dilution volume V with the cross section q. v_g is the so-called dilution rate. If radioactive tracer solutions are used in model and field tests (for example, $NH_4{}^{82}Br$, $Na^{131}I$), the variation of the pulse rate n as a function of the deconcentration can be logged with a Geiger-Müller counting tube or a scintillation counter in a dilution volume sealed off by packers in the filter tube of the measuring well. In the case of a circular cylindrical measuring probe with a radius r_0 in a filter tube section of the same height with an inside radius of r_1, there results for the dilution rate v_g:

$$(2) \qquad v_g = \frac{-\pi(r_1^2 - r_0^2)}{2r_1 t} \ln \frac{n}{n_0}.$$

Fig. 3, a model test having been used as example, contains the graphic representation of Eq. (2) from which v_g can be determined. Regarding the question as to how large the logging tool may be in relation to the filter tube diameter, it was possible to demonstrate for a test range of $0 \leq r_0/r_1 \leq 0.92$, that Eq. (2) holds valid.

In order to be able to convert the dilution rate to the filtration velocity of the aquifer surrounding the measuring well, the effects of the well construction

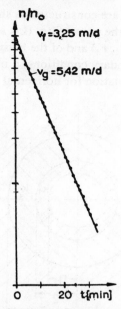

Fig. 3. Example of deconcentration curve for a 2-inch steel filter tube

on the groundwater flow field, as well as disturbances through the measuring process proper, must be allowed for. Thus the measuring well produces a distortion of the parallel flow lines in the undisturbed aquifer, which, assuming a pure potential flow, is generally open to calculation. This field distortion is described by the relation α between the width attained asymptotically of a tracer cloud emanating from the filter tube and the inside diameter of this tube (Fig. 4).

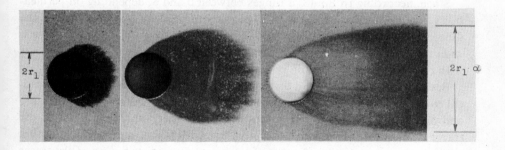

Fig. 4. Photographs of stages at different time intervals of the effluence of a $KMnO_4$-solution from a filter tube in the quasi-two-dimensional flow model.
Aquifer: coarse sand, 0.5 ... 1.5 mm ($k_2 = 0.35$ cm/sec)

In general, measuring wells are constructed as shown in Fig. 5, the permeabilities of the filter tube (k_1), the gravel filter (k_2) and the aquifer (k_3), as well as the radii of the filter tube (r_1, r_2) and of the boring (r_3) determining the flow line field. Under specific boundary conditions, α in this case results from the integration of the potential equation for horizontal flow as (see [2, 5, 6, 11]):

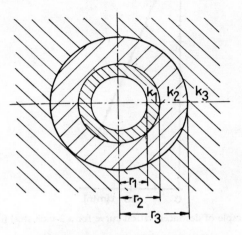

Fig. 5. Scheme of a well construction

$$\alpha = \cfrac{8}{\left(1 + \frac{k_3}{k_2}\right)\left\{1 + \left(\frac{r_1}{r_2}\right)^2 + \frac{k_2}{k_1}\left[1 - \left(\frac{r_1}{r_2}\right)^2\right]\right\} + \left(1 - \frac{k_3}{k_2}\right)\left\{\left(\frac{r_1}{r_3}\right)^2 + \left(\frac{r_2}{r_3}\right)^2 + \frac{k_2}{k_1}\left[\left(\frac{r_1}{r_3}\right)^2 - \left(\frac{r_2}{r_3}\right)^2\right]\right\}}$$

(3)

Since, according to Eq. (3) α ranges between 0 and 8, depending on permeability conditions, a contraction or expansion of the flow line field toward the dilution volume may arise (Fig. 6). Table II reveals the good agreement over a wide range of well construction parameters, of the results of the model tests with the calculations based on potential theory.

Besides this horizontal field distortion determined by potential theory, vertical currents in the filter tube and its vicinity, caused by differences in pressure and density, may influence the dilution rate. Further, inhomogeneities of the tracer concentration in the measuring volume will result in erroneous dilution rates, as the supposition in Eq. (1) of a homogeneous distribution of the tracer is not fulfilled. On the basis of numerous tests with the three-dimensional models, where the dilution rate v_g was determined by Eq. (2), α by Eq. (3), and the filtration velocity $v_f = Q/F$ from the flow Q and the flow cross section F, it

a)   b)

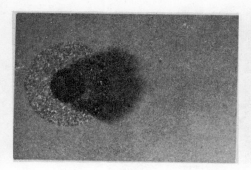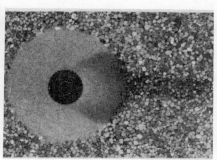

Fig. 6. Examples of model tests with dye tracers to study the flow through a filter tube
($k_1 = 1.0$ cm/sec): horizontal section through a simulated aquifer
a) for $k_3 < k_2$; $k_2 = 3.35$ cm/sec, $k_3 = 0.35$ cm/sec
b) for $k_3 > k_2$; $k_2 = 0.35$ cm/sec, $k_3 = 6.00$ cm/sec

TABLE II

Comparison of a-values calculated from Eq. (3) and measured on the two-dimensional
flow model (see Fig. 4) with 2″-filter tubes and different well constructions ($v_f = 11$ m/d)

| Parameters of well-constructions and aquifers | | | | | | a-values | |
r_1(mm)	r_2(mm)	r_3(mm)	k_1 (cm/sec)	k_2 (cm/sec)	k_3 (cm/sec)	calculated from Eq.(3)	measured (see Fig. 4)
26	29	—	1.31	0.35	—	2.16	2.16
26	29	—	1.31	3.35	—	1.75	1.77
26	29	—	1.31	6.00	—	1.49	1.48
26	29	50	1.31	0.35	3.35	0.52	0.52
26	29	50	1.31	3.35	0.35	2.72	2.66
26	29	75	1.31	0.35	3.35	0.44	0.49
26	29	75	1.31	3.35	0.35	2.84	2.84
26	29	100	1.31	0.35	3.35	0.41	0.44
26	29	100	1.31	3.35	0.35	3.02	2.99
26	29	100	1.31	3.35	6.00	1.26	1.26
26	29	100	1.31	6.00	0.35	2.80	2.80
26	29	100	1.31	6.00	3.35	1.78	1.88

became possible to develop a logging tool (Fig. 7) that does not disturb the
dilution process, meaning that

$$v_f = v_g/\alpha \qquad (4)$$

is satisfied over a wide range of test conditions. Table III shows the consistency
of the performance of the two- and three-dimensional models compared with
each other, and under different test conditions (see [2, 4, 5,] for details).

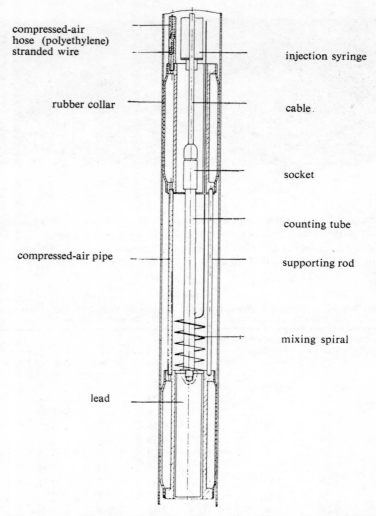

compressed-air
hose (polyethylene)
stranded wire

rubber collar

compressed-air pipe

lead

injection syringe

cable

socket

counting tube

supporting rod

mixing spiral

Fig. 7. Flow rate measuring probe with inflatable rubber seals (not to scale). Several probes
were constructed for different filter tube diameters (minimum diameter: 1.5″)

The models also offered the opportunity of checking the applicability of the
dilution method in the case of very small filtration velocities which are indeed
of interest when dealing with an unpumped aquifer. It appeared that in the
range of a filtration velocity of $v_f < 0.3$ m/d dilutions of the tracer solution by
molecular diffusion and convection must be allowed for. From Table VI can be
learned that a measurement of the filtration velocity by the dilution method is

TABLE III

Comparison of results obtained on both types of flow models with the same well-constructions and aquifers

filter tube	gravel filter	r_3(mm)	aquifer	three-dimensional model $v_f = Q/F$(m/d)	v_g (m/d)	v_f/v_g	quasi-two-dimensional model a
-inch.	—	—	sand 0.5 ⋯ 1.5 mm	3.53	7.60	2.16	2.16
r_1=26 mm	—	—	gravel 2 ⋯ 4 mm	3.56	6.25	1.76	1.77
r_2=29 mm)	gravel 2 ⋯ 4 mm	100	sand 0.5 ⋯ 1.5 mm	3.52	10.60	3.01	2.99
	gravel 4 ⋯ 6 mm	75	sand 0.5 ⋯ 1.5 mm	3.53	1.56	0.44	0.47
2-inch	—	—	sand 0.5 ⋯ 1.5 mm	3.49	7.25	2.08	2.06
(r_1 = 54 mm	—	—	gravel 2 ⋯ 4 mm	3.57	6.84	1.92	1.96
r_2 = 57 mm)							

TABLE IV

Influence of molecular tracer diffusion(v_d) and convection(v_k) for $v_f < 0.3$ m/d on the measurement result when using an inflatable packer according to Fig. 7, 2-inch filter tube:
aquifer: coarse sand, 0.5 ⋯ 1.5 mm
a-value: $a=2.72$
initial tracer concentration in dilution volume: $c_0 = 5 \cdot 10^{-5}$ mol $(NH_4Br)/1$

v_f(mm/d)$=Q/F$	v_g(mm/d)	v_g/v_f	v_f(mm/d)$=v_g-v_d-v_k$ over α
244.0	666.0	2.73	239.0
103.0	288.0	2.80	99.0
100.0	278.0	2.78	96.0
92.0	261.0	2.84	89.0
80.0	235.0	2.93	83.0
68.0	206.0	3.03	69.0
63.0	193.0	3.06	64.0
63.0	193.0	3.06	64.0
63.0	191.0	3.03	63.0
63.0	193.0	3.06	64.0
55.0	170.0	3.09	56.0
42.0	133.0	3.16	42.0
34.0	110.0	3.25	34.0
25.0	84.0	3.36	24.0
22.0	79.0	3.59	22.0
19.0	70.8	3.73	19.5
18.0	69.2	3.86	18.5
14.5	60.5	4.17	15.5
12.4	58.1	4.69	14.5
11.0	55.0	5.00	13.5
6.0	38.6	6.22	7.3
4.8	31,6	6.60	4.8
3.0	24.0	8.00	2.0
1.5	22.2	14.80	1.5
0.0	18.5$=v_d+v_k$	—	—

possible down to the lowest set velocity values ($v_f = 1.5$ mm/d) if the dilution rates v_d and v_k caused by diffusion and convection are deducted:

$$(5) \qquad\qquad v_f = (v_g. - v_d - v_k)/\alpha.$$

v_d and v_k can be determined from the behavior of the tracer in standing water ($v_f = 0$), the other test conditions remaining the same. Fig. 8 contains α-values calculated from measurements with and without correction according to Eq. (5).

Finally, an investigation was undertaken on turbulence effects in water currents approaching narrow filter slots, or where high heads are involved (in pumped aquifers, for example), which result in a decrease of the value of α.

3. MODEL TESTS TO MEASURE THE DIRECTION OF GROUNDWATER CURRENTS IN A BOREHOLE

Principle of the method (see [2, 6, 8] for details):

The water in a filter tube is labeled using a radioactive tracer, and the direction of the outflowing water then determined with a collimated revolvable radiation detector.

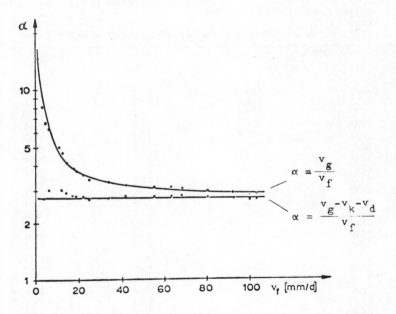

Fig. 8. Influence of molecular diffusion and convection: α-values calculated from measurements with and without correction according to Eq. (5)
4-inch filter tube ($r_1 = 54$ mm, $r_2 = 57$ mm, perforation: 24 %)
tracer concentration: $c_0 = 10^{-4}$ Mol/1
aquifer: coarse sand, 0.5 ... 1.5 mm

It was possible on the quasi-two-dimensional model to clearly follow all phases of the variation with time of the emanating tracer cloud as a function of the well construction and of the method of injection (compare, for example, Fig. 4).

Comparative experiments on three-dimensional models with different tracers revealed that it is preferable, in contrast to filtration velocity measurements, to use tracer solutions for direction measurements that become fixed in the surrounding gravel filter or aquifer outside the filter tube (for instance, $^{198}AuCl_3$ — or $^{51}CrCl_3$-solutions) because tracer distribution during the measurement then remains more or less stable. As regards recording, the most satisfactory energy range of the emitted gamma radiation was found to be $0.4 - 0.7$ MeV.

Besides the method of injection, the borehole construction and the tracer solution employed, the design of the logging tool governs the pattern of the pulse-rate distribution measured during a revolution in the filter tube, i.e. the so-called direction diagram (see Fig. 9) The measuring probe represented schematically in Fig. 10 was designed turning to advantage the results of numerous tests on the three-dimensional models; its performance in field tests has been good so far.

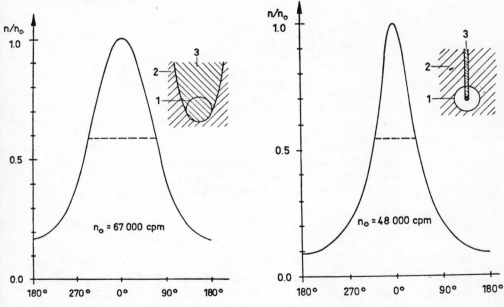

Fig. 9. Direction diagrams obtained 24 h after injection of 50 μCi of ^{198} Au. The different injection methods employed are described in the text.
1 well screen (diameter, 4 inch), 2 aquifer: gravel, 2–4 mm, 3 tracer fixed at the aquifer, n=counting rate, n_0=maximum counting rate

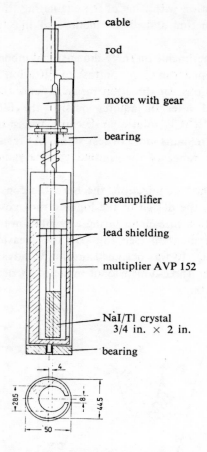

Fig. 10. Scintillation-counter probe for measuring the direction of ground water
flow with radioactive tracers.
Dimensions, mm; length of probe, 600 mm

4. OTHER APPLICATIONS OF THE MODELS

In addition to the application results of the models just described, which have
been confirmed in the meantime by a large number of field tests, studies are
being carried out at present on the dispersion of tracers in porous aquifers
(supplementing column experiments), together with comparative measurements of the velocity and point-to-point travel rate of the groundwater as well
as porosity measurements using gamma and neutron logging tools. Moreover,
the models are being employed for exploratory measurements on vertical
flows in boreholes whose evaluation and explanation are apt so shed new light

on hydrological phenomena (see [2, 4]). Implementing this line of investigation, the 4-m-high three-dimensional model referred to in Table I is being erected at present, the goal being to extend the studies on layered aquifers conducted so far.

ACKNOWLEDGMENT

Part of these investigations was sponsored by the International Atomic Energy Agency Vienna (Research Contract No. 259/RB).

REFERENCES

1. BOROWCZYK, M., MAIRHOFER, J. AND ZUBER, A. (1965) Laboratory investigation on the determination of filtration velocity by means of radioisotopes, *Atomenergie 10*, 51.

2. DROST, W., KLOTZ, D., KOCH, A., MOSER, H., NEUMAIER, F. AND RAUERT, W. (1968) Point dilution methods of investigating ground water flow by means of radioisotopes, *Water Resources Research 4*, 125.

3. GUIZERIX, J., CALMELS, P., COROMPT, P., GAILLARD, B. AND MAESTRINI, A. (1967) "Appareil utilisant une source de neutrons pour la mesure des vitesses horizontales de filtration", *Steirische Beiträge zur Hydrogeologie*, Jahrgang 1966/67, Graz.

4. HALEVY, E., MOSER, H., ZELLHOFER, O. AND ZUBER, A. (1967), Borehole dilution techniques: a critical review—in: Isotopes in Hydrology, *IAEA, Wien*.

5. KOCH, A., KLOTZ, D. AND MOSER, H. (1967) Anwendung radioaktiver Isotope in der Hydrologie VI. Einfluß des Filterrohres auf die Messung der Filtergeschwingigkeit nach dem Verdünnungsverfahren, *Atomkernenergie 12*, 361.

6. KRÄTZSCHMAR, H. (1966) Beitrag zur Bestimmung der Grundwasserfließgeschwindigkeit und -richtung, *Diss. Techn. Univ. Dresden*.

7. KROLIKOWSKI, CZ. (1965) The influence of measuring probes on the examination of underground waters in boreholes and piezometers, *Atomkernenergie 10*, 57.

8. MAIRHOFER, J. (1963) Bestimmung der Strömungsrichtung des Grundwassers in einem einzigen Bohrloch mit Hilfe radioaktiver Elemente, *Atompraxis 9*, 2.

9. MOSER, H., NEUMAIER, F. AND RAUERT, W. (1957) Die Anwendung radioaktiver Isotope in der Hydrologie: Ein Verfahren zur Ermittlung der Ergiebigkeit von Grundwasserströmungen, *Atomkernenergie 2*, 225.

10. OGILVI, N. A. (1958) Elektroličeskij metod opredelnija skorostej filtracii, *Bjull. O.N.T.I.* No. 4, Gosgeoltehizdat.

11. RÖSLER, R. (1967) Der Einfluß einer Bohrung auf die Grundwasserströmung, *Z. angew. Geologie, 13*, 351.

INSTITUT FÜR RADIOHYDROMETRIE DER GESELLSCHAFT
FÜR STRAHLENFORSCHUNG MBH,
D 8 MÜNCHEN 2, LUISENSTRASSE 37

6

Surface phenomena in flow through porous media

ETUDE DES PHENOMENES INTERFACIAUX DANS DIFFERENTS MODELES ANALOGIQUES DE MILIEU POREUX

C. Thirriot et J. M. Aribert

1. INTRODUCTION

L'écoulement de fluides dans un milieu poreux naturel est tellement complexe qu'il déroute vite l'homme. Celui-ci cependant a besoin d'une explication et d'une représentation des mécanismes d'écoulement. Une explication pour répondre à la curiosité scientifique des lois de la nature. Une représentation dans le souci pragmatique de pouvoir agir sur la nature afin qu'elle serve mieux l'homme.

La forme la plus vraie d'appréhension de la réalité est évidemment l'observation directe. Mais elle est aussi souvent la plus difficile à interpréter. Pour faciliter la compréhension, on peut envisager d'utiliser des images: en recherche scientifique, c'est l'analogie. Bien sur, le phénomène alors étudié ne représente plus exactement le phénomène naturel mais il permet d'en mieux saisir l'aspect fondamental, les aspects saillants.

Les modèles analogiques de milieux poreux interviennent aussi bien au niveau de l'explication qu'au stade de la représentation. Excellent outil pédagogique, ils peuvent fonctionner comme un microscope pour l'étude explicative des phénomènes interfaciaux lors de l'écoulement de plusieurs fluides dans un tube cylindrique ou non représentant une succession de pores par exemple. Comme instrument de prévision, les modèles analogiques tels les modèles Hele-Shaw jouent le rôle de simulateur, ou calculateur assurant la résolution globale d'un modèle mathématique comme l'équation de Laplace qui régit après simplification aussi bien le phénomène naturel que l'écoulement en modèle analogique.

Dans ce qui suit, nous voudrions rapporter l'expérience de notre Laboratoire dans l'utilisation des modèles analogiques au double point de vue de l'explication des phénomènes interfaciaux et de la simulation des problèmes d'écoulement multiphasiques. Plusieurs personnes ont collaboré de manière décisive à ces études dont M. Raoul Lopardo, Ingénieur de l'Université de La Plata, Mm. Esteban, Geli et Monferran, Techniciens au Laboratoire.

2. RÉPARTITION DE VITESSE AU VOISINAGE DE L'INTERFACE

La forme de la surface de séparation entre fluides non miscibles est encore un paramètre global, essentiel par sa relation avec la pression capillaire. Si l'on passe à une observation microscopique plus fine, mais encore sans atteindre l'échelle moléculaire on doit se poser la question de la distribution des vitesses au voisinage de l'interface. La connaissance de cette distribution a un double intérêt: d'abord, elle aide à mieux comprendre le mécanisme du remplacement d'un fluide par un autre, ensuite elle permet d'évaluer avec plus d'exactitude la dissipation d'énergie par frottement visqueux au voisinage du ménisque, détail très important pour juger de la validité de l'analogie entre l'écoulement laminaire entre plaques à faces parallèles et l'écoulement en milieu poreux suivant la loi de Darcy.

Evidemment le problème le plus général doit être posé en supposant a priori indéterminée la forme du ménisque dans un chenal de géométrie quelconque. Tel quel, le problème nous paraît difficilement soluble surtout par voie analytique. Nous simplifierons donc son exposé et nous examinerons le cas d'un ménisque plan et d'un canal de section constante. Un cas particulier important est celui du canal Hele Shaw où, compte tenu de la faible épaisseur ($2h$) de l'entrefer, on peut admettre avoir affaire à un écoulement à deux dimensions suivant le plan normal aux plaques comme l'indique la figure 1. Lorsque l'écoulement est très graduellement varié, loin du ménisque, les répartitions de vitesse dans le fluide 2 déplacé et dans le fluide 1 injecté sont paraboliques. Compte tenu de la réalité expérimentale du déplacement du ménisque sans déformation sensible, on ne peut admettre au voisinage de l'interface une telle répartition de vitesse à composantes uniquement longitudinales. Intuitivement, on peut imaginer que les molécules du fluide 2 vont décoller normalement des parois et venir vers le milieu du canal. Les molécules du fluide 1 amenées à grande vitesse dans l'axe du canal vont être réparties vers les parois en remplacement des molécules du fluide 2 chassées au voisinage du ménisque. Ce schéma très simple a été confirmé par un calcul analytique qui donne aussi la réponse à la question posée par l'incompatibilité des sens opposés de circulation des molécules du fluide 1 et du fluide 2 au voisinage du ménisque.

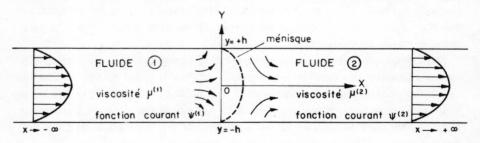

Fig. 1.

MODÈLE MATHÉMATIQUE SIMPLIFIÉ

Dans l'étude théorique, nous supposerons les fluides incompressibles et constant le débit $2Uh$ par unité de largeur du modèle Hele Shaw. Les coordonnées (X, Y) seront comptées par rapport à un plan normal aux plaques et lié à la position du ménisque. Soient $p^{(i)}(X, Y)$, $u^{(i)}(X, Y)$, $v^{(i)}(X, Y)$ la pression et les composantes de la vitesse absolue ($i = 1$ ou 2 suivant que l'on considère un point dans le fluide injecté ou le fluide déplacé).

A nombre de Reynolds faible, les équations générales d'écoulement de Navier-Stokes se réduisent aux équations de Stokes des fluides visqueux en régime permanent:

$$(5) \qquad \frac{\partial p^{(i)}}{\partial X} = \mu^{(i)} \left(\frac{\partial^2 u^{(i)}}{\partial^2 X} + \frac{\partial^2 u^{(i)}}{\partial^2 Y} \right)$$

$$(6) \qquad \frac{\partial p^{(i)}}{\partial Y} = \mu^{(i)} \left(\frac{\partial^2 v^{(i)}}{\partial X^2} + \frac{\partial^2 v^{(i)}}{\partial Y^2} \right).$$

Comme l'avait déjà indiqué R. Berker [2], nous remarquons qu'avec l'approximation de Stokes, les champs de vitesses absolues $u(X, Y)$ et $v(X, Y)$ exprimées dans le repère mobile satisfont à un modèle mathématique identique à celui obtenu par rapport à un repère fixe.

(Si l'influence de la pesanteur n'était pas négligeable, on pourrait en tenir compte de manière approchée en faisant jouer à $p^{(i)}$ le rôle d'une pression piézométrique évaluée en prenant une valeur moyenne des masses spécifique $\rho^{(1)}$ et $\rho^{(2)}$).

En tenant compte de l'équation de continuité des fluides incompressibles

$$(7) \qquad \frac{\partial u^{(i)}}{\partial X} + \frac{\partial v^{(i)}}{\partial Y} = 0$$

on peut introduire des fonctions courant $\psi^{(i)}$ donnant les vitesses par les relations

$$(8) \qquad u^{(i)} = \frac{\partial \psi^{(i)}}{\partial Y} \qquad v^{(i)} = - \frac{\partial \psi^{(i)}}{\partial X}.$$

Ainsi, les équations résultantes sont ramenées à des équations biharmoniques de la forme:

$$(10) \qquad \Delta\Delta\psi^{(i)} = 0.$$

Pour définir complètement le problème, précisons les conditions aux limites:

1. sur les parois (sauf aux points d'attachement du ménisque), la vitesse des molécules est nulle:

$$(11) \qquad u^{(i)} = 0, \quad (12) \quad v^{(i)} = 0 \quad \text{pour} \quad Y = \pm h, \, \forall X.$$

2. sur l'interface, la composante longitudinale $u^{(i)}$ de la vitesse est continue et uniforme:

$$(13) \qquad \frac{\partial \psi^{(1)}}{\partial Y} = \frac{\partial \psi^{(2)}}{\partial Y} = U, \quad X = 0, \ Y \in [-h, +h].$$

3. sur l'interface, il y a continuité des composantes tangentielles:

$$(14) \qquad \frac{\partial \psi^{(1)}}{\partial X} = \frac{\partial \psi^{(2)}}{\partial X}, \quad X = 0, \quad Y \in [-h + h].$$

4. enfin, il doit y avoir égalité des contraintes tangentielles de part et d'autre de l'interface

$$(15) \quad \mu^{(1)}\left[\frac{\partial^2 \psi^{(1)}}{\partial Y^{(2)}} - \frac{\partial^2 \psi^{(1)}}{\partial X^2}\right] = \mu^{(2)}\left[\frac{\partial \psi^{(2)}}{\partial Y^{(2)}} - \frac{\partial^2 \psi^{(2)}}{\partial X^2}\right], \quad X = 0, \quad Y \in [-h, h].$$

Comme le ménisque est supposé de forme connue, il n'y a pas lieu de faire intervenir de conditions sur les valeurs des contraintes normales.

5. très loin du ménisque, les vitesses sont parallèles et suivent une distribution parabolique:

$$(16) \qquad \psi^{(i)} \to \frac{3}{2} U Y \left(1 - \frac{Y^2}{3h^2}\right) \quad \text{si } x \to -\infty$$

$$(17) \qquad \psi^{(2)} \to \frac{3}{2} U Y \left(1 - \frac{Y^2}{3h^2}\right) \quad \text{si } x \to +\infty.$$

ÉTAPES ESSENTIELLES DE LA RÉSOLUTION

Nous construirons la solution sur un espace de fonctions solutions de l'équation biharmonique et qui restent finies dans une bande semi-infinie. Pour alléger les écritures, nous utiliserons désormais des grandeurs réduites $x = X/h$ et $y = Y/h$. Tenant compte de la solution asymptotique loin du ménisque, nous poserons:

$$\frac{\psi^{(i)}}{Uh} = \frac{3}{2} Y \left(1 - \frac{Y^2}{3}\right) + \frac{\psi^{(i)*}}{Uh}$$

tel que:

$$(20) \quad \psi^{*(1)} = \sum_{n=1}^{\infty} \left\{ A_n^{(1)} e^{P_n x} \sin P_n y + B_n^{(1)} x e^{P_n x} \sin P_n y + \right.$$

$$\left. + \int_0^{\infty} C_n^{(1)}(y \operatorname{ch} ly - \coth l \operatorname{sh} ly) \sin lx dl \right\}$$

et

$$(21) \quad \psi^{*(2)} = \sum_{n=1}^{\infty} A_n^{(2)} e^{-P_n x} \sin P_n y + B_n^{(2)} x e^{-P_n x} \sin P_n y +$$

$$+ \int_0^{\infty} C_n^{(2)}(y \operatorname{ch} ly - \coth l \operatorname{sh} ly) \sin lx dl \right\}.$$

Les coefficients des formes linéaires ci-dessus seront déterminés par la vérification des conditions aux limites.

La condition (14) sera assurée si l'on pose $P_n = n\pi$ avec n entier.

La condition (11), soit $u^{(i)} = \partial\psi^{(i)}/\partial y = 0$ pour $y = \pm h$ va permettre de déterminer $C_n^{(1)}$ comme fonction linéaire de $A_n^{(1)}$ et $B_n^{(1)}$.

Compte tenu des relations

$$e^{-P_n x} = \frac{2}{\pi} \int_0^\infty \frac{l}{P_n^2 + l^2} \sin lx \, dl \quad \text{et} \quad P_n x e^{-P_n x} = \frac{4P_n^2}{\pi} \int_0^\infty \frac{l}{(P_n^2 + l^2)^2} \sin lx \, dl$$

la relation (11) dans le fluide (1) qui était:

$$(-1)^n P_n A_n^{(1)} e^{-P_n x} + (-1)^n P_n B_n^{(1)} x e^{+P_n x} + \int_0^\infty C_n^{(1)} \left(\text{ch } l - \frac{l}{\text{sh } l} \right) \sin lx \, dl = 0$$

donne

$$(22) \qquad C_n^{(1)} = \frac{(-1)^n 2nl}{(n^2\pi^2 + l^2)(\text{ch } l - l/\text{sh } l)} \left[A_n^{(1)} - \frac{2n\pi}{n^2\pi^2 + l^2} B_n^{(1)} \right].$$

De manière analogue, on obtient:

$$(23) \qquad C_n^{(2)} = \frac{(-1)^{n+1} 2nl}{(n^2\pi^2 + l^2)(\text{ch } l - l/\text{sh } l)} \left[A_n^{(2)} + \frac{2n\pi}{n^2\pi^2 + l^2} B_n^{(2)} \right].$$

La relation (13) conduit à:

$$\left(\frac{\partial\psi^{*(1)}}{\partial y} \right)_{x=0} = \left(\frac{\partial\psi^{*(2)}}{\partial y} \right)_{x=0} = -\frac{Uh}{2}(1 - 3y^2) \quad y \in [-1, +1]$$

soit

$$\sum_{n=1}^\infty n\pi A_n \cos n\pi y = \sum_{n=1}^\infty n\pi A_n^{(2)} \cos n\pi y = -\frac{Uh}{2}(1 - 3y^2)$$

$$= \sum_{n=1}^\infty (-1)^n \frac{6Uh}{n^2\pi^2} \cos n\pi y$$

d'ou par identification

$$(24) \qquad A_n^{(1)} = A_n^{(2)} = (-1)^n \frac{6Uh}{n^3\pi^3}.$$

En posant $M = \mu^{(2)}/\mu^{(1)}$, la condition d'égalité des contraintes tangentielles à l'interface donne:

$$\sum_{n=1}^\infty 2n\pi\{n\pi A_n^{(1)} + B_n^{(1)} - M(n\pi A_n^{(2)} - B_n^{(2)})\} \sin n\pi y = (M-1)3yUh$$

$$= (M-1) \sum_{n=1}^\infty (-1)^{n+1} \frac{6Uh}{n\pi} \sin n\pi y.$$

Soit, puisque $A_n^{(1)} = A_n^{(2)}$

$$(25) \qquad B_{(n)}^{(1)} + MB_n^{(2)} = (M-1) \left[n\pi A_n^{(1)} + (-1)^{n+1} \frac{3Uh}{n^2\pi^2} \right] = (-1)^n \frac{3(M-1)Uh}{n^2\pi^2}$$

La dernière relation nécessaire sera fournie par là condition

$$(14) \qquad \frac{\partial \psi^{(1)}}{\partial x} = \frac{\partial \psi^{(2)}}{\partial x} , \quad X = 0 .$$

Il vient:

$$\sum_{n=1}^{\infty} \int_{0}^{\infty} (C_n^{(2)} - C_n^{(1)})(y \operatorname{ch} ly - \coth l \operatorname{sh} ly) l dl = \sum_{n=1}^{\infty} \left[(B_n^{(1)} - B_n^{(2)}) + 2n\pi A_n^{(1)} \right] \cdot$$
$$\cdot \sin n\pi y.$$

Comme précédemment, nous userons de développement en série de Fourier pour $y \in]-1, 1[$

$$y \operatorname{ch} ly - \coth l \operatorname{sh} ly = \sum_{k=1}^{\infty} f_k(l) \sin k\pi y$$

avec

$$f_k(l) = (-1) \frac{k 4 k\pi l \operatorname{sh} l}{(l^2 + k^2\pi^2)^2} .$$

D'où, pour l'équation (14):

$$\sum_{k=1}^{\infty} \left\{ \sum_{n=1}^{\infty} \int_{0}^{\infty} (C_n^{(2)} - C_n^{(1)}) f_k(l) l dl \right\} \sin k\pi y = \sum_{n=1}^{\infty} \left\{ B_n^{(1)} - B_n^{(2)} + (-1)^n \frac{12Uh}{n^2\pi^2} \right\}$$
$$\cdot \sin n\pi y.$$

En permutant les indices k et n on voit que l'équation ci-dessus est satisfaite si:

$$(26) \qquad B_k^{(1)} - B_k^{(2)} + (-1)^k \frac{12Uh}{k^2\pi^2} = \sum_{n=1}^{\infty} \int_{0}^{\infty} (C_n^{(2)} - C_n^{(1)}) f_k(l) l dl .$$

Compte tenu des relations linéaires (22) et (23) entre coefficients déjà obtenues, on dispose d'une équation résultante ne faisant intervenir que les coefficients d'une seule famille par exemple la différence $B_n^{(1)} - B_n^{(2)} = D_n$. En fait, la détermination de D_n sera compliquée par la présence d'une série d'ordre infinie. Nous passons sur les détails du calcul approché mi-analytique, mi-numérique qui conduit de manière relativement économique à l'obtention des coefficients. Remarquons cependant que la tabulation de D_n, étape la plus laborieuse, est indépendante de la valeur de M et donc ne doit être effectuée qu'une seule fois.

RÉSULTATS CONCERNANT LA DISTRIBUTION DES VITESSES

A l'aide des relations (20) et (21) et des tables de coefficients nous avons déterminé la répartition de la fonction courant pour trois valeurs du contraste M de viscosité 1, 2 et 5. Les résultats obtenus sont rassemblés sur les planches 2, 3 et 4. Les courbes sont paramétrées suivant les valeurs réduites de la fonction courant.

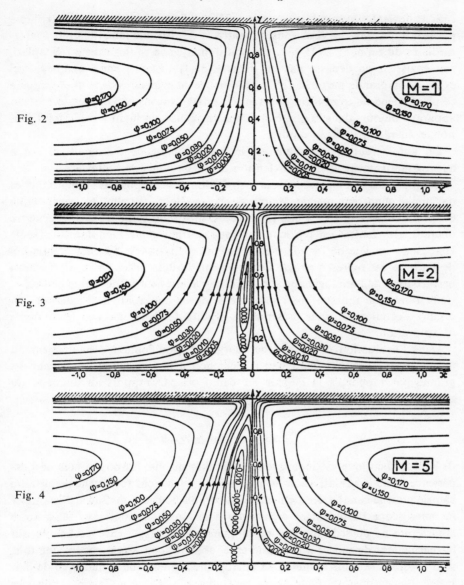

Fig. 2

Fig. 3

Fig. 4

L'écoulement va de gauche à droite. Le fluide injecté est le moins visqueux. Pour $M = 1$ (fluides de même viscosité) les lignes de courant sont parfaitement symétriques par rapport au ménisque plan mais il s'agit là d'un écoulement asymptotique. La vitesse tangentielle le long du ménisque est nulle. Lorsque M croît, on voit apparaître un tourbillon laminaire qui assure la compatibilité des sens de circulation des fluides 1 et 2 au voisinage du ménisque. La vitesse tangentielle devient importante.

Si M devient très grand, le rouleau ne s'amplifie pas beacoup. Ceci s'explique par le fait que dans les équations, le contraste de viscosité intervient par l'intermédiaire du rapport $M/M + 1$. On voit donc que ce rouleau reste très aplati, son épaisseur ne dépasse pas un dixième de la distance entre plaques; ceci explique en grande partie la difficulté à mettre en évidence par la photographie ce phénomène de tourbillon pourtant essentiel. Notons encore que le rouleau apparaît toujours dans la phase la moins visqueuse, vérifiant ainsi la loi d'économie énergétique de la nature.

Un autre résultat important est la zone d'influence du ménisque qui a approximativement la même extension des deux cotés de l'interface.

Les planches 5 et 6 présentant la distribution des composantes longitudinales de vitesse montrent que le régime de Poiseuille à distribution parabolique est pratiquement rétabli à une distance du ménisque égale à l'épaisseur entre plaques. Ce résultat est particulièrement rassurant pour l'emploi du modèle Hele Shaw en vue de l'étude d'écoulements diphasiques. En effet, l'équation de Laplace est représentée par l'écoulement de Poiseuille. Donc, il n'y aura perturbation de l'analogie que sur une distance inférieure à deux fois l'entrefer. Cette remarque indique aussi une raison de la limitation de l'épaisseur du chenal d'écoulement pour préserver la validité de l'analogie sur le domaine d'exploration.

Dans l'évaluation des pertes de charge, la modification d'écoulement au voisinage du ménisque aura peu d'incidence si la longeur d'écoulement est grande par rapport à la largeur du canal ou par rapport au diamètre du capillaire.

ÉTUDE EXPÉRIMENTALE

Si l'étude théorique est independante des dimensions, les possibilités de l'expérience et de la visualisation du phénomène dépendent fortement de la taille choisie pour le canal d'écoulement. Bien sur, on pourrait envisager d'examiner au microscope l'écoulement diphasique dans un tube capillaire. Mais notre laboratoire ne possède pas encore le matériel et l'expérience technique qui lui permettent de mener à bien une telle observation. Aussi, encore une fois, sommes-nous revenus à une étude analogique sur un tube cylindrique de diamètre 10 centimètres. Pour éviter tout effet d'inertie nous avons choisi les viscosités et les vitesses de façon que les nombres de Reynolds restent compris dans la gamme 3 à 20. Les liquides utilisés sont:

— d'une part un mélange d'huile de mayoline et d'huile de silicone permettant de faire varier la viscosité de 0.25 à 0.5 poise.

— d'autre part une solution à l'eau distillée de pluracol pouvant atteindre 3 poises pour une concentration de l'ordre de 40% de Pluracol, la masse spécifique restant voisine de 1 g/cm^3.

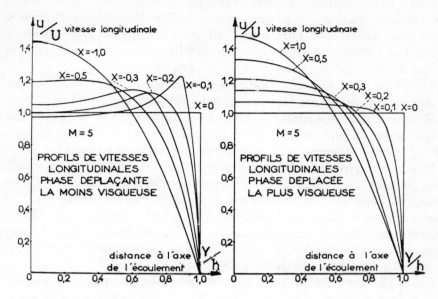

Fig. 5

Fig. 6

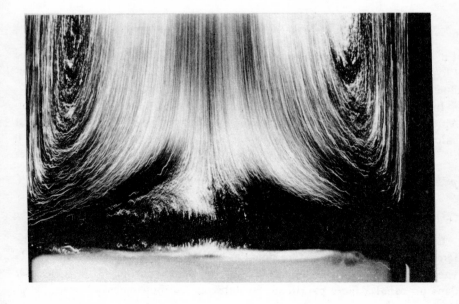

La vitesse de déplacement du ménisque est de l'ordre du centimètre par seconde. Pour visualiser les lignes de courant, nous avons introduit dans les deux phases visqueuses, des paillettes d'aluminium en suspension, suffisamment fines pour éviter tous phénomènes gênants d'orientation des particules. L'éclairage a été réalisé dans une section diamétrale du tube sur une épaisseur de 2 à 3 mm et sur une hauteur de 40 cm environ.

Evidemment, un récipient parallélépipédique entoure le tube sur la zone d'observation pour éviter tout effet optique parasite.

Seuls les clichés pris par un appareil se déplaçant à la vitesse du ménisque permettaient une interprétation nette. Les prises de vues cinématographiques sont d'ailleurs nettement plus instructives que les photographies.

Bien que la géométrie de l'écoulement soit différente de celle considérée dans le calcul, les résultats expérimentaux confirment bien les prévisions théoriques aussi bien sur l'existence du rouleau, la longueur de la zone d'influence que les mécanismes d'alimentation et de nettoyage des parois. Sur la photographie jointe et qui correspond à $M = 10$, on distingue bien dans la phase la moins visqueuse le tourbillon laminaire qui peut être fort affecté par la présence d'impuretés ou de bulles d'air.

Des essais ont été réalisés aussi sur un écoulement à trois phases successives. On observe dans le bouchon intermédiaire l'écoulement tourbillonnaire alliant les rouleaux dus aux deux interfaces.

3. VARIATION DE FORME DE L'INTERFACE

En extrapolant assez témérairement, les résultats théoriques obtenus avec l'hypothèse d'un ménisque plan, on peut obtenir des informations sur l'influence de la vitesse moyenne d'écoulement sur la forme de l'interface.

Indiquons brièvement les hypothèses intervenant dans le raisonnement. Après avoir montré que $P^{(i)}$ est le conjugué harmonique de $\mu^{(i)} \Delta \psi^{(i)}$ on peut calculer la différence de pression $p_1 - p_2$ à la traversée de l'interface en $Y = 0$

$$p_1 - p_2 = \frac{2}{h^2} \sum_{n=1}^{\infty} \left[n\pi(\mu^{(1)}B_n^{(1)} - \mu^{(2)}B_n^{(2)}) - \int_0^{\infty} (\mu^{(1)}C_n^{(1)} - \mu^{(2)}C_n^{(2)})l\,dl \right] + C^{te}.$$

(27)

Comme $B_n^{(1)}$ et $C_n^{(1)}$ sont proportionels à Uh, on en déduit que la différence de pression (ou encore pression capillaire) dépend linéairement de la vitesse: (28) $p_1 - p_2 = c^{te} + m(U\mu^{(1)}/h)$, m dépendant de $M = \mu_2/\mu_1$. Or dans un canal à faces parallèles, en l'absence de mouvement:

$$(28) \qquad p_{cs} = p_{1s} - p_{2s} = \frac{\sigma}{R_s} = \frac{\sigma \cos \theta_s}{h}$$

où σ est la tension superficielle, R_s le rayon de courbure du ménisque, θ l'angle de raccordement du ménisque et de la paroi.

En écoulement, admettons que la pression capillaire vérifie encore une relation de la forme (28):

$$p_{cd} = \frac{\sigma}{R_d} = \frac{\sigma \cos\theta_d}{h} ;$$

d'où, par comparaison avec la formule (27)

$$(29) \qquad \cos\theta_d = \cos\theta_s + m\frac{\mu^{(1)}U}{\sigma}$$

Pour un tube capillaire, en poursuivant l'extrapolation compte tenu de l'expression de la pression capillaire, on obtiendrait:

$$\cos\theta_d = \cos\theta_s + m_t\frac{\mu^{(1)}U}{2\sigma}$$

Cette relation extrapolée (moyennant plusieurs hypothèses criticables, il est vrai) coïncide avec elle établie expérimentalement par Rose et Heins [7].

Des expériences ont été effectuées dans notre Laboratoire sur le déformation du ménisque dans un capillaire de 3 mm de diamètre, séparant de l'huile de mayoline d'une solution de napiol à 35%. Les résultats des enregistrements photographiques rapides du ménisque en déplacement nous ont permis de tracer la courbe de la figure 7. La relation linéaire (29) est valable tant que la vitesse est inférieure à 0.1 cm/s. Au-delà l'effet hydrodynamique est si important que l'angle de contact est nécessairement limité à 180°. Il apparaît alors sur les parois du tube un film résiduel d'huile de mayoline dont l'épaisseur quasi constante après le passage du ménisque, augmente avec la vitesse moyenne d'injection.

Si l'on considère l'expression donnant $m(M)$, on vérifie aisément que:

$$(30) \qquad m\left(\frac{1}{M}\right) = \frac{1}{M}m(M) .$$

La tabulation de la fonction $m(M)$ conduit à un graphe dont on peut donner la formule de lissage suivante:

$$(31) \qquad m = -25(1 - M)^3 \qquad M \in [0,1]$$
et

$$m = -25\frac{(M - 1)^3}{M^2} \qquad M > 1 .$$

La relation (30) et le résultat de la tabulation montrent bien que le coefficient de U dans la formule (29) est toujours négatif. Ceci confirme encore les résultats expérimentaux de Rose et Heins [7], Siegel [9], Legrand et Rense [10], et Chaudhari [11].

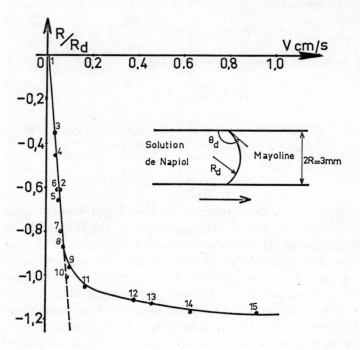

Fig. 7.

Pour l'extension des résultats obtenus pour l'écoulement entre plaques parallèles à l'écoulement dans un tube, nous proposons d'utiliser l'affinité sur le rapport $\Delta P/P_c$ de la variation de pression dynamique à la pression capillaire.

$$\left(\frac{\Delta P}{P_c}\right) \text{ plaques } = \frac{3}{4}\left(\frac{\Delta P}{P_c}\right) \text{ tube }.$$

Compte tenu de la relation (31), on obtient pour le coefficient m_t dans les tubes:

$$m_t = -33(1-M)^3, \qquad M \in [0,1]$$

Pour l'écoulement d'un couple liquide gaz dans un capillaire, on obtient donc $m_t(0) = -33$.

Des mesures effectuées par Rose et Heins sur le couple air-nujol à 22C° font apparaître une valeur expérimentale égale à -34. Il s'agit peut être d'une circonstance heureuse, mais tout de même encourageante.

Les prévisions théoriques afférentes aux couples de fluides de même viscosité pour lesquels $m(1) = 0$, sont aussi bien confirmées par les expériences de Chittenden et Spinney sur le couple eau distillée/cyclohexane pur présentant la même viscosité 0,843 cp à 2% près.

4. CONCLUSION

Ce qui précède n'est qu'une vue très partielle du problème puisqu'elle n'intéresse que l'aspect hydrodynamique. Il serait essentiel d'examiner aussi l'aspect physico-chimique et les bilans thermodynamiques.

Plusieurs autres études sont en cours dans notre Laboratoire portant aussi bien sur la critique de l'analogie Hele Shaw que sur les mécanismes de phénomènes capillaires dans les tubes non cylindriques en particulier. Les résultats déjà obtenus permettront de mieux comprendre les mouvements de l'eau en milieu non saturé et le remplacement d'un fluide par un autre dans la matrice poreuse pétrolifère.

BIBLIOGRAPHIE

1. BATAILLE, JEAN (1966), Etude de l'écoulement au voisinage du ménisque séparant deux fluides se déplaçant lentement entre deux plaques planes parallèles ou dans un tube capillaire, *C.R. Acad. des Sci., Paris*, † *262* (4. avril).

2. BERKER, RATIP (1963), *Encyclopedia of Physics*. Fluid dynamics, page 200.

3. BRETHERTON, F. P. (1961), The motion of long bubles in tubes, *J. Fluid Mechanics*, *10*, 166.

4. ARIBERT, J. M. (1967), Etude théorique et expérimentale de l'influence du contraste des viscosités sur le déplacement d'un ménisque entre deux plaques parallèles. Rapport interne IMF.

5. ARIBERT, J. M. (1967), Quelques résultats numériques sur l'écoulement au voisinage d'un ménisque plan, *C.R. Acad. des Sci. Paris*, † *265*.

6. ROSE, W. (1963), Aspects des processus de mouillage dans les solides poreux, *Revue I F P XVIII, 11*, 1571.

7. ROSE, W. AND R. W. HEINS (1962), Moving interfaces and contact angle rate dependency, *J. of Colloïd Science, 17*.

8. YARNOLD G. D. AND B. J. MASON (1949), A theory of the angle of contact, *Proc. Physic. Soc. Sect., B 62*, 121–125.

9. SIEGEL, R. (1961), Transient capillary rise in reduced and zero gravity fields, *J. Appl. Mechan., 83*, 165–170.

10. LEGRAND, E. J. AND W. A. RENSE (1945), Data on rate of capillary rise, *J. Appl. Physics, 16*, 843–846.

11. CHAUDHARI, N. (1962), *Capillary forces acting in manometer systems*, M.S. Thesis. University of Illinois.

12. CHITTENDEN AND SPINNEY (1966), An experimental study of the surface composition effect on two phase flow in a glass capillary tube, *J. of Colloid and Interface Science*, *22*, 250–256.

13. ARIBERT, J. M. (1967), Sur quelques modifications du critère de stabilité des digitations en canal Hele Shaw. Rapport interne IMF CT 30.

14. ARIBERT, J. M. ET C. THIRRIOT (1968), Contribution à l'étude des instabilités interfaciales en modèle Hele Shaw, *IFP 15* 845.

INSTITUT DE MÉCANIQUE DES FLUIDES,
2, RUE CHARLES-CAMICHEL,
31, TOULOUSE, FRANCE

SORPTION IN FLOW THROUGH POROUS MEDIA

INTRODUCTION

In one dimensional flow of a dissolved solute through a saturated homogeneous porous media, in which the solute sorbs on the granular material, the concentration of that solute species in the liquid phase $C(Z,t)$, and in the solid phase, $\bar{X}(Z,t)$, will vary with time, t, and distance, Z. It is the objective herein to outline the fundamental principles relevant to the solutions in time and space for both solid and liquid phases, $C(Z,t)$ and $\bar{X}(Z,t)$.

Such solutions are useful in designing and in specifying operation of ion-exchange and adsorption columns, and in evaluating the movement characteristics of a ground water contaminant.

MATHEMATICAL DESCRIPTION

The elements of the overall problem are readily delineated by a conventional mass balance analysis on a column slice of infinitesimal thickness, which results in the differential equation

$$(1) \qquad \frac{\partial C}{\partial t} = -\bar{v}\,\frac{\partial C}{\partial Z} + D\,\frac{\partial^2 C}{\partial Z^2} - \rho\,\frac{1-P}{P}\,\frac{\partial \bar{X}}{\partial t}$$

$$\begin{pmatrix}\text{net rate of}\\\text{change of sorbate}\\\text{conc.}\end{pmatrix} = \begin{pmatrix}\text{net}\\\text{convection}\\\text{rate}\end{pmatrix} + \begin{pmatrix}\text{net}\\\text{dispersion}\\\text{rate}\end{pmatrix} - \begin{pmatrix}\text{sorption}\\\text{rate}\end{pmatrix}$$

where:

C = concentration of sorbate species in liquid phase (gm/ml)

t = time from a convenient reference point, such as initial introduction of the sorbate (min.)

Z = distance from beginning of flow path (cm)

$\bar{v}$ = interstitial flow velocity (cm/min.)

D = coefficient of disperion for the porous media at velocity v (cm^2/min.)

ρ = density of the individual granular particles comprising the porous media (gm^3/cm)

P = porosity of the porous media (expressed as a decimal fraction)

$\bar{X}$ = concentration of sorbate species in the solid phase (gm sorbate/gm sorbent

384

Eq. 1 is intuitively meaningful, if each of the several groupings of variables is expressed in terms of its respective physical significance, as written below Eq. 1. Thus at any given location on the flow path, the rate of change of sorbate concentration depends upon the transport rate to that location by convection and dispersion, and the rate of removal by the sorbent. The conditions change, of course, in time and space, and the solution $C(Z, t)$ is two dimensional, being expressed as $C(Z)_t$, the concentration profile at a fixed time, or $C(t)_z$, known as the "breakthrough curve," at a given distance, Z.

The variables $\bar{v}$, $(D/\bar{v})$, ρ, and P must be evaluated by measurement, which is straightforward. The term D/v is the measurable property of the porous media (see Rifai, Kaufman and Todd, 1956). The kinetic term, $\partial \bar{X}/\partial t$, is the most profound mathematically, and also is the most difficult term to evaluate.

KINETICS

Three kinetic mechanisms are presented in the literature (Helferrich, 1962; Hiester and Vermeullen, 1952]; these are: (1) chemical reaction kinetics, (2) liquid phase diffusion, and (3) solid phase diffusion. It is generally conceded that (1) is not rate limiting (Hiester and Vermeullen, 1952). Either or both of the other two mechanisms may be rate limiting, however, depending upon chemical and flow conditions; it is very difficult to discern the conditions controlling the respective kinetic mechanisms. Obtaining a *practical* rate law for either category of diffusion is also difficult. Fick's first law describes either solid or liquid phase diffusion, but is difficult to apply in its differential form.

Hiester and Vermeullen (1952) have presented approximations to Fick's first law (flux density = diffusion coefficient × grad. concentration) for both liquid and solid phase diffusion which, when combined with the mass action equilibrium equations, results in expressions for each mechanism having a form identical to the equation for second-order reaction kinetics for the sorption reaction. Keinath and Weber (1968) have applied the methods of Hiester and Vermeullen to predict breakthrough curves for fluidized beds; their experimental results agree well with predicted breakthrough curves.

As pointed out by Helferrich (1962), many rate equations are suggested in the literature, and may apply quite well under the conditions in the systems investigated, but fail under other conditions. For the work reported herein, the experimental data ($\bar{X}$ vs t) from batch sorption tests were found to be best described by the equation:

(2) $$\left(\frac{\partial \bar{X}}{\partial t}\right)_p = \bar{D}C(\bar{X}^* - \bar{X})$$

where,

$(\partial \bar{X}/\partial t)_p$ = kinetic term when solid phase diffusion is rate controlling

$\bar{D}$ = diffusion coefficient in solid phase (gm-ml/gm-min.)

$\bar{X}^*$ = equilibrium concentration of sorbate in solid phase, corresponding to the liquid phase concentration, C (gm sorbate/gm sorbent)

Experimental data using two experimental systems (rhodamine-B dye and Dowex 50 resin and rhodamine B dye and Eau Claire sand), were fitted to several empirical kinetic equations, but Eq. 2 was the best fit. It can also be argued that Eq. 2 has some rational basis, having similarity with second order kinetics (the rate is proportional to the product of two reactants).

While a suitable kinetic equation having a broad application is desirable, in the absence of such an equation, or if it is difficult to apply, it may be more pragmatic to find an empirical equation which fits the data for the range of interest.

Application of Eq. 2 was somewhat more formidable than immediately apparent, however. It is necessary to evaluate the isotherm, $\bar{X}^*(C)_T$, for the particular sorbate-sorbent system in question. The Langmiur isotherm is both widely used and has a rational basis, and fits the experimental data, so was used in this work; the Langmiur isotherm is defined

$$(3) \qquad \frac{\bar{X}^*}{X_m} = \frac{\alpha C^*}{1 + \alpha C^*}$$

where,

X_m = ultimate capacity of sorbent (gm sorbate/gm sorbent)

α = Langmiur coefficient

C^* = equilibrium liquid phase concentration (gm/ml) (C is used for calculating $\bar{X}^*$, however)

This equation was experimentally defined for two temperatures 23°C and 21°C having respective Langmiur constants

$$\alpha = .96, \ X_m = 4000 \ \text{and} \ \alpha = 0.62, \ X_m = 2760$$

(the values given for α and X_m are for the rhodamine-B-Dowex 50 system and should be multiplied by 10^{-6} to obtain units in grams). This strong temperature sensitivity caused considerable instability in the $C(Z)_t$ measured profiles.

Another impediment in using Eq. 2 lies in evaluating the term $\bar{D}$, which unfortunately is not a constant. This term was found to have the empirical form

$$(4) \qquad \bar{D} = (.00417 \times 10^{-1.5\,\bar{X}/\bar{X}^*})C^{-1.02}$$

which also can be rationalized. The fit for this equation to the data is shown in Fig. 1, which is for the rhodamine B-Dowex 50 system. Eq. 4 was corroborated also for the rhodamine-B-Eau Claire sand system with more limited data,

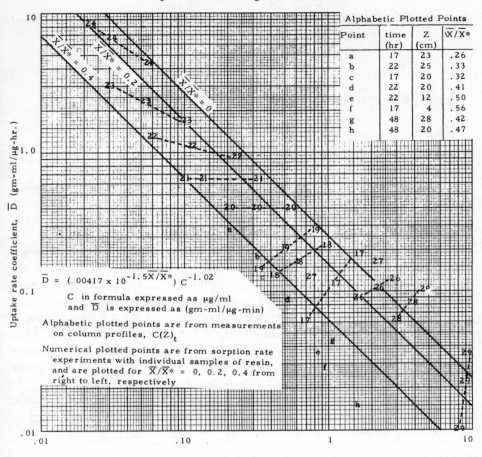

Fig. 1. Uptake rate coefficient, $\bar{D}$, for Dowex 50 resin as affeced by solution concentration and relative degree of resin saturation

the constants were different, of course. Alphabetic points shown are from measurements on column profiles, which provides an independent corroboration of the equation.

At this point it should be noted that Eqs. 2, 3, and 4 comprise the description of kinetics, as measured for the experimental system studied. Conditions for C and $\bar{X}$ change with time and space, which implies that Eq. 2 also changes both along the distance, Z, and with time.

CONVECTIVE DISPERSION

A fourth kinetic mechanism is convective-dispersion. This mechanism is rate limiting for the portion of the $C(Z)_t$ profile from the inflection point forward.

When this condition prevails the sorbate is sorbed as quickly as it is delivered to the sorbent neighborhood, and $\partial C/\partial t \approx 0$. Thus Eq. 1 becomes

(5)
$$\frac{\partial C}{\partial t} = 0 = -\bar{v}\frac{\partial C}{\partial Z} + D\frac{\partial^2 C}{\partial Z^2} - \rho\frac{1-P}{P}\frac{\partial \bar{X}}{\partial t}$$

Solving for $\partial \bar{X}/\partial t$ give

(6)
$$\left(\frac{\partial \bar{X}}{\partial t}\right)_{CD} = \frac{1}{\rho}\frac{P}{1-P}\left(-\bar{v}\frac{\partial C}{\partial Z} + D\frac{\partial^2 C}{\partial Z^2}\right)$$

The subscript "CD" refers to the convective-dispersion mechanism.

The task now is to find an expression for $C(Z)$ which will allow solution for Eq. 6. This function is unique for the particular porous media in question and is influenced also by flow velocity. Thus $C(Z)$ must be an experimentally determined function under conditions such that $(\partial X/\partial t)_{CD} < (\partial X/\partial t)_P$.

The nature of the $C(Z)$ function is reasoned as follows. Consider a single sorbate particle traveling with the interstitial stream. As the stream bifurcates and undergoes velocity changes, in a random manner, a single sorbate particle will have a certain fixed probability, unique for the particular porous media, of making a collision with the solid phase. Velocity of the fluid stream is relevant to the collision since this affects the zone of diffusion influence. Thus for a given porous medium and a given flow velocity, a single sorbate particle will have a probability of say 0.50 of making a collision with the solid phase within a certain distance of travel (call it the "half distance").

Consider now 100 sorbate particles in the fluid stream initially; 50 will remain after one half distance, 25 after the next half distance, and so on. This suggests a decay of the form

(7)
$$C(Z) = C_0' \cdot e^{-\lambda Z'}$$

where,

$C'_0 = C(Z'_0)$
$Z'_0 =$ the pivot point, where $(\partial \bar{X}/\partial t)_P = (\partial \bar{X}/\partial t)_{CD}$
$Z' = Z - Z'_0$
$\lambda =$ collision probability coefficient.

Taking derivatives on Eq. 7 gives

(8)
$$\frac{\partial C}{\partial Z} = -\lambda C'_0 \cdot e^{-\lambda Z'}$$

(9)
$$= \lambda \cdot C(Z)$$

and

(10)
$$\frac{\partial^2 C}{\partial Z^2} = \lambda^2 \cdot C(Z).$$

Substituting Eqs. 9 and 10 in Eq. 6 gives

$$(11) \qquad \left(\frac{\partial \bar{X}}{\partial t}\right)_{CD} = \frac{1}{\rho} \frac{P}{1-P}(\bar{v} + D \cdot \lambda) \cdot \lambda \cdot C(Z).$$

The terms $\bar{v}$, D, and $C(Z)$ in Eq. 11 are indicative of the transport rate at the position Z; the term λ is the collision probability coefficient. This coefficient is determined experimentally by Eq. 7. Fig. 2 illustrates the existence of Eq. 7 for a measured $C(Z)_t$ curve. Fig. 3 shows the effect of flow rate (veloctiy) upon the collision probability coefficient, λ.

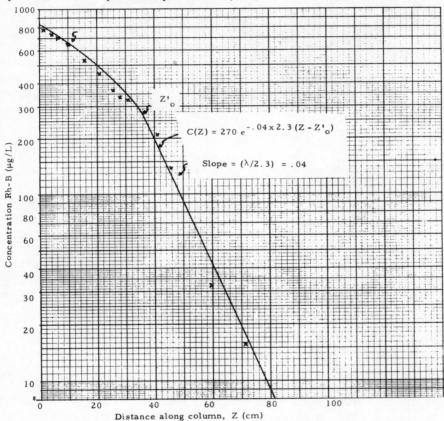

Fig. 2. Illustration of decay equation for measured concentration profile at $t = 30$ min. for Eau Claire sand and rhodamine B dye

CALCULATIONS

Either Eq. 2, or Eq. 11 may be rate controlling. The computer must test each of these equations at each Z increment along the column, and for each new time increment. The equation having the smaller value for $(\partial \bar{X}/\partial t)$ governs,

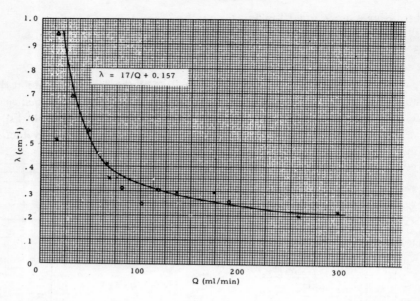

Fig. 3. Sorbate – sorbent collision probability, λ, as affected by flow rate in 1.5″ dia. column of 16–20 mesh Dowex 50 resin

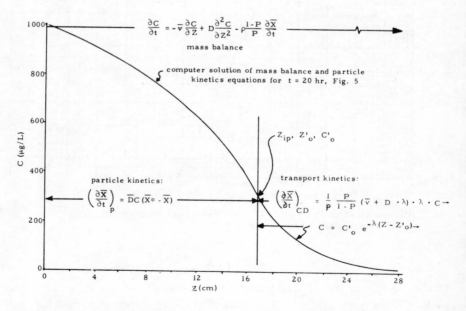

Fig. 5. Illustration of model components showing zones of influence of governing equations

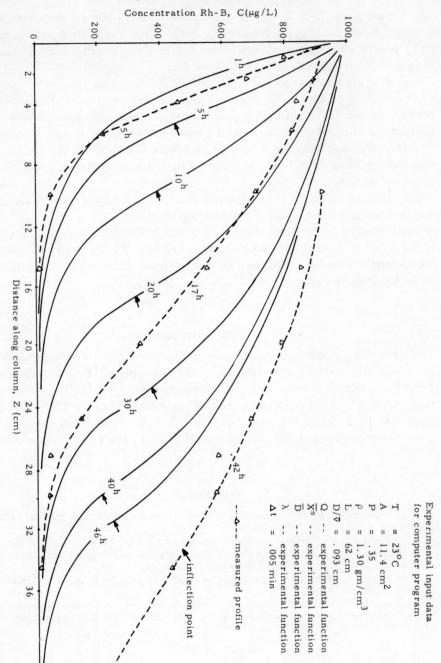

Fig. 4. Concentration profiles of Rh–B in liquid phase a various time intervals for feed concentration $C_0 = 1000$ and Dowex 50 resin — calculated from computer program

and is used in the finite difference calculation of Eq. 1 for a given time and distance. It has been proven (Hendricks, 1965) that the inflection point in the $C(Z)_t$ profile is the transition point from Eq. 2 to Eq. 11. Fig. 5 summarizes the zones of applicability of Eqs. 1, 2, and 11.

Fig. 4 is a computer output for solutions of Eq. 1 in the form $C(Z)_{t_1}$, $C(Z)_{t_2}, \cdots C(Z)_{t_n}$. These solutions are compared also with several measured profiles, shown as dashed lines in Fig. 5; the match is only nominal due to fluctuations in room temperature which had a marked effect on the isotherm. The computer solutions did behave in an identical manner to the measured solutions, however, when the same variations in room temperature and flow rate were imposed.

Computer solutions to $\bar{X}(Z)_t$ were also found; this solution consists merely with a single measured solution was quite favorable.

Further work is now underway to investigate kinetics more exhaustively under a variety of chemical and physical conditions. Work is also proceeding on bacterial adsorption which will be related to the mechanisms of transport of bacteria through granular porous media.

ACKNOWLEDGMENT

This investigation was supported, in part, by a research fellowship (number 1–F1–WF 24125–01) from the Division of Water Supply and Pollution Control U.S. Public Health Service during the academic year 1964–65. This report is based upon a Ph.D. dissertation completed in June 1965 at the University of Iowa. Dr. H. S. Smith was major professor for the work; his criticisms are gratefully acknowledged. The assistance of Dr. G. F. Lee is also appreciated.

REFERENCES

1. HENDRICKS, D. W. (1965), *Sorption in flow through porous media*, unpublished Ph.D. dissertation, Univ. of Iowa, Iowa City.

2. HELFFERICH, F. (1962), *Ion Exchange*, McGraw-Hill, New York.

3. HIESTER, N. K. AND T. VERMEULEN (1952), Saturation performance of Ion-exchange and Adsorption columns, Chemical Engineering Progress, *48*, 505–516.

4. KEINATH, T. M. AND W. J. JR. (1968), *A predictive model for the design of fluid-bed adsorbers*, J. WPCF, May.

5. RIFAI, M. N. E., W. J. KAUFMAN AND D. K. TODD (1956), *Dispersion Phenomena in Laminar flow through porous media*, Report No. 2, r.E.R. Series 90, Sanitary Engin. Research. Lab., Univ. of Calif., Berkeley.

UTAH WATER RESEARCH LABORATORY,
UTAH STATE UNIVERSITY,
LOGAN, UTAH 84321